# Children of Prometheus

## A History of Science and Technology

Benjamin Franklin Discovering Electricity, *by Benjamin West (c. 1816-1817). Philadelphia Museum of Art/Corbis/Magma*

# Children of Prometheus

## A History of Science and Technology

Second Edition

**James MacLachlan**

Wall & Emerson, Inc.
Toronto, Ontario • Dayton, Ohio

## *The Author*

James H. MacLachlan is a writer living in Toronto. Formerly he was Professor of History at Ryerson Polytechnic University; before that he was Assistant Professor at the Institute for the History and Philosophy of Science and Technology, University of Toronto; and before that was Head of Science at W.A. Porter Collegiate Institute in Toronto. He holds a B.A.Sc. in Engineering Physics from the University of Toronto and a Ph.D. in the History of Science from Harvard University.



Orders for this book or requests for permission to make copies of any part of this work should be sent to:

Wall & Emerson, Inc.
Six O'Connor Drive
Toronto, Ontario, Canada M4K 2K1

Telephone: (416) 467-8685
Fax: (416) 352-5368
E-mail: wall@wallbooks.com
Web site: www.wallbooks.com

Cover design and picture research: Alexander Wall

Cover picture: *Benjamin Franklin Discovering Electricity*, by Benjamin West. Philadelphia Museum of Art/Corbis/Magma

**National Library of Canada Cataloguing in Publication Data**

MacLachlan, James H., 1928-
Children of Prometheus : a history of science and technology

2nd ed.
Includes bibliographical references and index.
ISBN 0-921332-51-3

1. Science and civilization. 2. Science—History.
3. Technology—History. I. Title

CB478.M14 2002 509 C2001-901711-1

Printed in Canada.

# Table of Contents

# Preface

This book is based on the course of 48 one-hour programs in the history of science and technology that Phyllis Rose and I presented on radio in 1986-87. The director of CJRT-FM Open College in Toronto, Margaret Norquay, provided us with continuous and unflagging support. Although Phyllis and I shared the microphone (and an office), I have been ungracious enough to write the book by myself. Instead of using her scripts in the history of technology, I have derived the relevant chapters from my lectures in the history department at Ryerson Polytechnical Institute, Toronto.

Most of the other chapters are based on the scripts I wrote for the Open College course. As a result they largely preserve the conversational tone so necessary for lifting lifeless words off the page so they can be poured into a blind microphone. Whether you'll find that style helps your eyes to pour these ideas into your head remains to be seen.

For such skills as I developed in writing for radio, I'm grateful to Diane Rotstein, who edited and produced the series. She edited my scripts so thoroughly that I often wondered which words were still mine; yet so gently I never doubted that the ideas were all mine. Even when they weren't.

I have taken the opportunity provided by this paperback edition to add several discussions of relations between science and society. You will find them in Chapters 9, 12, 16, and 21. In other parts of the text various corrections and alterations have been made. I am grateful to Ronald B. Thomson for numerous suggestions for improvement. He will wish I had taken them more to heart.

*James MacLachlan,*
*Toronto, July 2001*

*Prometheus ("Forethought") steals fire from the gods.*

# Foreword

Science and technology are activities we humans use to deal with the world around us—to understand it and adapt it to our needs and desires. We study the history of these activities in order to understand how we've gotten to where we are today. We do that because we're fascinated by our origins. We might also do it because we hope that understanding our past can assist us on our path into an unknown future, but we needn't suppose historians to be somehow better equipped to deal with the future than other folk.

A fascination with human origins has been a feature of human thought for many centuries. My title, *Children of Prometheus,* illustrates one aspect of the history of that fascination. Prometheus is the Titan credited in Greek mythology with having stolen fire from the gods on Mount Olympus for the benefit of humankind. The story was elaborated by the Greek dramatist Aeschylus (525-456 BCE) in his play *Prometheus Bound.* For Aeschylus, Prometheus did much more than just provide fire to humans: "all the technical skills of mortals are from Prometheus" (line 506).

That line is the summation of a sixty-line passage in which Aeschylus had Prometheus describe the broad range of his gifts to mankind: he gave them mind and reason; taught them brick-laying and carpentry; invented astronomy, arithmetic, and writing for them; and provided them with the skills of harnessing horses and the wind. We might note in passing that Aeschylus used the Greek word *tekne* for his summary of all those skills. *Tekne* (the root of our word "technology") is usually translated as "art, skill, or craft" and contrasted with "knowledge" (*episteme*), for which we usually say "science." Yet Aeschylus includes within the term the skills of astronomy and arithmetic.

So, I find in Aeschylus ancient support for the notion of including both science and technology within a single book. Science and technology are by no means identical, but they *are* related.

To call us children of *Prometheus* also signals that I'll deal mostly with "western" (or "European") science and technology. That is not to say that one could not find fascination in Chinese science or African technology; they just don't loom large in the story I'm telling. Readers wanting to learn more about non-western science and technology would find good introductions in Colin Ronan's *Science: Its History and*

*Development among the World's Cultures* and Arnold Pacey's *Technology in World Civilization: A Thousand Year History*. These and other suggestions for further reading are listed at the end of this Foreword and following the introductions to each of the parts of the book. The culture of the children of Confucius is not very prominent in the western world, though, as we'll see, Chinese influences in the west are far from negligible (see especially Chapter 8). There is also some influence in the west by the children of Israel. And we can find in their *Genesis 2* and *3*, the Hebrew story of the source of science and technology: the serpent talked Adam and Eve into eating the fruit of the tree of knowledge (assume that includes science). They stitched fig-leaves together to cover their nakedness. As Yahweh drove them out of the Garden of Eden for their disobedience, he made them tunics of animal skins. As for other technologies, we're only told that Yahweh left Adam to win his food from the accursed ground and "gain your bread by the sweat of your brow."

My story is intended to elaborate on steps in the developments by which our ancestors learned by sweating and beating their brows to survive and eventually prosper in a hostile world. The Prometheus story is too easy. To achieve the items listed by Aeschylus took mortals many thousands of years.

When people use the words "science" and "technology," they're likely to attach current meanings to them—modern science with its panoply of high-powered computers and little men in white coats scurrying about in antiseptic laboratories; and high tech, with more computers, exploding rocket engines, and fearsome reactors. These are images many people find uncomfortable: "If God had intended us to fly, He'd have given us wings," as my old grandmother used to say.

We can't avoid being part of the world of sci-tech—it's all around us every day. The science that brought us nuclear bombs also brought us chest X-rays; while we can kill 100,000 in Hiroshima, we can also eliminate tuberculosis. The technology that brings us acid rain also allows us to fly back to Jamaica for our sister's wedding. You can think of examples like this endlessly.

A historical treatment of science and technology is intended to help us to grasp the awful truth—"We have seen the enemy, and it is us." Science and technology are human activities every bit as much as religion and politics are.

### *Science and Technology*

For the purposes of this study, I want to clarify how I'm using the terms "technology" and "science." Most of the time I want to emphasize that they're activities humans engage in. Occasionally, I'll slip into the traditional mode of treating them as products—as when I say a stone ax is a piece of technology. What I mean (but I'm too lazy to say it) is "a stone ax is the product of the activity of a technologist." I realize this is a somewhat peculiar usage, but I think you'll not find it obtrusive.

I also want to say that "science is what scientists do." That's not tautological, because if you want to probe that definition more deeply, you just have to watch scientists in action. And that's exactly what I intend to do in a historical sense—watch Galileo and Newton and the others transforming the way to do science. I could have kept closer to the traditional meaning of science as knowledge, by saying "watch Galileo and Newton making new science." I think both styles are appropriate enough, but I want to stress that it's the *new* making that was important, that those men were doing science—they were "sciencing."

Technology is a cluster of activities we humans have been using, developing, and elaborating from before our ancestors were fully human—from before our ancestors became *Homo sapiens,* according to the anthropologists. Technology consists of systems of skills in which we construct and use tools, implements, and materials for the benefit of human life by modifying or transforming natural materials. I stress "natural" because everything artificial (that is, "made by skill") had to start out being in or part of our natural environment.

Try out that definition on anything you please, all the way from paleolithic hunters stalking antelopes with chipped-flint clubs to you calculating with your silicon-chip calculator. It also includes the making of our houses and clothing; traveling under the power of wind, steam, jet, or rocket; designing materials and organizing workers to build an electric power network or transplant a human heart.

There is also a strong knowledge component in technology—knowing how to make or do or organize and how to deploy their results. Knowing, for example, which stones make good axes, where to find the best flint, where the deer and the antelope play, and how to creep upwind to get close enough to them to use your good ax.

I reject the simple, common distinction between technology as doing and science as knowing. The way I draw the categories, technology must include *knowing* how to do and science must include *doing* how to know. Some people want to think that all technology is applied science, and they're willing to attribute science to stone age creatures—that classifying rocks according to usefulness was at least proto-science. Classifying is certainly a scientific activity, but there's a lot more to it than that.

Science is a cluster of activities humans use to understand the structure and operation of the natural world. The stress here is on *understanding*, which to define adequately would raise a whole other set of problems. I think of us "understanding" something when we have a coherent and satisfying picture of it in our heads that makes sense to us. It seems to imply a regularity of behavior in things that may sometimes say more about our heads than about the things.

Take a simple example. You're watching the gang playing pool at the local billiards parlor. You understand that when the (white) cue ball hits a red ball, the two balls then go along paths that you soon begin to predict. If the cue hits the white ball dead center, the ball goes in the direction the cue was pointing. If the white ball hits the red ball squarely, the white ball stops dead and the red ball continues along the same direction at almost the same speed. When the balls hit the side cushions they're usually deflected in an understandable way.

If this is your first visit, you may soon be shocked to see that the balls sometimes deviate from the simple behavior described above. Your friend may let you in on the secret of "english"—there are ways to impart various spins to the cue ball. Pretty soon, you're an old pro at watching and figuring out how the shark must have stroked the ball that time. After a little bit of instruction on the rules of scoring, you should be ready to say you understand pool. You have a coherent and satisfying picture in your head about your observations (experiences) at the pool table.

Now, some science teacher comes along and tells you that if you want to understand the behavior of gases (the relations among volume, pressure, and temperature, say), you should imagine that gases are made of molecules that are like tiny (perfectly) hard billiard balls bouncing around in three dimensions. They collide with one another and with the walls of a container. Mathematicians have worked out what pressure you should expect when you increase the temperature of a gas in a cylinder, or decrease its volume by pushing in on a piston. Their system of equations comprises the kinetic theory of gases.

When deductions from the theory are compared with experimental measurements of related changes in volume, pressure, and temperature, they match—often very closely. Then we say that we *understand* the gas laws. That's because we've imagined gases behaving according to the rules of collisions that we've found working for billiard balls. We've matched the picture in our heads of behavior at the pool table with a wholly imaginary picture of bouncing gas molecules. With luck, both pictures will be coherent and satisfying.

This example is intended to point up the major features of the cluster of activities we call science: *imagining* gas molecules as tiny hard balls, making mathematical *deductions* from the imagined assumptions or hypotheses, and *matching* those deductions to experimental *measurements*. Imagining, deducing, measuring, and matching—they form the cluster of activities we call science.

Now, we face a quandary! Should we confine "science" only to that cluster as a whole, or can we think of the four individual activities each as being scientific? My answer is simple-minded: any one of the activities in complete isolation from any other is not scientific; but an *intention* to perform it in conjunction with the others makes it scientific. If you find that unsatisfactory, just think about it a bit and match it with the activities of scientists described in the rest of the book.

One historian of science once defined science as "the abstract understanding of nature." I object to that because he seemed to be stressing the imaginative element at the expense of the others. Probably, he didn't intend that. But his definition does seem to imply that scientists like Einstein, who indulge in abstract speculations, are somehow more scientific than the ones who concentrate on experimenting and measuring, like Michael Faraday, the "prince of experimenters." I'm convinced that all the activities I've mentioned are equally important because it's their combined cluster that makes science.

You should find it helpful to look at science this way in order to better understand the place of theory. We have theories in science because they provide understandings we cannot get from direct experience. Science helps us to understand our experience by postulating structures and behaviors (like molecules, quarks, and quasars) at levels that are forever inaccessible to direct observation. In some sense, it's that inaccessibility that makes us need science for our understanding of nature.

That notion leads off in another direction: I choose to use the word "science" (unmodified) only for enterprises that are consciously part of the cluster of activities that includes speculating, deducing, and measuring. "Science" has also been used for a long time to mean "organized knowledge." And so you will often see various modifiers of "science," such as "Greek," "Chinese," "Arabic," "medieval," "modern," etc. These bodies of knowledge have been intellectually useful to their developers, by providing them with satisfying, coherent pictures of the world. But they are not *science* as I'm defining it. I will mention them only occasionally, as they are seen to impinge on science (in my sense). This results in my concentrating largely on the science of western Europe. And nowadays, the Chinese and Arabs mostly do the same. That's because western science is unique.

This notion was once very clearly expressed by Albert Einstein. In 1953 he received a letter enquiring why he thought the Chinese had not developed the kind of science I'm talking about. Einstein's reply was succinct, and perhaps startling:

> Dear Sir,
>
> Development of Western Science is based on two great achievements, the invention of the formal logical system (in Euclidean geometry) by the Greek philosophers, and the discovery of the possibility to find out causal relationship by systematic experiment (Renaissance). In my opinion one has not to be astonished that the Chinese sages have not made these steps. The astonishing thing is that these discoveries were made at all.
>
> Sincerely yours,
>
> Albert Einstein
> April 23, 1953.

What this means is that cultures and civilizations can get along perfectly well without science—though they could not even be human (I claim) without technology.

I have one final item to clarify in this section. It has to do with the connections between science and technology. As you'll see, science and technology have travelled relatively distinct paths for most of human history. Some authors have claimed that they didn't have much to do with each other until the nineteenth century—until industries like electrical and chemical manufacturing took off. I think that's going too far. For two main reasons.

First of all, even if we think of science as knowing and technology as doing, there are many examples before 1800 of people trying to make something of what they understood. The science of astronomy (Chapter 4) had an influence on the technology (*technique* if you prefer) of astrology. In fact, I think it's likely that astronomy was developed in order to serve astrology. (There's more to it than that, because astronomy in the hands of Ptolemy became far more precise than most astrologers needed. Ptolemy as scientist was just curious about the planetary motions. But Ptolemy was also an astrologer.) The alchemists had theories about the structure of matter (Chapters 4 and 6) and engaged in the technology that tried to change lead into gold, as well as succeeding in changing wine into spirits. You'll find other examples in other parts of the book—before 1800.

For, I believe, human beings do not act within the confined limits of the neat categories scholars try to fit them with. Any time in history that anyone thought they had knowledge that *could* be applied, they'd try to apply it. Until 1700, the knowledge they had wasn't secure enough to lead to useful results very often. That didn't keep them from trying. In the eighteenth century the theory of phlogiston (Chapter 15) was applied to the making of steel. We believe the phlogiston theory was wrong—that doesn't mean that it couldn't be usefully applied. After all, in some senses, phlogiston was almost perfectly upside down from our oxygen theory. If it was applied upside down it might work. The test (which I've not made) is to see if metallurgical practice was enhanced by applying the phlogiston theory.

Nor should you think that just because some technology today is science-based, therefore it all is. Technologists are continually confronted with the task of manipulating intractable materials into products they can find a market for. In that process, they're willing to use whatever comes to hand. If it's good, sound theoretical science, fine. But they won't abandon all hope if there is no science.

My favorite example comes from within a science-based industry—electroplating. It's quite true the industry would not exist had it not been for the scientific investigations of Volta, Davy, Faraday, and others (Chapters 15 and 17)—laboratory investigations with small vats and ample time. Technicians in industry have large vats and short deadlines and economic considerations. They don't have time to worry about what's scientific or what the explanation is for any particular phenomenon. They have the robust task of turning out chrome-plated bumpers at so many per hour. And the plating has got to be permanent. As a result, the technicians have learned by experience that the job goes better if they throw various chemicals into the vat along with the steel bumper and the chromium rods, etc. They have no better reason (or explanation) for some of those chemicals than that they work—they make the product better or cheaper. No theory—just experience.

So, don't be misled by the dictionary definition of technology as applied science.

However, on the other side of the coin, there are some laboratories today where you'd have a hard time telling the engineers from the scientists. They both make measurements. They both put their results through computers. They both have speculations or theories or hypotheses about how to manipulate their data. By the standard definitions the only difference between them is one of motivation. The engineers want to figure out how to improve the product; the scientists want to understand. Curiously enough, doing one often leads to the other. But, as my friend John Abrams used to say, "You can't observe motivations."

We historians are leery about anecdotal evidence, about asking people what they think they're doing that for. We've been lied to before, and besides, the people may not even know, or they may have mixed motives. So, for much of the twentieth century, distinguishing science from technology is very difficult. As a result, most separations of the two are pretty arbitrary—if you find them in this book, realize that I know better.

Now, you might think there must be some extreme cases, pure science without technology or vice versa. For example, arguing about whether pandas are more closely related to raccoons or bears or whether Supernova Shelton (1987A) will require a revision of theory seem pretty remote from the technology of practical life. But the history of pandas may inform us about the environment and how to save them from extinction, which could have certain intriguing political consequences (no, I'm not looking forward to a Panda War). What may come from the supernova seems remote enough, but in the 1940s when the Manhattan Project was looking for someone who knew about thermonuclear reactions, they found an astrophysicist who'd been working on the theory of energy generation in stars. As Benjamin Franklin and Michael Faraday said, "What's the use of a new-born babe?"

We don't do science *because* it will be useful— but it may be. And we don't do technology because we'll learn something—but we might. As one historian has put it, science benefitted more from the steam engine (in thermodynamics) than the reverse. And the science of aerodynamics grew out of the startling discovery by the Wright brothers that heavier than air flight was possible. Only two months before (in October 1903) Professor Simon Newcomb, a notable American scientist, had written eloquently about the futility of flying machines. Two brief quotes will suffice:

> ...there is no mechanical combination, and no way of applying force, which will give to the aeroplanes the flexibility and rapidity of movement belonging to the wings of a bird....

> If, therefore, we are ever to have aerial navigation with our present knowledge of natural capabilities, it is to the airship floating in the air, rather than the flying machine resting on the air, to which we are to look.

See all those airships floating across the sky!

## *History*

I try to do history like a scientist. History is about the actions of humans in the past. If we had an absolutely complete record of all the activities of all past humans, then writing history might seem to be merely a matter of *selecting* the juicy bits.

However, even with such a complete record, historians would still lack direct knowledge of the motivations and passions of the historical actors. For example, although we possess abundant documentation for the Great World War, there is still much contention among historians about its causes.

Writing history provides a pattern of human behaviors that helps us to understand the events of the past, and how we got to be the way we are.

So, historians have to speculate—they have to supply imaginative stories about what motivated Caesar to go to Britain in 55 BCE, or what Lee Harvey Oswald had against President Kennedy in 1963. That would be so, even if historians had all the evidence of actions—who did what when—which of course they don't. So, they reconstruct. Take a trivial example: you won't imagine that Caesar's boats had sternpost rudders if you're sure they weren't invented till hundreds of years later. Or, if you're making a movie of Roman times you don't dare put stirrups on the horses for the same reason. Historical reconstructions have to be plausible.

Plausible reconstructions and invented episodes sometimes make history books sound more like novels—both the glory and shame of historians. Glory when they do it well, and shame when they try to avoid it. What are the criteria for doing history well? As in science, nothing in a historical reconstruction should conflict directly with the evidence. Evidence can be both direct—as in dating Caesar's British venture—and indirect, as with the stirrups. Plausibility comes from the historian's insight into human nature. Can a person actually have been expected to have behaved that way for those reasons? It's because novelists spend more effort plumbing the depths of human nature than most historians that sometimes a historical novel rings truer than the history book.

Historians and novelists are more limited in their speculations than scientists are. Scientists can invent weird entities like electrons and genes (Chapters 19 and 21), whereas historians must confine themselves to what people can realistically be expected to do. Frequently, they make the implicit hypothesis that human nature has not changed drastically throughout historical time. Anthropologists make essentially the same hypothesis when they try to imagine life at our ancestors' campsites a million years ago.

One example in the history of science is my main claim to fame in the field. In the *Two New Sciences*, Galileo described what he said he saw when he submerged the tiny mouth of a bottle of water beneath the surface of wine in a glass. He said he saw the wine wisping up into the water (like smoke from a cigar sitting on an ashtray—that's the quickest way I can give you Galileo's precise word image—he didn't mention cigars). And soon he also saw a perfectly clear band of water forming at the bottom of the wineglass.

Now, historian of science Alexandre Koyré had doubts about the veracity of most of Galileo's descriptions of experimental observations. He wrote quite scathingly about this particular one. Being French, he knew how readily water mixes with wine—this separation described by Galileo was surely implausible. Another imaginary experiment! When Stillman Drake was translating Galileo's *Two New Sciences* in 1972, he asked me what I thought of the passage. Between us, we decided I should try it. With an after-shave bottle full of water and a glass of wine, I soon was treated to exactly the sight Galileo described. So much for implausibility!

Sometimes we can test plausibility by working through the presumed steps of our historical subjects. One intriguing study of Newton's optical writings has made some inferences about the quality of his vision. When I have followed some of Newton's arithmetic in detail, I've found him occasionally making exactly the kind of numerical slip (typographical error) that I often make. Sometimes it seems possible for us poor mortals to get inside the heads of our heroes, but of course, it's pretty chancy.

Getting inside the lives of historical characters is not easy to do—and it's where novelists excel. But, there's a really significant problem in that. Good novels are seen as "true to life"—they ring bells in our lives: "Hey, our Bill is just like that." That's okay for contemporary novels—we all live in much the same world. But, how does it work for historical novels (or for science fiction)? How do we know how people behaved three hundred or three thousand years ago?

For example, it's hard to find many atheists in the seventeenth century. How deeply into the head of Newton or Shakespeare can a modern historian get if he's an atheist? If you don't have religious motivations, how can you understand them in others. I think it's not impossible, but it may be very difficult.

In any case, we cannot be sure so we make the hypothesis that people are pretty much alike, even though we know they aren't. Even without switching time periods, think how few bells might ring in the head of a Shiite Muslim villager in Iran reading a novel by Margaret Atwood. We know there are many differences among cultures in our modern world. I think there are enough differences between a young professional living in her high-rise on Toronto's Yonge Street with her DVD player, her cellphone, her high speed internet access and her personal digital assistant and a !Kung bushman in Botswana that we could say they live in different worlds.

Certainly they live in different worlds technologically; they also have different world views—perhaps the bushman still has tribal and elemental gods, while Ms. North America's TV tells her about the jet stream and radio-telescope searches for extra-terrestrial life. Such extremes encourage me to surmise that we'd better assume *some* significant differences in the historical past.

So, I'd like you to share an imaginative exploration of previous worlds with me, assuming that their inhabitants had essentially the same genetic structure as we, but also being willing to assume that they had sufficiently different cultural structures that we may not actually be able to understand their behavior and motivations in all their intricate detail. We'll use all the evidence we can to make our reconstruction look plausible, but we'll remember that one person's plausibility is another person's nonsense.

I want you to avoid thinking that history tells you directly what the past was like or that science tells you directly what the world is like. The past was, and nature is; we've invented history and science to help us make sense of the past and of nature. History and science as "rhetorics of conclusions" seem to be inventions of teachers intent on keeping inquisitive pupils in line: "This is what the experts tell us. Maybe someday you'll get to be an expert. Until then, shut up and listen."

Too often in schools, it feels like: read the books, pass the exams, then forget it. Maybe schools can't avoid being like that—lots of folk have tried to change things, to little avail. I share Joseph Schwab's contention that authority does not consist in possessing information (any dumb computer can do better at storing, if not retrieving, information); rather, authority consists of being a competent inquirer. Historians and scientists are inquirers into the past and nature. In learning about them, you should be inquiring too.

Finally, I should make a couple of technical points. Historians are forever chopping up the past into more or less digestible bits. Yet, we know history is continuous, even when it looks otherwise. What we don't know—why historians have to speculate—is which events were in fact significant precursors of succeeding events. I have given medieval "science" short shrift because I doubt it had much influence on the science of the Renaissance. I may be wrong. Certainly, many medieval historians disagree with me. So, when we chop up the past, we look for "watersheds" in history, even though we know full well that most of the time, one day was pretty much like the next for the people living at the time. You know, don't dare ask "What year did the Renaissance start?"

I have divided history into four periods: the first (ancient) starts way back and continues down to the end of the Roman Empire—some time around 450. After the "decline and fall" of Rome, Europe (in the south and west at least) lapsed back into a peasant existence entirely innocent of the glory that was Rome (with a few exceptions, such as Rome itself).

I start medieval times (the middle ages) about 500. Medieval does not have any connotation of barbaric or "dark" for me." A twelfth-century renaissance brought many changes to western Europe, which deserve notice. Wherever that might have led was nipped in the bud by the Black Death (1348-1351)—at least *those* are firm dates. The Renaissance by no means started the next day-it should be expected to take a while to recover from having your population cut by a third or more (more in some places than in others).

I begin the Renaissance about 1400 (the last year of the fourteenth century). Century years are convenient for history merely in the same way the hour is the time to start a new radio program: "Next news at 11:00." Then, I let the Renaissance run till 1700. Other historians are inclined to stop it at 1600, and to call 1600-1800, "early modern." I didn't want to have five divisions, and besides 1800 falls right in the middle of the Industrial Revolution (according to me). Indeed, it may be because of that arbitrary 1800 that you'll find people starting the Industrial Revolution then. Too much industrialization happened between 1700 and 1800 for me to let them get away with that. In addition, I find many aspects of the Renaissance taking three hundred years to catch on widely.

So, I'm not too wide of other people's mark in starting the modern period at 1700—they just qualified the first century of it as "early." I omit the qualification. Beyond today, it's anybody's guess when a new "period" starts. Some observers have already labelled us "post-modern," whatever that's supposed to mean.

Because I do think periodization is arbitrary, you'll find a few topics treated in the "wrong" places, simply because they seemed to fit better where I've put them. Gunpowder and cannon (Chapter 7) are one example; Mendel's genetics (Chapter 21), another.

Another prejudice I should expose is my large concern for origins. Since history is mostly about change (things get pretty dull when nothing *new* happens), I'm often looking for when new things started. Of course, I try to be cautious about claiming that "the earliest found" is actually "the first." 'Tis a caution I'd like to urge on others—in science as well as history.

Origin-itis is not unique to me. There have been times when I've looked in the sources to find out what life was like at a certain time to discover that my historical colleagues tell me only what was new then. A student once wanted some advice about sources for an essay on medical practice in the fourteenth century. All I could find in the books I looked in were a few paragraphs about a couple of the innovations of the time. Was I to assume that in everything else the physicians of the time followed the practices of earlier centuries? So I read back through the thirteenth, twelfth, eleventh, fifth, fourth, second centuries. Could I trust that our fourteenth century doctors put all that together? Not likely! Perhaps they read most of their medicine in manuscripts of Hippocrates and Galen; perhaps they got their remedies out of medieval herbals; perhaps they used large doses of folk medicine; who knows? For times before printing by movable type, the kind of picture my student wanted was bound to be pretty hit or miss.

So, even when I want to trace something on after its origins, I'm limited by what historians have already found in the traces of the past or found interesting for the particular story they decided to tell. Be aware that stories you might wish I'd tell are missing, either because I wasn't interested or because no historian has been or because there's just a big hole in the record.

Finally, for history, a note about centuries. The first century of our era began in year 1 and ended in year 100. The 1900s comprised the twentieth century, except for 1900, which was the last year of the nineteenth century, plus 2000, which was the last year of the twentieth century and of the second millenium. Thus, when you read "seventeenth century," please remember that's the 1600s. This is not a problem for Italians, who call the twentieth century *Il Novecento*, because they're thinking "nineteen hundreds." Actually 1900 is *mille novecento,* but Italians assume the "thousand" without stating it. Thus for Italians, the Renaissance of the fifteenth century is *Il Quattrocento* (literally "the four hundreds").

### *Units and Symbols*

From time to time on the following pages you'll see some abbreviations for units. The following list provides their meanings. I have been metric (*Système internationale*) most of the time though I seldom attribute metric units to people who lived before 1800 when I translate their own words. In case you're not metricized yet, I give rough equivalents for the main units.

The *meter* was supposed to be exactly 1/40,000,000 of the earth's polar circumference, but the guys who measured it first got it a little wrong. Today the meter is defined precisely by a particular number of wavelengths of a specific color of light. A meter is a little longer than a yard (almost 10%).

The beauty of the metric system is that it makes all multiples and fractions in factors of 10. A decimeter is one-tenth of a meter (0.1 m). A centimeter is one hundredth of a meter, 1 cm = 0.01 m (within about 0.5% of two-fifths of an inch); a millimeter is 1 mm = 0.001 m.

Going the other way, we skip to the kilometer, 1 km = 1000 m (short of two-thirds of a mile by about 4.5%).

The unit of volume is the liter, where one liter is a cubic decimeter. A liter is close to a quart (13% short of an Imperial quart and about 5% more than a U.S. quart).

The kilogram was supposed to be the mass of a liter of water, but they got that a little wrong too. A kilogram is now the mass of a particular lump of metal kept under glass in Paris. A kilogram is about 10% heavier than two pounds. For shipping weights, the metric unit is the tonne (1 t = 1000 kg—10% heavier than a U.S. or "short" ton, and about 2% less than a British or "long" ton).

The metric unit of energy is the joule. It's named after James Joule, with the symbol J. Although the names of all units are written in lower case letters, their symbols are capitalized when they're derived from the name of a person: so the newton of force is symbolized N.

This principle caused some interesting confusion a few years ago over the liter (*litre* in France, England, and Canada). The Metric Commission prefers to use a lower case script $\ell$ for the symbol, but since that would cause some typographical difficulties, they allowed L also to be used. (They couldn't use l, because it's too hard to distinguish from 1.) But that makes it look as if the liter is named for a person. So a physicist at the University of Waterloo invented a Sieur de Litre, an obscure French chemist of the eighteenth century, and published his brief biography in the April issue of a newsletter for chemistry teachers. Though he's had some fun with it, he sometimes wishes he hadn't.

I think the physicist may have been encouraged by how the Celsius scale got its name. For many years it had been the centigrade scale, but using C for its symbol. At some stage in metric deliberations, someone said we could keep that capital C and follow our new rules for symbols if we could find someone to name this scale after. Searching into the history of thermometry, they found a name beginning with C, Anders Celsius (1701-1744), a Swedish astronomer. He'd even suggested dividing the range between the freezing and boiling temperatures of water into 100 parts. But, for some obscure reason he put zero for the boiling point and 100 for freezing. That was close enough, and saved them the trouble of inventing someone. Nowadays scientists mostly use the absolute scale of kelvins (named after the man who had something to do with it, William Thomson, Lord Kelvin), where (approximately) 0° C = 273 K (no degree symbol).

The metric unit of power is the watt, 1 W = 1 J/s. The s is the symbol for "second" of time, which the Metric Commission decided to leave in the traditional base sixty: 60 s = 1 min, 60 min = 1 h, 24 h = 1 d, 365.24 d = 1 a (French *an* for year).

An electric mixer may run at 400 W, while a toaster takes 900 W (or 0.9 kW). When you buy one kilowatt-hour of electricity from the power company for a nickel or a dime, you're getting 3.6 million joules of energy—a 100 W lamp running for 10 h, the mixer for 2.5 h, the toaster for 1.1 h.

You can see that the symbols for the multipliers, like k for kilo- and c for centi-, can be attached to the symbols for any of the units. When you get into high-powered physics, you also get high (and low) powered multipliers. For a million, mega- (M); a billion, giga- (G); a trillion, tera- (T), etc.; and a millionth, micro- ($\mu$, the Greek letter *mu*); a billionth, nano- (n); a trillionth, pico- (p), etc.

The metric units for electricity are our old friends ampere (A) for current (not amperage) and volt (V) for potential difference (or voltage), named for A.M. Ampère and Alessandro Volta. The hertz is the new unit for cycles per second (1 Hz = 1/s), named after H. Hertz.

If you ever get the urge to write any of these symbols, you should realize that upper and lower case are crucial:

1 mA = one milliampere (a thousandth of an ampere)

1 Ma = one megayear (a million years, or a thousand millennia).

You can figure out the meanings of 1 ma, and 1 MA.

### *For Further Reading*

*Here, and at the end of each of the major divisions of the book, I provide lists of books for further reading. They include books that I used as sources, as well as others that cover narrower areas more deeply or go into matters that I have omitted or touched very lightly. You should not consider the lists to be exhaustive. Most of the books were still "in print" when the lists were compiled.*

*Books that encompass more than one Part are listed only with the first Part to which they are relevant. The list that follows here includes mostly general books in the history of science and technology, as well as several on the philosophy of science.*

Aeschylus. *Prometheus Bound & Other Plays*. Trans. Philip Vellacott. Penguin Books, 1961.

Alioto, Anthony M. *A History of Western Science*, 2nd ed. Englewood Cliffs, NJ: Prentice-Hall, 1993.

Basalla, George. *The Evolution of Technology.* Cambridge, MA: Cambridge University Press, 1988.

Bronowski, Jacob. *The Ascent of Man*. Boston: Little, Brown, 1979.

Burke, James. *Connections*. Boston: Little, Brown, 1979.

Burke, James. *The Day the Universe Changed*. Boston: Little, Brown, 1985

Cardwell, Donald. *The Norton History of Technology.* New York: Norton, 1995.

Cohen, I. Bernard. *Revolution in Science*. Cambridge, MA: Harvard University Press, 1985.

Daumas, Maurice, ed. *A History of Technology and Invention: Progress Through the Ages*, 3 volumes. New York: Crown Publishers, 1969-1979.

Hacking, Ian. *Representing and Intervening: Introductory Topics in the Philosophy of Natural Science*. New York: Cambridge University Press, 1983.

Jaffe, Bernard. *Crucibles: The Story of Chemistry: From Ancient Alchemy to Nuclear Fission*, 4th rev. ed. New York: Dover, 1976.

Kragh, Helge. *An Introduction to the Historiography of Science*. Cambridge: Cambridge University Press, 1987.

Kranzberg, Melvin & Carroll W. Pursell. *Technology in Western Civilization*, 2 volumes. Oxford: Oxford University Press, 1967.

Kuhn, Thomas S. *The Structure of Scientific Revolutions*, 3rd ed. Chicago: University of Chicago Press, 1996.

Motz, Lloyd and Jefferson Hane Weaver. *The Story of Physics*. New York: Avon Books, 1989.

Pevsner, Nikolaus. *Outline of European Architecture*. New York: Penguin Books, 1950.

Pacey, Arnold. *The Maze of Ingenuity: Ideas and Idealism in the Development of Technology*. Cambridge, MA: MIT Press, 1976.

________ *Technology in World Civilization: A Thousand Year History*. Cambridge, MA: MIT Press, 1991

Price, Derek de Solla. *Science since Babylon*. New Haven: Yale University Press, 1975.

________. *Little Science, Big Science, and Beyond*. New York: Columbia University Press, 1986.

Ronan, Colin A. *Science: Its History and Development among the World's Cultures*. Facts on File, 1985.

Schneer, Cecil J. *Mind and Matter: Man's Changing Concepts of the Material World*. New York: Grove, 1969.

Spielberg, Nathan and Bryon D. Anderson, *Seven Ideas that Shook the Universe*, 2nd ed. New York: Wiley, 1995.

Staudenmeier, John M. *Technology's Storytellers: Reweaving the Human Fabric*. Cambridge, MA: MIT Press, 1985.

Toulmin, Stephen. *Foresight and Understanding: An Enquiry into the Aims of Science*. Harper & Row, 1961.

Wall, Byron E. ed. *Science in Society: Classical and Contemporary Readings*. Toronto: Wall & Emerson, 1989.

The Universe. *A 16th century engraving portraying the Aristotelian-Ptolemaic system.*

# Children of Prometheus

A History of Science and Technology

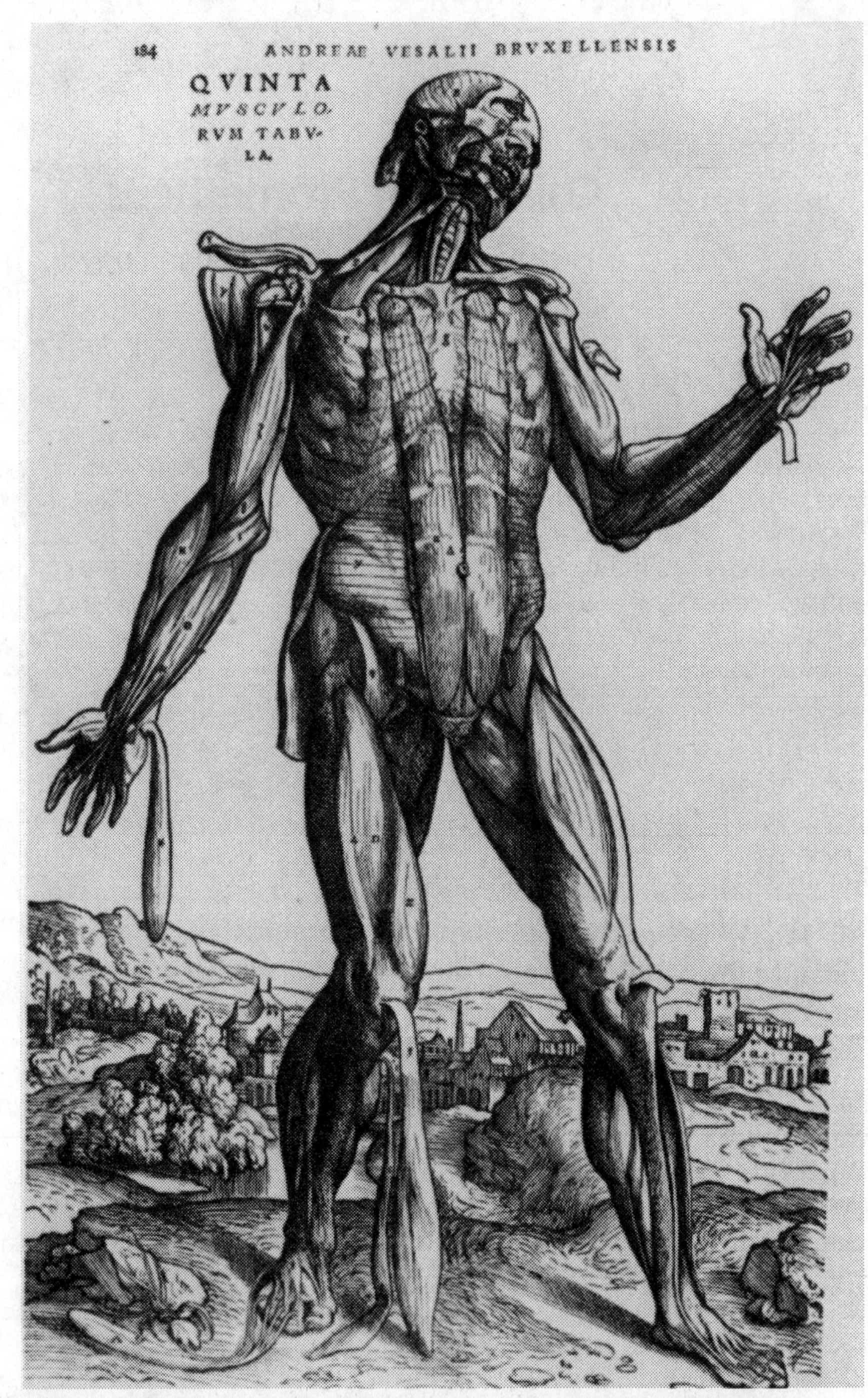


*This illustration of the structure of human musculature combines Renaissance art with scientific anatomy from Andreas esalius,* On the Construction of the Human Body *(1 3).*

# Prologue

## *Change of Pace*

Ten million years ago there were no humans on earth. Creatures ancestral to us were roaming through African forests. Then the first of our ancestors that walked on their hind legs appeared about five million years ago. Anthropologists date their finds of the earliest crude stone tools to about two million years ago. By about one million years ago they can identify such a variety of fabricated stone tools in different sites to substantiate claims of the existence of several distinct cultures among our ancestors. Soon after 100,000 years ago, our own species, *Homo sapiens,* was emerging from the ancestral stock, occupying sites in Africa and ranging across Eurasia from France in the northwest to Java in the southeast.

Already we can see a quickening pace of change. From then on, changes in human life styles occurred at ever more dazzling rates. By 20,000 years ago, humans occupied all the continents. Their hunting-gathering lifestyle had enough staying power that it still exists (though probably modified over twenty millennia) in some of the more remote regions of the earth today—among the Inuit of North America (at least until the recent introduction of rifles and snowmobiles), the !Kung of Botswana, and the Aboriginal peoples of Australia.

Fifteen thousand years ago there were no farms and no cities. Yet by 5000 BCE, settled agriculture was flourishing along the banks of several rivers in Asia. By 3000 BCE emerging cities can be found near the southeast corner of the Mediterranean. These cities depended on agricultural surpluses in surrounding farm villages, won from the soil by people using techniques of cultivation and herding, tools of stone, and vessels of pottery, who dressed in woven fabrics, and lived in homes of mud brick. In the cities, artisans fabricated tools, vessels, and buildings in stone and metal. Priests administered the cities and the surrounding countryside, and used holy vessels and graven images of bronze. Merchants traded a wide variety of goods with distant lands by ships and pack animals. Soldiers with bronze armor and weapons developed their skills in battle. Civilization!

History (based on written records) begins at 3000 BCE as priests and merchants devised ways of writing on clay, stone, and papyrus. These written records give us our first solid glimpses of our ancestors' mental processes—in calculating, praying, organizing, and soon in star-gazing.

For the next 3500 years, the history of the Middle East is a tangled tale of the rise and fall of empires: Egyptian, Sumerian, Akkadian, Hittite, Assyrian, Median, Chaldean, Persian, down to the rise of European domination of the area with Alexander (around 300 BCE) and the C sars (after 100 BCE).

During those 3500 years, artisans perfected their skills in building and fabricating a wealth of structures (from Egyptian pyramids to Roman aqueducts) with implements and techniques ranging from stone masonry to blacksmithing. This period also saw the emergence of astronomy as a science, geometry and logic as instruments of thought, and the wide-ranging philosophical system of Aristotle.

After about 500, our focus blurs, then shifts to Mecca in the east (after 700) and Paris in the west (after 1100). For a thousand years we watch the gradual emergence of the nations that would put a distinctly European stamp on the course of world history.

Medieval Europe emerged out of the remnants of Roman colonies and the vigor of invading tribesmen from the northeast. Over that thousand years Europeans improved on Roman techniques, invented some of their own, and borrowed others from the Muslims and Chinese. The Europeans used them to develop local political economies based on wealth and power, and in the sixteenth century Europeans conquered the world.

After the invention of printing with movable type, about 1450, a new intellectual vigor spread across Europe. Scholars disseminated Greek learning (in Latin dress) and built the scientific revolution on it. In less than 150 years thereafter, isolated scholars transformed themselves into a scientific community, with societies and journals.

By 1700 European mathematicians and experimenters had the model of Isaac Newton (and the giants upon whose shoulders he stood) to forge tools of analysis of the gamut of nature from distant stars to invisible atoms. They probed and analyzed and speculated with full confidence in the power of human minds to understand the mysteries of nature's bounty.

That confidence spilled over into the technological vigor we call the Industrial Revolution. Before the eighteenth century, humans were limited to the power they could extract from animal muscles, flowing water, and blowing wind. At the end of the eighteenth century, transportation of goods and troops was still limited to the pace of horses and sailing ships. Then, they began to exploit the expansive power of steam. The nineteenth century brought railways, steamships, and automobiles. The twentieth added piston aircraft, jets, and rockets. And electrical communication sends messages whizzing across continents and oceans in the twinkling of an eye.

Technology and science have combined to speed the pace of human commerce, as they continue to dazzle us with their own ever-accelerating rate of change.

### *Making the World Our Own*

We humans live in a world largely of our own making. Indeed it is our capacity to transform our environment that marks us as human. From the first pebble tools of two million years ago to the sophisticated computers of today, we've been using the materials of the earth to create tools and machines to make the world the way we want it, instead of the way we found it.

We've made the world the way we want it—over and over again. As we look back over the past we should be aware that our ancestors have lived in a variety of worlds. Our world is different from the world of the Elizabethan adventurers who conquered the new world of the Americas. And their world was different from the classical world of Greece and Rome, which in its turn was different from the Egypt of the age of pyramids. Before them were the first farmers, and before them the stone age hunters.

Life-styles in every age have been determined by the technology of the time: by the way humans used the materials available to them, by the plasticity or intractability of those materials, by the ingenuity humans could bring to bear on the problems they faced. Humans in every age have been limited by the hard realities of what was available, and by what they could recognize as *being* available. Until 1824 silicon dioxide was just sand on the sea shore. Now that same sand is the source for the chips that drive our computer age. Yet who among us today could recognize a good piece of flint to make a hand ax? Ten thousand years ago, our neolithic ancestors valued high grade flint deposits the way we value deposits of iron or petroleum today.

We humans live in the material world created by our technology. But we also live in a mental world created by our science. On a summer night, far from city lights, gaze up at the star-studded black velvet sky. With Captain Kirk at your elbow you can dream of distant galaxies, of spending a lifetime in warp drive "going where no one has gone before." Although the universe is not infinite, it's at least "rather large."

But with Icarus at your side you might dream of a feathered flight to the moon, with the vault of heaven just beyond it. Or with John Donne beside you, you might think of "a little world, cunningly made." Our view of the universe is conditioned by the science of the time we live in. We live in worlds we've made with our science as well as our technology.

In order to think of these various worlds of our own making we have to imagine ourselves as part of an international and historical human community. That is, you and I, right now, are two humans out of a collection of perhaps 60 billion who have ever lived. If by the effort of your imagination you can participate in the making of the worlds of the past, then you'll be in tune with my theme of examining the worlds we've made for ourselves.

This will require you to imagine yourself gathering or hunting with the paleolithics, farming with a wooden hoe made with a stone adze, hauling stones for the pyramids, philosophizing with the Greeks, battling with the Romans, praying with medieval monks, sailing with Vasco da Gama, calculating with Kepler, looking through Galileo's telescope, tending a spinning machine in Lancashire, pounding a telegraph key in Wichita, working on the assembly line in Detroit, living in space-station Mir with the cosmonauts, and dreaming about the future by yourself.

### *In a Mind's Eye*

In that magical moment of twilight, with the grey-blue sky pausing before plunging into an abyss of blackness, Swe Ansa stepped out of the shelter of her hillside cavern, walked past the glowing coals of the campfire and settled into the seat of the lookout rock. The children and dogs had settled down for the night, while the women chatted and groomed.

She looked across the tree tops to the plain beyond. Somewhere out there her menfolk would be camping down for the night. She hoped they'd been having some luck with their hunting. Antelope were getting scarce, and they'd not been doing well lately. They've been gone for five and two suns. Swe Ansa wondered if they might find a better site to which to move the family.

Meanwhile, she'd been directing the women in collecting berries, roots, and grains nearby, but their larder was not as full as it had been the last time the great hunter appeared in the southern sky.

Swe Ansa's gaze turned skyward. Shining eyes peeped out from the depths of heaven's cave. To the southeast she saw the form of the mighty hunter spirit. Her eyes moistened as she thought of that other mighty hunter the family had put to rest only five and three suns ago.

He lay there now, in the back of that other cave high up the hillside. Swe Ansa mourned for her soulmate, the father of her children. She put out of her mind the image of that grey, crouched figure, strewn with flowers now wilted. She tried instead to recapture scenes of joy—stories by the campfire, swimming in the streams of their youth, running in the sun with the shouting children.

Soon the cry of a distant leopard brought her up short. Swe Ansa moved to stir up the fire and added another few sticks of wood. The hard reality was, their food supply was low. How could they survive?

She settled again on the lookout rock, and looked toward the southeast. The waning moon, just risen above the horizon, was casting its pale glow across the plain. Somehow she was briefly comforted by the certainty of the moon spirit's regular returns. So much more certain than the luck of the hunters. When would they return?

Swe Ansa plucked some seed pods from a few golden stalks of grain growing in a clump beside her. As her worries crept about through the memory traces in her head she idly rolled the pods between her palms. As the husks fell away, she slipped the small handful of hard kernels into her mouth and chewed them slowly. How would the family survive the impending winter without a sufficient store of food?

Swe Ansa looked again at the moon and wondered.

*Plato (left) and Aristotle (right) detail from the* School of Athens *by Raphael. Vatican Musuem*

# PART I

# *Ancient Times*

CONTENTS

Human technology and science have a long history. Before we get to them, we begin with the history of humankind itself.

When we measure time in the units of geological eras, hundreds of millions of years each, we find that planet earth has supported life forms of some type or other for about half of its five billion year existence. No trace of anything like the human form appeared until very recently. Vertebrate fish swam in the ancient Ordovician seas 350 million years ago. Early amphibians were crawling along Devonian shores 250 million years ago. Early reptiles were stalking through the Permian vegetation 200 million years ago. Then, for almost 150 million years, reptiles covered the earth with the multitude of forms that we see in museums' dinosaur galleries today. Part way through that Mesozoic era, much smaller mammalian forms emerged to scurry about in the underbrush, ever vigilant, keeping just beyond the grasp of hungry *Tyrannosaurus rex* and his cousins.

About 60 million years ago, the dinosaurs died out; whether suddenly, for some cosmic reason, or more gradually, for some other reason, is still a question for research by paleontologists. With the demise of the dominant dinosaurs, the world was now open to the mammals. By about 45 million years ago, the mammals had radiated into the numerous niches we find them still occupying today—antelopes and horses on the plains, cats and other carnivores in the forests, along with ancestors of rabbits, rats, and squirrels, and of monkeys, apes, and us. Mammals also found the friendly seas teeming with the bony descendants of ancient fish as the ancestors of whales, dolphins, and walruses slipped slowly into the water.

That briefly sets the stage for Chapter 1: the earth full of living forms, exquisitely adapted to survive in niches ranging from arctic tundra to tropical jungles, from high mountains to deep seas. Precisely adapted by evolutionary forces (we may say) to follow specialized lifestyles in particular niches.

Into this world, our hominid ancestors made their first tentative steps—puny creatures walking on their hind legs with heads held high. With dexterous hands and growing brains they fashioned tools of bone, stone, and wood. They gradually turned the apparent disadvantage of their lack of specialization into the supreme advantage of learning how to adapt themselves with their tools to a wide variety of niches throughout the world.

About ten thousand years ago, fully human, our *Homo sapiens* ancestors in certain parts of Asia invented agriculture (Chapter 2). Until then, the impact of humans on the world was not large. Our pre-agricultural ancestors roamed where they needed in order to find food to live on. Occasionally they'd have found regions lush enough in vegetation or fish that they could live a semi-settled life. Although some anthropologists attribute the extinction of some mammals to neolithic over-hunting, their total impact on the earth was small. Agriculture changed that!

The process was slow at first—hunters didn't all become farmers in one or two generations. Over the course of five thousand years (250 generations) farming techniques gradually modified a few species of plants and animals to provide our ancestors with more assured supplies of food.

The settled life of the early farmers made possible local increases of population density previously unknown. A neolithic hunting family (an extended family of a few adult couples and their children—25 or 30 in all) covered a range of about 10 km from their campsite. In a fully agricultural economy, the same range could feed as many as 300. Whether in scattered hamlets or homesteads or in more centralized villages, neolithic farmers needed more social organization than neolithic hunter-gatherers. Much evidence points to spiritual or religious bases for the techniques of creating group solidarity. Families became tribes with chiefs and shamans in charge of the symbols of group identification—especially the totem animals they identified themselves with.

With cities and civilization (Chapter 3) larger populations became concentrated in small areas, with growing specialization in handicrafts such as potting, tool making, carpentry, masonry, weaving, and metallurgy; and in the organizational skills of priests, merchants, and soldiers. The invention of writing about 3000 BCE gave voice for the first time to what till then was the mute testimony of surviving artifacts for historical investigators.

Agriculture and craft products and specialization led to trade and warfare and empires. Empires were founded on the technologies of stone tools, weapons of bronze, and the labor of artisans who built walls and houses, chariots and ships. From 3000 BCE to 1000 BCE, our historical record (especially in the Middle East) recounts the rising and waning fortunes of a score of imperial powers, all based on essentially these same techniques.

After about 1000 BCE, the strength of iron was added to the armory of empires and the tool kits of artisans (Chapter 4). By now we also have written records in which Hebrews and Greeks depict their origins, with a growing awareness of themselves and of their place in the scheme of the world. The mythologies of other and earlier peoples portray humans searching for their natural place in a world populated by spirits and creatures much like themselves. In different ways the Hebrews and Greeks depopulated the spirit world and separated themselves from nature.

Nature "out there" became an object for humans to think about and to control, as their separation from nature filtered through their consciousness. After about 600 BCE, the Greeks developed logic and philosophy. This was no more sudden than many other human developments. Controlling nature still had "magical" components as logic and observation were combined with incantations and spells in astrology and alchemy—sometimes called "fossil sciences."

Astronomy developed within astrology as the technique for discovering the rules that governed the heavens. "Discovering" worked best when the Greeks used their mathematical logic to combine speculative hypotheses about the stars in their courses with careful observations. Over a period of 500 years they created astronomy as the first science.

While mathematicians probed the heavens, Alexander of Macedonia gave the Greeks sway over the eastern Mediterranean and southwest Asia all the way to India. The astronomer Ptolemy created his grand synthesis of the heavens in Alexandria at the mouth of the Nile in Egypt. But by then (AD 150) the dominant power in the Mediterranean was Roman.

Spreading out from Italy, while Alexander was roaming eastward, the Romans established their dominance in the western Mediterranean, and then swept over much of Alexander's empire in the couple of hundred years after his death in 321 BCE. Though dominant in the Mediterranean until after AD 400, the Romans never held the regions farther east, which fell to the Parthians.

At its peak, the Roman Empire extended from Britain, France, and Spain in the west over all the rest of Europe south of the Danube, to Turkey and Palestine in the east, and to the whole northern coast of Africa. Their language, literature, laws, customs, and Christianity provided a solid foundation on which European civilization was built.

Those Roman achievements were themselves founded on a technology of iron weapons and tools, and stone buildings, along with outstanding organizational skills. Roman power was strong and vigorous. Its techniques were honed to perfection over seven centuries. Yet, throughout that time, the more contemplative side of Roman life was largely Greek—philosophy and mathematics for example. Even Roman mythology seems to have been mostly a matter of translating Greek names into Latin. And when Rome was Christianized after 300, the Hebrew scriptures and the New Testament came in Greek.

So, when Europeans in the middle ages and Renaissance began to elaborate philosophy, astronomy, and mathematics, they used Greek models, though they sometimes came to them in Arabic script. Another, newer culture had done more to preserve Greek learning than the Romans had (see Part II).

General histories describe the latter days of Rome—degeneration from within and barbarian invasions from without. It's not a story I have to tell. Such gloss of civilization as the Romans had laid upon the far-flung reaches of their empire soon turned to dross.

## *In a Mind's Eye*

He'd slept fitfully. Now, well before dawn, Junius Publius was wide awake. Might as well get up. Perhaps the salt air would brace his spirits. He slipped out of the encampment with a nod to the guards and walked the short distance to the cliff overlooking the port.

The news from Rome was disturbing; there in the harbor he could see the galley already moving on the outflowing tide, taking the commandant's reply back to Rome. Soon, another cohort would be sent north to meet the invading Vandals, but Junius wondered if it wouldn't be better to keep them here to defend Rome's oldest outpost in Gaul.

Being only the legion's architect, he knew he had no say. He had no right even to be consulted. And even while his work force was contracting, he was expected to hurry the fortifications—well, at least to repair the most serious breaches. Narbo had been so far from the center of action for so long that the ravages of time and local peasants had taken their toll. He clenched his teeth at the thought of the stone hut his legionnaires had torn down, not a thousand paces from the wall the peasants had mined for their own benefit.

Junius looked back toward the vineyards and grain fields the peasants tended. With the mood they were in, that cohort might be better deployed to strengthen the tax collectors. Then, with another grimace of disgust he recalled the filth those peasants lived in, even in a stone hut—occupied more by animals than by people—if you dared dignify them as people. Rome might rule the seas and the cities, but superstition and lust ruled the countryside. Why they weren't even Christians, though God knows, the bishop had tried often enough.

Junius shook off the thought and clutched his cloak more tightly. His gaze turned seaward. The galley had cleared the harbor entrance and even now was hoisting her sail. Back to Rome. Mother Rome. Despite her current troubles he longed to be back in the streets of his youth. At least they were clean streets, washed down regularly with water flowing from the aqueducts—aqueducts he'd learned to build and then had spent long years merely repairing. Why was it the schools couldn't prepare us for the life we'd actually live?

Another shudder, another shake of his head. So many memories kept welling into his troubled mind. He looked up at the sky. No help there. The priests really frowned on astrology, though we all know our zodiac.

Above Rome, far to the east, Venus shone a cold light he didn't find comforting. Nearby, Junius made out Regulus in Leo, just rising. Then he scanned across the ecliptic past Cancer, Gemini, and Taurus, with Orion below them, to Pisces and Aquarius. At least he knew Aquarius was there beyond the full moon just setting behind the hills to the west.

Looking back toward the south, he found Mars and Jupiter. As his eyes defocussed, he seemed to feel a soft Roman breeze on his cheek as he recalled his astronomy lessons. Vitruvius's book had been tough going for budding young architects. Sons of patrician families often rebelled at having to learn so many subjects. He'd shared his chums' resentment, quietly, and then topped his class.

Junius recalled Vitruvius's list of the studies an architect needed:

> Let him be educated, skilful with the pencil, instructed in geometry, know much history, have followed the philosophers with attention, understand music, have some

> knowledge of medicine, know the opinions of the jurists, and be acquainted with astronomy and the theory of the heavens.... Men have no right to profess themselves architects hastily, without having climbed from boyhood the steps of these studies and thus, nursed by the knowledge of many arts and sciences, having reached the heights of the holy ground of architecture.

Junius grinned at the thought, and straightened up proudly.

Even though he'd never been sure why architects needed to know more than how to tell directions from the stars, he'd become fascinated in spite of himself. The planets and stars were old friends now. That comforted him a bit.

He looked back to the west as the moon disappeared behind the trees. How big she seemed. He felt like reaching out to touch her. Was she a touchable thing? He knew enough not to call her the goddess Diana, but what then? There was another Diana, more remote in both time and space, and less touchable—a double shudder. Concentrate on the moon!

He recalled Plutarch's essay *On the face of the moon*. The moon could be earthlike—riven with gorges and peaked with crags. At least that was one account Plutarch seemed to credit for Diana's face. (Diana's face!)

Then he laughed out loud, in spite of himself, as he recalled the fantastic voyage to the moon recounted by Lucian. Thoughts of ludicrous lunar legions clothed in vegetables helped him shake off his gloom.

As the first grey streaks lightened the eastern sky, he heard the armorers and blacksmiths beginning their daily clamor. They'd be busy equipping that cohort, and he'd be busy too. Junius turned back toward camp with lightened pace.

## *For Further Reading*

Clagett, Marshall. *Greek Science in Antiquity.* New York: Collier, 1963.

Clark, Grahame. *World Prehistory in New Perspective.* Cambridge: Cambridge University Press, 1977.

de Camp, L. Sprague. *The Ancient Engineers.* Cambridge, MA: The M.I.T. Press, 1963.

Cohen, M. R. and Drabkin, I. E., eds. *A Source Book in Greek Science.* Cambridge: Cambridge University Press, 1958.

Dreyer, J. L. E. *A History of Astronomy from Thales to Kepler.* New York: Dover, 1953.

Farrington, Benjamin. *Greek Science.* Penguin Books.

Gingerich, Owen. *The Eye of Heaven: Ptolemy, Copernicus, Kepler.* New York: American Institute of Physics, 1993.

Gribbin, John and Mary. *Children of the Ice: Climate and Human Origins.* Oxford: Basil Blackwell, 1990.

Hadingham, Evan. *Circles and Standing Stones.* London: Heinmann, 1975.

Hodges, Henry. *Technology in the Ancient World.* New York: Knopf, 1970.

Johanson, Donald & Maitland Edey. Lucy: *The Beginnings of Mankind*. Warner Books, 1982.

Kirk, G.S., and J.E. Raven. *The Presocratic Philosophers: A Critical History with a Selection of Texts*. Cambridge: Cambridge University Press, 1963.

Kitto, H. D. F. *The Greeks*. Baltimore: Penguin, 1951.

Leakey, Richard & Roger Lewin. *Origins: What New Discoveries Reveal about the Emergence of our Species and its Possible Future*. Dutton, 1982.

Lindberg, David C. *The Beginnings of Western Science: The European Scientific Tradition in Philosophical, Religious, and Institutional Context, 600 B.C. to A.D. 1450*. Chicago: University of Chicago Press, 1992.

Lloyd, G. E. R. *Early Greek Science: Thales to Aristotle*. New York: Norton, 1970.

________. *Magic, Reason and Experience: Studies in the Origin and Development of Greek Science*. New York: Cambridge University Press, 1979.

Lucian, of Samosata. *Satirical Sketches / Lucian.* Trans. and with introduction by Paul Turner. Bloomington: Indiana University Press, 1990.

Mendelsohn, Kurt. *The Riddle of the Pyramids.* London: Thames and Hudson, 1974.

Neugebauer, Otto. *The Exact Sciences in Antuquity,* 2nd ed. New York: Dover, 1969.

Renfrew, Colin. *Before Civilization: The Radiocarbon Revolution and Prehistoric Europe.* London: Cape, 1973.

Saggs, H. W. F. *Civilization before Greece and Rome*. New Haven, Yale University Press, 1989

Sambursky, S. *The Physical World of the Greeks*. London: Routledge and Kegan Paul, 1956.

Stahl, William H. *Roman Science: Origin, Developments and Influence to the Later Middle Ages*. Madison, WI: University of Wisconsin Press, 1962.

Starr, Chester G. *A History of the Ancient World,* 4th ed. New York: Oxford University Press, 1991.

Tester, S. Jim. *A History of Western Astrology.* Woodbridge, Suffolk: Boydell Press, 1987.

Wenke, Robert J. *Patterns in Prehistory: Humankind's First Three Million Years*, 3rd. ed. New York: Oxford University Press, 1990.

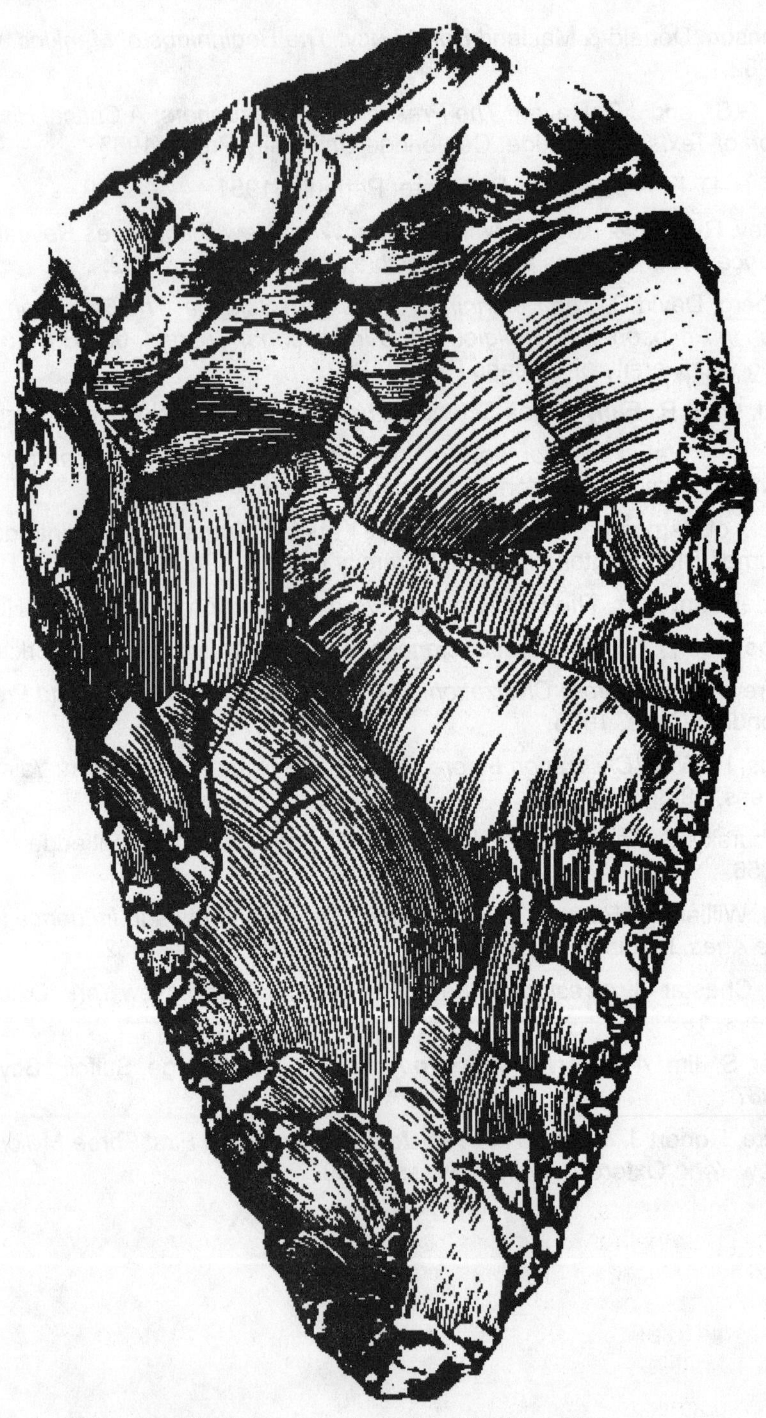

*Stone hand ax used by Homo erectus about half a million years ago (approximately this size).*

## CHAPTER ONE

# The World of Early Humans

Five million years ago, the ancestors of humans were still living in a completely natural world. Yet the evidence from anthropology suggests that a clear genetic line already distinguished our ancestors from those of our nearest cousins, the chimpanzees. By one million years ago, our ancestors were making and using quite refined stone tools. The human revolution was well under way.

And yet, in the world of a million years ago, human life-styles were still bound to nature. Our ancestors of that time foraged for their food. They had stone tools for pounding, chopping, scraping, and cutting, which they used on the seeds and fruits of plants, and on the skins and flesh of scavenged animals.

From one million down to about fifteen thousand years ago, the earth experienced a series of ice ages. Glacial sheets periodically covered large tracts of the northern continents. During the ice ages our ancestors began to tend campfires and developed a more elaborate stone tool kit, which extended beyond the earlier tools to tipping projectiles like spears and arrows. When our ancestors emerged from the last ice age, they were fully human—genetically indistinguishable from ourselves. These humans were still in a balance with nature, but they were now much better equipped to withstand attacks by large carnivorous animals like tigers and leopards.

Let's now consider in more detail the stages in the development of our human ancestors from the beginnings down to the end of the last ice age.

## *The Origin of the Genus* **Homo**

We belong to the group of mammals called **primates**. The primates include all the monkeys and apes, as well as smaller tree-dwelling animals like lemurs and tarsiers. The primates are distinguished from other mammals by having grasping hands (with finger nails instead of claws) and stereoscopic vision. The apes, with whom we share the largest heritage, separated from other members of the primate line about 25 million years ago. The fossil ape that seems to be most directly in our line of descent is *Ramapithecus*, dated at about 12 million years ago.

*Ramapithecus* is the first in the group called **hominids**—meaning human-like, that is, with a few features that distinguish it from its ape cousins. Fossils of this creature have been found in China, Hungary, India, and Kenya, indicating that it must have been well adapted to a wide range of environments. Its main distinguishing feature is a tooth pattern that suggests a diet containing both plant seeds and animal flesh. Its front teeth are suited for slicing and its back teeth for grinding and chewing. *Ramapithecus* does not possess the large canine teeth other animals use for tearing flesh.

From ten to about five million years ago, the fossil record is empty. Then from five to two million years ago, sites in eastern Africa reveal several distinct lineages of hominids that may be descendents of *Ramapithecus*. All display one major anatomical feature that distinguishes them from the apes—the ability to walk regularly upright on their hind legs—**bipedalism**.

What had happened? Because of the large gap in the fossil record, we can only surmise a sequence of developments during that interval.

Anthropologists conjecture a process of responses to environmental change beginning with *Ramapithecus*. At about 15 million years ago, geological evidence reveals a decline in rainfall and a gradual replacement of large forests by grasslands or savannahs. In Africa, up to that time, ecological niches in the forests had been relatively clearly defined. Among the primates, the monkeys and smaller creatures dominated in the upper branches and the trees. The larger apes were closer to the ground, swinging through the lower branches and spending some of their time on the ground.

With climatic changes competition for food must have become intense. The savannahs were the domain of the larger herbivores, like antelopes, and predators of the cat family—not a safe place for tree dwellers. Hominids and apes less able to compete within the forests were pushed to the edges. With a decline in their traditional food supply and the predation of cats, their populations declined drastically.

Some of the groups were able to adapt to the environment at the edges of the forests. They were the ancestors of today's gorillas and chimpanzees. Those great apes dominated the forest edges to the detriment of their cousin hominids. These latter were forced out of the protection of the forest cover.

We have to surmise that the hominids were able to maintain a precarious existence by managing to make short, quick trips across the intervening savannah to the next clump of trees. Often finding that wooded area already occupied, they'd have to move on. Confronted by sharp-toothed, quick-footed cats on the savannah, many of these early hominid waifs perished.

Yet, *some* must have survived! With a tooth system that enabled them to consume a varied diet of plant seeds and roots, and small animals, they fitted into a number of marginal niches. They also must have been able to protect enough infants to ensure the continuation of their species. Their grasping hands would allow them to carry their infants, if they could manage to travel the required distance on their hind legs. We needn't suppose these earliest hominids to have been regularly bipedal, just capable of covering a few hundred meters at a time without having to put hands on the ground. We conjecture that this incipient bipedalism enabled the early hominids to get across forbidding territory to the next clump of trees—*not* that it was an adaptation designed for invading the savannnah (see box).

Since full bipedalism had emerged by about four to five million years ago, we can surmise what must have happened in the intervening time. Hominid anatomy changed in the direction of increasing bipedalism. Now, apes and monkeys have grasping feet as well as hands. As hominid feet became better adapted to walking, they became less well adapted for grasping. That meant infants couldn't hold on to their mother's fur as well, and carrying became even more important. So there is a feedback interaction: better walking requires better carrying and *provides for it* by removing the need to use the hands for walking.

Fully bipedal hominids, presumed to be descendants of *Ramapithecus*, are found by four million years ago. The first ones are now classified in the genus *Australopithecus*, meaning "southern ape." Being now fully adapted to walking upright all the time, these hominids had several advantages over their ancestors. The main one was the ability to carry things. Not only infants, but also food. For seeds, they'd need containers, which we can only guess at, since solid evidence is lacking. But they could easily carry off pieces of meat from a scavenged animal.

Some anthropologists have derived a couple of consequences from that carrying ability. First, scavenged meat could be carried away from where the carcass lay to a safer place—a "campsite." Second, bringing chunks of meat back to camp implies that it would be shared with other members of the group. These features imply the existence of families, not necessarily monogamous, but more clearly differentiated than a troop of baboons.

Walking on their hind legs freed *Australopithecus's* hands for carrying, and for ever-improving dexterity. Since many animals carry things in their mouths, creatures who carry in their hands can use their mouths for other things. Monkeys chatter a lot. Anthropologists suppose a gradual increase in the sophistication of vocal communication among hominids over a few million years.

The dexterity of hominid hands implies an ability to use tools, like sticks for digging or animal leg bones for clubs. However, there is no clear evidence for such tools at the time of early *Australopithecus*. In fact, solid evidence for tool using comes only with evidence for **tool making**!

Anthropologists have found stone tools in east Africa that date to as early as two million years ago. These tools consist of either flakes or cores of rocks that give evidence of having been worked by hominid hands. The hominids struck a rock to be worked with a hammer stone. The fracture of the rock created sharp edges that could be used for cutting or scraping. The pieces or **flakes** that were knocked off could be used immediately. Sometimes, the **core** that remained was worked further to produce a fist-sized **hand ax**.

## *Romer's Rule*

Paleontologist A.S. Romer devoted his career to uncovering the details of the evolution of vertebrate animals. In 1959 he published the fruits of his research in *The Vertebrate Story.* In his discussion of the development of amphibians from lobe-finned lungfish about 400 million years ago, Romer argued that bony, somewhat muscular fore-fins enabled some lungfish to survive when they were stranded in pools cut off from a larger body of water. Their survival depended on managing to cross a narrow neck of land, *in order to get back to the water*—not to invade the land. With enough repetitions of such activities, the amphibian descendants of those lungfish did eventually come to spend a large portion of their lives on the land. Over many generations a variety of genetic changes enabled them to breathe air directly, and to move about on the land; though as you may know, most amphibians even today have to lay their eggs in water.

The point of this explanation of amphibian evolution is that the genetic changes involved were not "intentionally" progressive—not "designed" to invade the land. Rather, we should say that:

> The initial survival value of a favorable innovation is *conservative*, in that it renders possible the maintenance of a traditional way of life in the face of changed circumstances.

This was the phrasing given in 1964 by anthropologists C.F. Hockett and R. Ascher to what they called *Romer's rule*.

Hockett and Ascher then applied this principle to the evolution of bipedalism in our hominid ancestors. By analogy with Romer's story of lungfish surviving by being able to regain the water, they proposed that the early steps to bipedalism enabled the hominids to regain the trees. And just as the amphibians were bound to the water for parts of their lives, the hominids regularly nested in or near trees for several million years after achieving full bipedalism.

Romer's rule is not a new principle in biology; rather it is a sharpening of the principle of natural selection developed by Charles Darwin (see Chapter 16). It is a useful rule to keep in mind, because you may be able to find applications of it for developments of human activities in more recent times. Keep it in mind as you read about the growth of science and technology. See if you can find examples in which a change begins by undertaking to preserve a tradition in the face of changed circumstances, and yet eventually results in a radical transformation of how things are thought or done. (For an example to start on, consider the appearance of the earliest printed books about 1450—made to look as much like manuscripts as the printers could manage.)

Recent experiments have shown that this simple technique can be used to produce an effective kit of stone tools. And further experiments have shown those stone tools capable of being used to skin animals and cut up their meat. Thus, by about two million years ago, hominids had taken a decisive step beyond the capacities of any other animal. They were making tools that could extend their ability to deal with their environment.

Now, these earliest stone tools have been found by anthropologists in association with hominids having larger brain cavities than *Australopithecus*. That has led them to identify the earliest tool makers with our own genus, *Homo*. There is just no evidence of tool making among *Australopithecus*.

Since the fossil record is pretty sparse, even after four million years ago, anthropologists cannot be sure of the lines of descent among these hominids. Most consider the earliest *Australopithecus* to be ancestral both to the later *Australopithecus* and to *Homo habilis*, the first member of our own genus. However, it is now quite certain that some species of *Australopithecus* and *Homo habilis* were both foraging in Kenya at about two million years ago. So, clearly, some of the later *Australopithecus* cannot be in the line of descent to modern humans. By one million years ago, all the *Australopithecus* species had become extinct.

## *The World of Paleolithic Hominids*

A million years ago, then, the pathway to human life was firmly established. Our ancestors of that era were equipped with a stone tool kit that enabled them to survive and prosper in a hostile environment. *Homo habilis* was succeeded by *Homo erectus*. And during the next half million years, *Homo erectus* spread out from Africa into southwest Asia, southern and western Europe, and southern Asia, all the way to the Pacific coast and as far south as Java.

The period from a million to about 25 thousand years ago is labelled **paleolithic**—the old stone age. Even at the start, there is evidence of what must be called **culture**, that is, a collection of regular habits that distinguish one hominid group from another. The evidence is based on the finding of *several* major types of stone tools from about the same time, but from different regions. In addition to the simple flakes and cores described above, the paleolithics soon developed more elaborate techniques for shaping stone tools. They shaped the core by striking it repeatedly with the hammer stone until they had a conveniently shaped hand ax. Differently shaped axes have been found in different geographical areas. (See example on page 16.)

Since these different tool types persisted for hundreds of thousands of years, we presume they represent different cultures—distinctive sets of instructions which parents in different groups passed to their children. It is this idea of transmission, of training, that suggests genuine culture. If a tool type persists for many generations, it must result from techniques by which the young learned traditions from their elders.

The evidence for the life-style of *Homo erectus* is very scanty. That's because fossilization is a process that preserves stone and bone much more readily than softer, more easily destructible materials. So, we have to be prepared to make many inferences from small amounts of evidence. One of the techniques of inference is simply to imagine the whole range of activities that must have accompanied the shapes and markings on the stones and bones that have survived.

Finding food, of course, continued to be a major preoccupation. For hundreds of thousands of years, *Homo erectus* probably lived on plant materials—fruits, seeds, and roots—and on meat that was scavenged from kills by predatory cats.

In addition to stone tools, we can only infer a variety of other tools and implements, made of bone, wood, skins, etc., which have left no trace in the fossil record. Some evidence suggests that early camp sites tended to be near bodies of water—lakes, water holes, and the like. Our ancestors probably retreated to nearby trees when a lion visited their water hole. Later, during the ice ages, they also inhabited convenient caves.

From about one million years ago, *Homo erectus* began to spread out from east Africa into Asia and Europe. With their stone tools and the ability to act together in the face of dangerous predators, the grasslands no longer needed to be feared. With their ability to carry implements and supplies, our ancestors were able to roam the countryside. We can think of small groups of five to ten families living a nomadic existence, traveling along to wherever adequate food could be obtained.

From about half a million years ago there is evidence of *Homo erectus* all across the Eurasian land mass, as well as in Africa. They were now using fire—whether only for warmth and protection or also for cooking is not clear. What is clear, is a gradual increase in the richness of their cultures as they responded to the pressures of the various environments that confronted them.

There is also a gradual increase in cranial capacity. Roughly, we can put ape brains at about 400 mL, *Australopithecus* at 500 mL, *Homo habilis* at 700 mL, *Homo erectus* at about 1000 mL, compared to modern humans (*Homo sapiens*) averaging around 1400 mL.

When we map the functions of the surface of the cerebral cortex of modern humans, we find relatively larger portions devoted to the mouth and hands than to other parts. This implies that our enlarged brains have evolved in the directions that supported both language and increasingly sophisticated hand manipulations.

Some anthropologists consider that hominid vocal signals gradually developed into language slowly over a couple of million years, though true language may date back to no more than a hundred thousand years ago. Our ability to express complex ideas with an immense variety of combinations of about forty simple sounds (in any one language) depends on a complex interaction of structures in our brains, mouths, and throats. They include what may be called neurological "hard wiring," especially in the region of the left hemisphere known as Broca's area; the considerable flexibility of our mouths and lips; and the fine control of sound-making by our glottis and pharynx. All of these are physiological changes appropriate for making clear, distinguishable sounds. Over the course of a hundred thousand generations, such changes would be incorporated into human physiology if they aided in survival; by a feedback mechanism in which better vocalizations produce better communications, making cooperative activities such as hunting and tool making more successful. Such features must have improved our ancestors' abilities to cope, else they'd not have been maintained.

## *The World of Neolithic Humans*

Our own species, *Homo sapiens*, emerged about 100,000 years ago, or a little later. From now on we can talk about **human** culture and human technology. And by 15,000 years ago, we find such a sophisticated kit of stone tools that we can call it **neolithic**—the new stone age. Anthropologists have spent the last hundred years making quite an elaborate scheme for various developments through the paleolithic era. They've identified about a dozen distinct cultures. These cultures gradually increased their range as the glaciers of the last ice age receded about 15,000 years ago.

*Early Neolithic stone tools*

Our neolithic ancestors occupied a variety of environmental niches. Many of them were near bodies of water—rivers, lakes, or the sea. In the later part of the period there is evidence of temporary structures for housing. They're mostly crude huts of various sorts, not involving enough investment of time and materials to be considered permanent. The simplest consisted of bent-over branches, sometimes anchored by stones. Some were covered over with mud to keep out the elements. By 15,000 years ago there are sites at the seashore built more substantially on pilings.

Though the popular media may have overdone the idea of "cave men," some of our ancestors were cave dwellers. There is one cave site in northern Iraq dated at about 50,000 years ago that shows clear signs of a burial ceremony, complete with flowers on the grave. Slim clues like this help us to round out the picture of humans' growing self-awareness—a sense of the mysteries of life and death. Cave paintings from as early as 20,000 years ago also suggest the same thing.

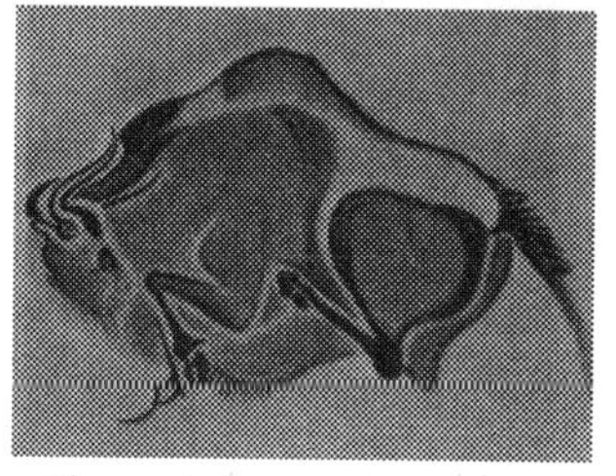

*Cave art.*

*Late Neolithic stone tools.*

The neolithic stone tool kit looks like a much more sophisticated version of the flake tools described earlier. Their structures imply their function clearly enough that we are tempted to call them knives, axes, hammers, picks, etc. These tools resulted from repeated working, such as chipping and polishing. They also represent the growth of knowledge about the kinds of stone material that were suitable for their intended purpose.

Indeed, by 15,000 years ago we have evidence of flint mines with nearby collections of flint pieces profuse enough that we might even identify the sites as tool-making factories.

The general life-style of neolithic humans is called **hunting and gathering**. Perhaps we may assume that a degree of sexual differentiation was occurring in the food gathering tasks. Small groups of men would roam away from camp in search of game, while the women stayed nearby gathering roots and seeds. However, that should not be taken to mean that men never gathered plant materials, nor that women never killed game. For hunting, the neolithics used such weapons as spears, throwing sticks, bows and arrows, and bolos.

We can also infer from today's variety of skin pigmentation that the lighter-skinned humans who lived in colder climates had learned to clothe themselves with the skins of animals. Some anthropologists have suggested that clothing may have originally developed not so much for protection from the elements, as for providing slings and bags for carrying tools and seeds. In the more recent neolithic sites, we find evidence of such bone implements as needles and fish hooks, which were probably shaped using stone knives, chisels, and awls.

Since communities of hunter-gatherers still exist in remote parts of the world today, we can assume that the neolithic life-style represented a completely adequate way of coping with the rigors of many different environments. The culture and technology of neolithic humans had staying power.

But they did not stay unchanged!

## *Conclusion*

The world of early humans should be contrasted with the world of their primate ancestors. The ancestral primates lived fully in the animal kingdom, roaming the forest floor, eating and being eaten. Their human descendants expanded into a variety of ecological niches and gradually established themselves in small family groups on all the continents.

Humans were able to change their life-style because of a series of interacting anatomical changes, extending from *Ramapithecus* at about 10,000,000 years ago, through *Australopithecus* at about 5,000,000, to early *Homo* from about 4,000,000 down to about 100,000 years ago, and ending up with *Homo sapiens* thereafter. The anatomical changes involved tooth and skull systems, and hip and foot systems. The result was that our ancestors changed the type of food they ate, the way they travelled, and how they procured their food.

Bipedalism freed their hands for carrying infants and food, and later for fabricating and using tools. Gradual changes in tool design gave early humans an increasing capacity to use those tools to gain their food.

In brief, improved brains and hands allowed humans to expand their skills with tools and language; to become generalized animals, rather than specialized ones. The evolutionary process works on animals to adapt them to particular niches. Stone tools and language allowed our ancestors to adapt *themselves* to a fantastically wide variety of different niches.

The enlarging range occupied by humans allowed their population to increase. Estimates have been made that before *H. erectus* expanded out of Africa their numbers did not exceed 100,000. When they extended across Eurasia, the population might have approached two million. By the end of the last ice age, with Australia and the Americas occupied, *H. sapiens* numbered perhaps three million. That represents a population density of about one neolithic human per ten square kilometers of habitable land. Such a population density is appropriate to the neolithic technology of hunting and gathering.

By the end of the last ice age what organization can we imagine for our hunter-gatherer neolithic humans? The family grouping already mentioned is presumed to be an "extended" family. That comprises perhaps a senior couple with their adult daughters or sons, their mates, and offspring. Evidence from modern ethnography suggests that the matings of sons and daughters were with members of the opposite sex from a neighboring group. Sometimes, the female would join the family of her mate, sometimes the reverse.

As a result, we can imagine a fairly broad territory occupied by a collection of ten to twenty extended families. Each family contained an average of about twenty-five members. The whole group of "cousins" comprised up to five hundred in number. Eventually, such groups can be thought of as tribes, sharing a common cultural heritage and a common dialect. For most of the year, each family roamed a relatively well defined range on its own. Perhaps once a year the whole tribe came together for festivities and for arranging matings. There is too little evidence for us to know whether the organization was any more complex than that—that is, whether there were chiefs or shamans, or whether any one family was superior to any other.

Such was the normal existence of neolithic humans when the earth emerged from the last ice age about 15,000 years ago. Our ancestors were making the best use they could of the materials in their environment. Some of those materials, stone, wood, hides, were transformed by human ingenuity into tools and implements never before seen in nature. Yet humans were still few in number and still living in balance with nature. Soon, the neolithics would be engaged in new activities by which they would transform the natural world and greatly increase their own numbers.

*A farmer in the savannahs of Africa using basic tools.*

## CHAPTER TWO

# The World of Early Farmers

From about 14,000 years ago, we find the gradual emergence of an entirely new life-style. Humans began to settle down in one place. And they began to control their food supply by the domestication of animals and plants—they tended herds of animals and crops of grain. They became farmers.

Humans added the technology of agriculture, the manipulation of plant and animal growth, to their technology of stone tools. By having more control over their food supply, they opened the way to an entirely different relation between the human species and its natural environment. Indeed, they began to transform their surroundings—and themselves.

### *The Setting*

Neolithic technology, culture, and organization were adequate to support a human population of about three million in environmental settings ranging from the arctic tundra, temperate forests, and tropical jungles, to the Australian outback. Their techniques for gathering plant material and for hunting and fishing were pretty refined. They had time left over from the search for food to make their tools, clothing, and shelter. We may suppose that they also had time for social activities. We should not suppose that they spent every waking moment in a never-ceasing struggle for bare subsistence. Even in harsh environments, recent "stone-age" hunter-gatherers like the Inuit of northern Canada are able to live off accumulated surpluses during certain times of the year, leaving them time for fun and games.

In more temperate regions in Africa, there are hunter-gatherer tribes today who manage well enough by spending only a couple of days a week in food gathering tasks. They may not have TV and rock concerts, but they do have their own kinds of leisure time activities.

So, if the neolithic life-style was that pleasant, why did it ever change? We may presume that it changed in response to changes in the environment. As the glaciers of the last ice age receded, the general climate became warmer and wetter. That resulted in increased plant growth. An increase in the supply of plant food meant that humans did not have to range as far for food as they had previously. In some particular regions, vegetation was luxuriant enough that they could stay in one place for an extended period of time. They could settle down.

Now, if less traveling is needed, a certain natural limit to population is removed. When a family is on the move much of the time, a woman cannot handle more than one infant at a time. This requires that births be spaced. Without techniques of artificial birth control, some other process is needed. Sometimes it was infanticide. Sometimes to the point that if twins were born, one would be killed. Some cultures practised abstinence from intercourse as long as the infant needed to be carried. When the family can settle down, such practices are less necessary.

Then, with more mouths to feed, more food must be found. There is evidence from about 16,000 years ago that the diet of some groups was extended to include a broader range of foods than had been usual up till then. This suggests that populations may have increased beyond the level appropriate to the initial improvement in vegetation.

For, we must realize that there is a negative feedback interaction between population and food supply. If the quality of the diet declines, death rates rise. However, if birth rates rise, an increased population can be sustained only by an enlargement in the food supply. One of the techniques of enlarging the food supply is to add new foods to your diet, which is apparently what happened first. Another technique is to find ways to increase the reliability of the supply of staple foods. The gradual domestication of animals and plants was under way by about 11,000 years ago.

### *The Agricultural Revolution*

As far as is known, the earliest steps toward domestication occurred in southwest Asia along an arc through Israel, western Syria, southern Turkey, northern Iraq, and western Iran. The process took at least a couple of thousand years. Somewhat later, probably independently, agriculture was also invented in the region of the Indus River in Pakistan and the Huang Ho River in central China. Much later, and again independently, maize agriculture was invented in Mexico. We shall concentrate on southwest Asia.

Let us imagine the following pre-agricultural scene along the lower slopes of the highlands to the east of the upper Tigris River. Consider the activities of a single extended family. Meadows of wild grains attract not only our human family but also grazing animals. The humans collect the grains and hunt the animals. The food supply is rich enough that our family makes a semi-permanent home in a nearby cave. They make their camp site on flat ground in front of the cave entrance. They gather sufficient grain and kill enough animals to provide a substantial diet.

To follow the steps to domestication we have to realize that there are small genetic variations within single species of animals and plants. Let's imagine an animal ancestral to modern sheep, and a plant ancestral to modern wheat. Initially, the sheep and wheat are adapted to a wild existence. Both are relatively scrawny, but vigorous. They have been propagating their own kind through many generations.

Occasionally, when an adult sheep has been killed by the humans for food, a lamb remains that's taken back to the camp for a pet. The lamb is fed and becomes somewhat tame. Others over the generations receive the same treatment; with the right genetic variations they may lose some of their wild instincts as the humans give them access to regular grazing fields. As small numbers of sheep grow into flocks, they provide the human family with an assured supply of mutton. Another genetic change shows up in a few sheep where the long, stringy outer hairs decline, and the soft woolly under hairs become more prominent.

Now, instead of simply using the skins for clothing, some inventive human uses a sharp stone knife to cut off the wool. Perhaps to make a softer bed for her infant. If the sheep that show this variation are segregated, succeeding generations can develop thicker, woollier coats. Eventually, the coat becomes so woolly that unshorn adult sheep cannot copulate. There's too much wool in the way.

Only with the intervention of human shearing is it possible to ensure the production of the next generation of the sheep. By this process, humans have grown dependent on the sheep for food and clothing, and the sheep have grown dependent on the humans for food and sex.

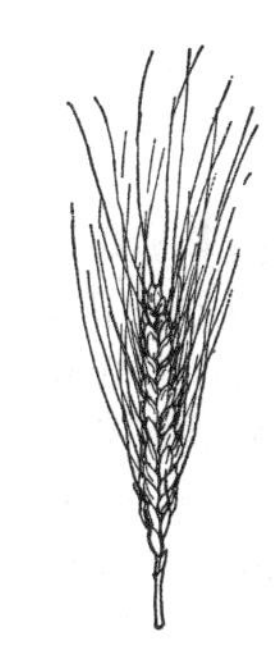

*Emmer.*

A similar story for cereal grains may be better founded. At least the steps in genetic modification are better known. Bread wheat is a hybrid grain having 42 chromosomes in its cell nuclei. It derived from earlier grains by a process of chromosomal addition—a process rather common in plants, though not in animals. An earlier grain called emmer had only 14 chromosomes. When it crossed with a related grain, the result was a hybrid having 28 chromosomes. Bread wheat resulted from a second crossing of that grain with another of 14 chromosomes.

A major difference between wild emmer and domesticated wheat is the way the seeds are attached to the stem of the plant. To propagate in the wild, the emmer seeds are loosely attached to the stem and have tufts to help them travel with the wind. Wheat seeds are much more tightly attached to the stem and lack most of the tufts. In the wild, wheat seeds would lose out to emmer seeds in the competition for fertile ground to germinate in. Wheat seeds just don't travel naturally.

*Bread wheat.*

These facts can provide the basis for speculation on how wheat arose. Imagine our family beside a meadow of emmer and other wild grasses. When the seeds ripen in the autumn, the humans go into the field and cut the stalks. Then they carry their harvest back to camp. Along the way, with shaking and jiggling, many of the seeds would drop off and be lost. Our family would have to be content with the seeds that remained by the time they got back to camp.

Now, we have to imagine that over the generations, members of our family would notice that some of the seeds that were scattered near the camp site, grew up there the following spring. Eventually, some one realized that the seeds could be planted deliberately. In this process, hybrid varieties with the seeds more tightly attached to the stem would have a greater chance of being the ones that got back to camp. And these were more prominent among the seeds that were planted than the more loosely attached seeds.

Through human intervention, natural accidents, and genetic hybridization, the bread wheat plant eventually emerged. Over many generations, our family developed a source of food they could depend on, which also depended on them. With the wheat seeds more tightly bound to the stem, more of them survived the trip back to camp, and our family obtained a higher yield. But being more tightly bound, the wheat seeds would be less efficient in self-propagation than their emmer ancestors. Wheat cannot survive in the wild.

Over a period of a few thousand years, our neolithic ancestors became gradually more dependent on domesticated varieties of animals and plants, which themselves were dependent on human action for their continued life. You might almost say that sheep, goats, cattle, wheat, barley, rice, and maize domesticated humans while the humans domesticated them.

We may note that the word *domesticate* comes from the Greek word *domos*, meaning "home." To be domestic means to be a homebody. And the coming of agriculture allowed humans to establish regular homes. In fact, the Greek word for "home" derives from an earlier Indo-European root word meaning "to build."

Such are the fruits of the agricultural revolution: a more stable food supply and a more settled existence. Evidence for early domesticated varieties comes from as far back as about 11,000 years ago. But you should not suppose that our ancestors' life-style was transformed overnight. It took about three thousand years, even in favorable locations, for the neolithic diet to be gradually changed from full dependence on wild plants and animals to a substantial dependence on domesticated varieties. And yet the transformation was virtually complete at many sites in southwest Asia by about 8,000 years ago. Notice that while paleolithic transformations in life-style fit time divisions of hundreds of thousands of years, the appropriate scale unit now is reduced to a thousand years. The pace of human change is quickening.

### *Early Farm Settlements*

The tending of herds and crops allowed our neolithic ancestors to settle down. And staying in one place gave them time for social and technological activities they couldn't indulge in while always on the move.

Among other things, they could build more substantial homes for themselves. There is a site in northern Iraq where a village stood about 9,000 years ago. It consisted of about 25 mud huts huddled together in a small space. The huts were built up in courses (layers) of mud, each being allowed to dry in the sun before the next one was laid down. Evidently, we have here a social group larger than a single extended family. Taking account of the overlap of generations, we can imagine 25 adult couples ranging from grandparents down to young adults with their first infant. Including children, the total population might have been about a hundred. A small "tribe" of cousins.

Within the huts there were crude ovens for baking and silos for storage. The tools of these neolithics included a variety of stone flakes for cutting, stabbing, etc. In addition, they had stone implements like sickles for cutting the grain stalks, and pestles and mortars for grinding the seeds. There is also evidence of plaited straw mats and numerous clay figurines of animals and women.

This is still a pre-pottery period, so archaeologists infer from the plaited mats that these neolithics also had plaited baskets for carrying. Also, since some stone tools are hafted, they infer wooden handles on stone blades. These are inferences because reed and wooden artifacts don't survive as long as bones. Occasionally, bone handles have been preserved.

Life-styles can only be inferred from the artifacts that remain. From the clay figurines and various decorations carved on bone, we know the neolithic farm villagers had spare time. These art forms demonstrate some sort of symbolism, and they are often assumed to be part of various ritual and religious activities. Certainly, it appears that they valued fertility and may well have used their art in symbolic ways to try to ensure it. Evidence of ceremonial burial of the dead also adds to our notion of their increasing religious sensibilities. However, this is not the place to try to delve into the psyche of pre-historic humanity.

## *Conclusion*

The world of early farmers represents a revolutionary change from the nomadic existence led by their ancestors during the ice ages. With their settled life-style they were able to congregate in larger permanent groups. Being settled, and with a more stable food supply, they could use their leisure time for technological advances. Not having to roam ceaselessly, they began to accumulate possessions.

Neolithic farmers had a place they could call home. They had a variety of stone and bone tools for converting their crops and herds into food, clothing, and implements. They had time for social, religious, and (we may presume) economic activities. Time to develop group solidarity, and time to think about their circumstances. Eventually, they had the time and facilities to specialize their production techniques and to begin to worry about protecting their surpluses from outsiders. We'll consider the gradual differentiation of members of the group into specialized roles—like priests and chiefs—in the next chapter.

Just after the last ice age (say 16,000 years ago), the total human population on earth was in the range of three million. Over the next seven thousand years, with the general climate getting somewhat warmer and wetter, certain groups became settled, and a few of them made the transition to agriculture. The size of settled groups increased, so that by 9,000 years ago, the total world population reached about four million.

While the hunter-gatherer life-style was maintained in many parts of the world for a long time, agriculture and its associated technologies gradually spread out from their original centers. In the next four thousand years, the total world population more than tripled. We're now close enough to historical times to start using the traditional dating system: years BCE ("Before the Common Era" – a term that is now preferred to the traditional "BC," "Before Christ." Likewise when it is necessary to distinguish later from earlier times, the abbreviation "CE" for the "Common Era" will be preferred to "AD"—"Anno Dominum"). So, at 7000 BC, the human population was about four million, and at 3000 BC, about 14 million and growing.

Villages became towns, and towns became cities. Civilization! Next chapter.

*An Assyrian bird-headed God.*

*A tomb painting showing Egyptian workers at their crafts.*

## CHAPTER THREE

# Early Civilizations

The word "civilization" is derived from a Greek word, *keimai,* meaning "to lie down." The Greeks derive other words from that root word, which include their words for putting to bed, a deep sleep (a coma), village, and cemetery. The Latins derived their word *civis* meaning "citizen" from the same Greek root, and then other words with the meanings of civil, civic, city, and civilization.

In present day English usage, civilization implies cities, with populations large enough to have the variety of shops, offices, and churches—providers of goods and services— that we associate with cities. In this sense, "uncivilized" should not be used as a derogatory term against cultures that do not have cities. To be less civilized, less "citified," does not make a person less human. On the other hand, "civilized" does imply a separation from nature, and it implies that cultures without cities, which are closer to nature, have not developed the social techniques required to live in daily contact with many strangers.

The world of civilization has brought our human species a long way from living in balance with nature. Even today, farmers are far more dependent on the whims of nature than city dwellers are. At 7000 BCE, all humans were either farmers or hunter-gatherers.

The earliest known city-based culture arose in southern Iraq, in Sumer. It was flourishing by 3000 BCE. By 2000 BCE, there were several civilizations stretching from the eastern end of the Mediterranean across Asia to China. The world's human population doubled in those thousand years.

So we wonder how our ancestors got from small farming villages in 7000 BCE to large cities after 3000 BCE. It cannot have been just a matter of simple enlargement, for the mark of a city is its complex organization. In the early farming villages we have to imagine that tasks were not clearly differentiated. Almost everybody could do everything. In cities, on the other hand, there are many specialized occupations—artisans, merchants, soldiers, clerks, priests, administrators.

In addition, city dwellers are dependent for their food and other raw materials on the farmers of the outlying land. A city rules over a territory. A city of a hundred thousand may have required the labors of as many as a million peasants in the countryside to keep it going.

In southern Iraq, this transformation from scattered villages to the Sumerian civilization spanned the centuries from about 7000 BCE to 3000 BCE. We shall concentrate on developments there, but occasionally draw inferences from evidence elsewhere.

## *Villages, Bricks, and Pots*

The central feature of Iraq is the plain of the two rivers, the Tigris and Euphrates. But the earliest farming communities were located beyond that plain to the north and east. As early as 7000 BCE, they consisted of small villages of perhaps 25 families, with homes made of packed, sun-dried mud.

Mud (or clay) provided the first building material that is shaped by being built up, rather than having parts cut or knocked off as with wood, bone, and stone implements. Very early in these villages there is evidence of clay figurines, mostly in the form of women or animals. Eventually, we must imagine, someone discovered that a clay figurine became much harder after being subjected to fire.

People may also have attributed the hardening of mud walls to the heat of the sun. Soon after 7000 BCE, some homes were built with rough, sun-dried mud bricks. So, we have evidence of a growing experience with the properties of clay.

Clay pots are clearly in evidence by 6000 BCE. They require more elaborate techniques than figurines and bricks, because they must be **fired** to be usable. At a temperature of about 850° C, the clay materials are fused. A chemical reaction produces a much more substantial product than can be achieved by air drying alone. Although some pots can be produced with an open fire, bonfires do not usually produce high enough temperatures. Thus, before 5000 BCE there is evidence of dwellings containing ovens for firing pottery, that is, **kilns**.

We may suppose that the earliest kilns were derived from ovens used for cooking food. However, before long, quite elaborate kilns were built. After all, 850° C is a lot hotter than is needed for cooking. Gradually, our neolithic villagers learned how to control the drafts and the atmospheres in their kilns in order to produce the temperatures needed for fine pottery.

Pots loom large in archeological field work because they have been so well preserved, and because the variations in pots from place to place allow various cultures to be identified. For after 5000 BCE pots became elaborately decorated, as well as finely shaped. Some people carved designs into the damp clay with stone tools. Others applied pastes of various materials onto the surface of a pot before firing, or painted it afterwards. Later, some applied glazes to the surface of a fired pot and then re-fired it. Before 4000 BCE, some kilns were reaching temperatures as high as 1200° C.

For a long time, pots and other forms were built up from coils of clay and shaped by hand. After 4000 BCE, the potter's wheel gradually came into use, which allowed much finer control of shapes.

Pottery provided containers for water and grain, which made life much easier. It also provided a new way of occupying time. The artistic tendency in humans seems to be almost as deep-seated as our need for food. Pottery provided many opportunities for its expression. Since some people were likely better than others at the craft, a degree of specialization must have begun. And, villages that were situated near ample supplies of high grade materials must eventually have been able to trade them for other supplies from neighboring villages.

*A Potter's wheel.*

## *Trade and Metallurgy*

Indeed, trade and barter are a very important feature of early human developments. Whenever one group (however small) has better access to some resource than another, they have an opportunity for trade. This could be, for example, bartering shells and fish from a seaside village for flint or pots from an inland village.

Since deposits of metallic ore are much less evenly distributed on earth than farm land is, metallurgy provided further opportunities for trade.

Metallurgy! Imagine how surprised the first neolithic farmer must have been when he hit a pebble with his stone hammer and instead of shattering, the pebble just bent. A nugget of native copper!

Copper does sometimes occur pure, in small nuggets. By banging away at it, our neolithic hero could form it into a small figurine. Such small copper pieces, including beads, are known from before 6000 BCE. The small amounts of native copper available gave people a chance to learn some of the properties of copper: that it was ductile; that it became brittle after repeated working; that it could be softened again in an open fire and reworked, perhaps to give a sharp edge.

Such small amounts of native copper would not have had much practical use. And the fact that copper melts at about 1100° C would not have been stumbled upon easily.

It is likely that the next step in metallurgy, the smelting of copper ores, occurred first in a pottery kiln. The ores of earliest interest were oxides and carbonates of copper. Some of them attracted attention because of their striking green or blue color. Ground into powder and mixed with water to form a paste, they were used as pigments for decorating bodies and pots.

Before 4000 BCE, potters had discovered that they could alter the finish on their pots by controlling the air flow in the kiln. Once they had their fire burning hotly, they could limit the draft of fresh air to produce a **reducing atmosphere**—one that is rich in carbon monoxide. In a reducing atmosphere, their pots were fired just as well, but not oxidized. The result was a dark grey or black pot, rather than the usual reddish color. And this reducing atmosphere was just what they needed to "reduce" copper ores, that is, to smelt the copper. The oxides or carbonates combined with the carbon monoxide and left the copper behind.

We must suppose this to have been an accidental discovery, say on a pot that had been decorated with a cupric pigment. The coloring of the pot turned out coppery-colored, instead of green or blue. Once some potters realized what was happening, they could start to produce copper metal in quantity. When they mixed powdered copper ore with powdered charcoal in a clay crucible, they could get relatively pure copper at the bottom of the crucible at a temperature of about 850° C. Some potters became metallurgists and were soon producing copper in quantity. They could work the metal by hammering it. But to get molten copper, they had to achieve temperatures of 1100° C.

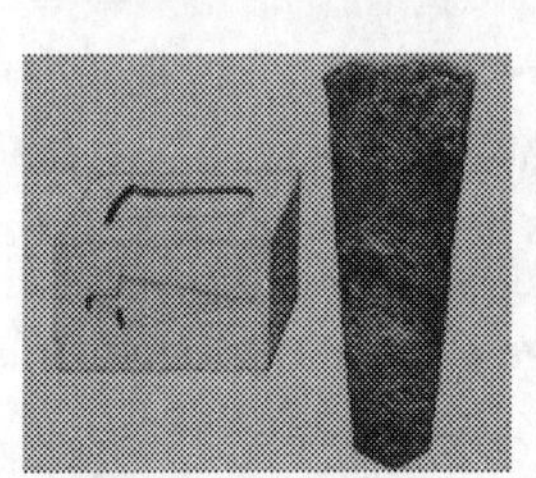

*Simple mold for making copper ax heads.*

By 4000 BCEE, metallurgists in southwest Asia and southeastern Europe had gotten their fires hot enough to be able to cast simple tools and weapons in copper. The simplest mold could be produced by depressing a stone ax in fine, damp sand and then pouring the molten copper into the depression. Over the next thousand years, they refined their techniques considerably and produced a variety of sophisticated molding techniques.

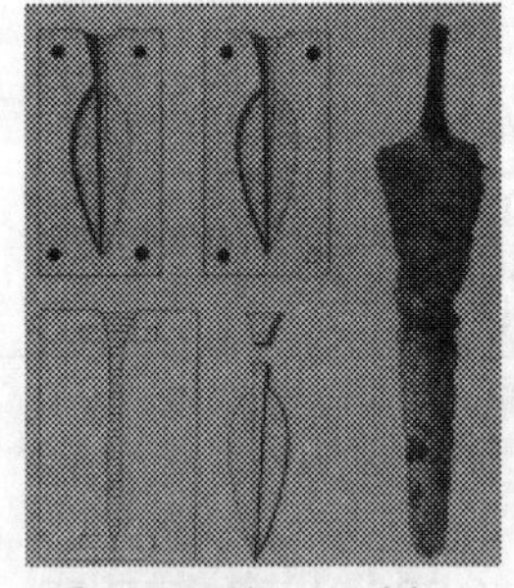

*Composite mold for casting.*

Since copper is a relatively soft metal, it does not hold an edge well. Copper makes a good enough hammer, but not a very good ax. Some hardening could be achieved by alternately hammering and heating at about 300° C. However, for many applications, stone tools were still superior to copper ones. And the technology of stone tools continued to improve, with techniques of fine polishing being added to the earlier flaking and chipping.

By about 3000 BCE, metallurgists had learned to mix copper with arsenic, lead, or tin to produce **bronze**. By adding 5 to 15% of one of those minerals to the copper, they could make much tougher tools. They were harder and held a sharp edge much better than copper. However, bronze was always expensive, and its use was largely limited to weapons of war and religious vessels. Cost is no object to soldiers and priests.

Ordinary artisans still used stone tools in most of their work. But the environment they worked in was changing. Some villages were getting big enough to be called towns.

Since resources are distributed unevenly over the earth's surface, some neolithic villages were better endowed with resources than others. As farming communities grew and spread, trade in specialized goods gradually increased. And villages that did well in that trade could expand.

## *Towns, Cemeteries, and Priests*

Through the period from 6000 BCE to 4000 BCE in southwest Asia, neolithic farming villages grew up along the **fertile crescent**: starting along the eastern coast of the Mediterranean (Israel, Lebanon, Syria), arcing north and east through southern Turkey and northern Iraq, then down along the eastern margin of Iran. Trade expanded in natural goods, as well as in pots and, later, metals. Sometimes a village, well situated on a trade route, could profit from that trade. For example, between sheep herders in the highlands, and fishermen on the coast.

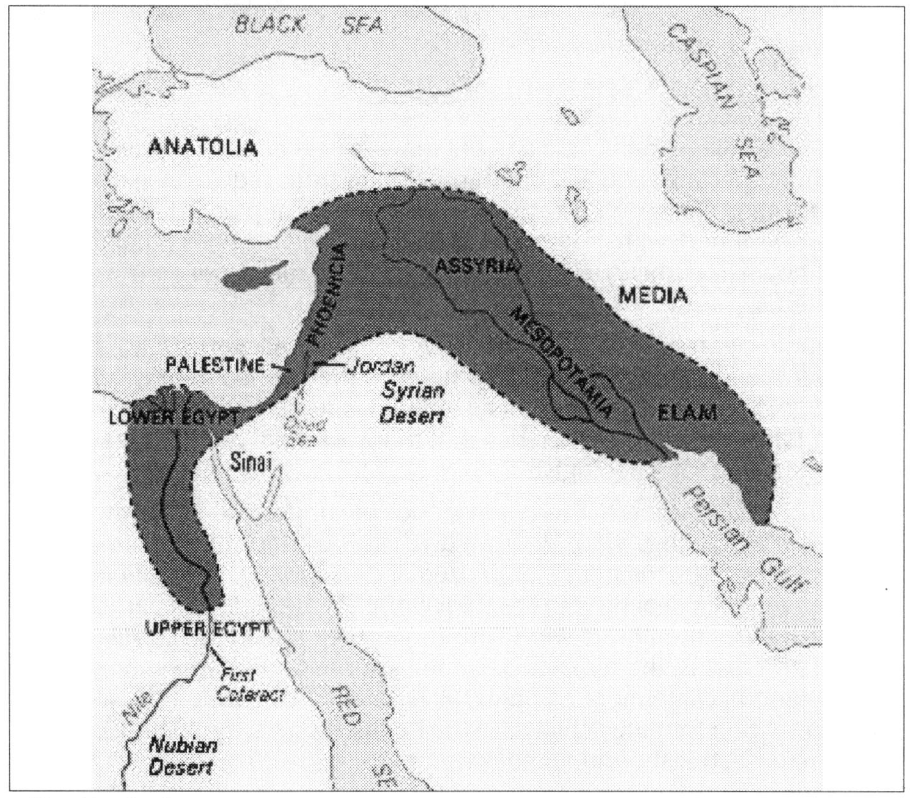


*The Fertile Crescent.*

*Sumerian god and worshippers.*

But for a village to become a town, more is needed than just wealth and population. Specialization and differentiation of crafts require an elaboration of organization. The evidence suggests that the focal point of organization was the priesthood, with its temples, cemeteries, and religious rituals. Yet too little is known from this pre-literate period for us to specify religious practices in any detail.

What is clear, however, is that these neolithic villagers took particular care over the burial of their dead. Sometimes they buried their dead within the village, in or between the huts; sometimes at a special site. They took enough pains with their cemeteries that many of them remain today, mute mysteries for our archaeologists.

In some early villages, there is evidence of shrines within the dwellings. Later, whole buildings were dedicated as shrines and mausoleums. From these early temples, the priests organized and controlled the villagers—using their superiority in dealing with the spirit world. As the wealth of particular villages increased, the priests were sure to get their share. So, as villages expanded into towns, the temples became larger and more magnificent. And organization became more complex, as whole families were now spending their time in non-farming activities in the towns—potters, metallurgists, builders, merchants, priests, and administrators.

Soon, the temples were the most prominent and the most permanent structures in the towns. Even in the Egypt of the pyramid age, while gigantic tombs were made of stone, the kings and priests lived in mud brick palaces.

## *Irrigation and the First Cities*

Along the fertile crescent, annual rainfall exceeded 30 cm. Within the crescent—in the Syrian desert and on the flood plain of the Tigris and Euphrates in Iraq—it was less. Instead of depending on rainfall, farmers might get water for their crops from the rivers. This was a somewhat precarious prospect since the rivers could vary over the year from raging floods to gentle meanders. But by giving nature a helping hand with a few irrigation canals, the farmers might gain greater security.

*Shadoofs. Irrigation devices to lift water from a river toan irrigation trench. A carving from Nineveh.*

The earliest known settlement on the alluvial plain is called Al'Ubaid. A few kilometers south of the Euphrates River in southern Iraq, Al'Ubaid was founded about 5000 BCE. Over the next 1500 years, Al'Ubaid became larger and more prosperous. We can only infer that the governing priests gradually improved their ability to mobilize the population to construct the dikes and canals needed to control the river waters.

By 3500 BCE, there were several sites in this ancient land of Sumer that were growing into cities with all the elaborate organization and redistribution of goods that implies. The priestly administrators had to arrange for the work to be done and for the non-farmers—the scribes, merchants, artisans, soldiers, and priests—to be fed, clothed, and housed.

Eventually, the most prominent of these cities was Uruk, 60 km from Al'Ubaid, on the opposite side of the river. About 3300 BCE, the outstanding feature of Uruk was the White Temple. The temple was almost square, about 20 m on a side, and built on a mud brick platform, about 70 m square and 13 m high. This was the administrative center of the sacred city of Uruk from which the priests controlled the religious, social, and economic life of the city and its surrounding farm villages.

*The Zigurat, or temple, at Ur, one of the major cities of Mesopotamia.*

The first civilization was under way in Sumer, later to be centered farther up the river at Babylon. Crucial to this creation of a new world was the human response to the challenge of the environment. The flood plain of the Euphrates could not have been occupied by our neolithic ancestors without the technological capacities to farm the land and control the river, the wealth brought by trade, and the organizational skills of the priests.

Those organization skills were greatly assisted by the development of writing, shortly before 3000 BCE. Indeed, it is hard to imagine the world of civilization without some way to record and communicate information. Thus we should not be surprised that the earliest writing found (in Sumer) are from temple account "books." Writing began as pictographic symbols on tablets of clay or limestone. In Egypt, hieroglyphics on stone were soon supplemented by writing with ink on papyrus.

*A wheeled cart, pulled by Onagers. A limestone relief from about 3000.*

Also, from about 3000 BCE, we have evidence of wheels being used—both for constructing pots by turning a lump of wet clay against hands or tools, and for transportation. Evidence for boats constructed from bundles of straw with turned-up ends dates back to the Al'Ubaid period—perhaps as early as 4000 BCE.

So, with the exception of good metal tools (which came only with iron after 1500 BCE), the early Sumerian cities possessed all the major features that we normally associate with a civilization—sufficient technology, transportation, trade, organization, and communications.

## *Egypt and the Pyramids*

In Egypt the Nile River provides an even more suitable alluvial plain for farming than in Sumer. Yet Egyptian cities seem to have arisen somewhat later than those in Sumer. We might even suppose that this resulted from the Nile being less of a challenge than the Euphrates—so, great organizational skills developed more slowly.

When Egyptian cities did arise, after 3000 BCE, they made rapid strides. The pyramids—those gigantic symbols of Egyptian civilization—suggest another role for technology, that of using a great building project to weld a nation together. For, in a period of 150 years, starting about 2700 BCE, the Egyptians transported 20 million tonnes of limestone across the Nile to build six great piles of stone.

The Nile River makes Egypt. The river floods its banks calmly and regularly every summer. The deposited silt makes a fertile strip along the valley of the Nile about ten kilometers wide and a thousand kilometers long. Just before the Nile reaches the Mediterranean it splits into several branches to form

*Boat construction in ancient Egypt.*

a rich delta area. Before 3000 BCE these two separate regions–the Valley along the Nile and the northern Delta–formed separate kingdoms.

The Delta region was divided into twenty provinces, each with about twenty thousand people. We may imagine a chiefdom for each province, with a king or superchief at the pinnacle of control. The king of the red crown was the king of the entire Delta region. Along with his priests in their temples, he resided in Buto, his capital city. There are very few archeological remains of this Delta civilization. We do know that the ordinary Delta people buried their dead among their mud and reed huts. We may assume that the king and high officials were buried according to their status.

The southern Valley region was a distinctively separate culture. It had roughly the same population and organization as the Delta, and the Valley king with his white crown lived in his capital city of Nekhen. In contrast to the Delta people, the Valley people buried their dead in a cemetery beside their villages. They lined a grave trench with linen and skins, and laid the dead in the trench in the fetal position, facing south with the tops of their heads pointing westward. These dead were well prepared for their trip into the afterlife–following the sun as it died in the west.

*Pharaoh Menkaure of the Fourth Dynasty and his queen.*

About 3000 BCE, the Valley people conquered the Delta people and put the new capital of the combined kingdom at Thinis in the Valley region. An indication of the uneasy unification of the two distinct cultures was the fact that from then on, the Egyptians always called their ruler or pharaoh, the Lord of the Two Lands; he wore a double crown, half red and half white. A couple of centuries later, the ruling pharaoh moved his capital to the edge of the delta, near Cairo. We may suppose he did it to have better, closer control over the people of the Delta.

Shortly after the new city was built, around 2800 BCE, the pharaoh's workmen started to build the first stone pyramid—a grave site more in the style of the Valley than of the Delta. They built it near the delta—perhaps another way to establish Valley dominance over the Delta. It took them ten to twenty years to build this first pyramid. It was situated away from the city to the south on the plateau overlooking the valley. Like all the later pyramids it was on the western or sunset side of the Nile.

The pyramid as a tomb combines two features of Egyptian religion. First, it shows great concern for the pharaoh's afterlife; and second, the pyramid shape, representing the rays of the sun, reflects the importance of the sun in their worship.

*The Step Pyramid at Sacarrah(sp).*

First, workers excavated a vertical tunnel, a shaft 30 m deep cut into existing solid rock. At the bottom of the shaft they cut out a tomb chamber. Then, covering the entrance to the shaft, they constructed a large stepped pyramid. In its final form the pyramid had six stone layers, each layer smaller than the one below it, reaching a total height of 60 m—taller than a modern twenty-story building. The bottom layer was almost square—about 130 m on a side—bigger than a football field. The total structure was composed of about a quarter million limestone blocks, each a cubic meter, weighing up to three tonnes. Around the pyramid they built a replica of the pharaoh's palace in the city. The palace in the city was made of mud brick, but the replica palace beside the pyramid was made of stone. Mud brick would do for this transitory life, but only stone would last for eternity.

A modest work force of perhaps ten thousand—half masons and half laborers—could have built this structure over a period of ten to twenty years. They could devote full time to the project and be fed from even a small agricultural surplus.

Over the next century, succeeding pharaohs continued to build pyramids as tombs. But none of these were as magnificent as that original stepped pyramid.

*The great pyramids of Giza, tombs of Khufu, Khafre, and Menkaure.*

Then, around 2700 BCE, at the start of a new dynasty, the Egyptians began a mammoth public works building project. They erected six more stone pyramids in less than a century and a half. All six were larger and better built than the original stepped pyramid—one was ten times larger. In all, these six pyramids used thirty times as many stone blocks as the stepped pyramid—more than 20 million tonnes of limestone.

Exactly how were the pyramids built? The Egyptians used limestone from nearby quarries for the core of a pyramid. For the outer layer, they used fine white limestone from Tura, a quarry on the eastern bank of the Nile. They cut the limestone blocks out of the rock using only tools of harder stone—hammers, chisels, and picks. They probably also used wooden wedges to crack the rock beneath the block being quarried. Laborers had to transport the enormous blocks to the pyramid site on barges across the river and then by earthen ramps up the slope and up the pyramid. Dragging and sliding. No pulleys or cranes, just human labor.

Calculations show that the task might have taken more than a hundred thousand laborers, working three months each year. There's general agreement that they were peasants recruited during the off season when the Nile flooded its banks. In addition, full-time specialists—stone masons—were needed. Perhaps there were ten thousand of them quarrying the blocks, and an equal number at the site organizing the blocks and chiselling them smooth so they'd fit together without mortar.

Why were these particular pyramids built? Were they just individual tombs for individual pharaohs? Here's where there's some disagreement. Traditional Egyptologists suppose that a new pyramid was built for each new pharaoh. But if a pharaoh reigned 40 years, and his pyramid could be completed in 20, the huge work crews would have to be sent home. And then have to be reorganized at the start of the next reign. The mammoth logistic problem of each new start up has led one analyst, Kurt Mendelssohn, to suggest that pyramid building may have been continuous. As construction neared the top of a pyramid, fewer laborers were needed. But instead of being sent home until time for the next pyramid, they were just ordered to a new site to start another pyramid right away. Independent of the reigns of the pharaohs. History's first make-work project!

This argument fits the numbers better too—the number of workers needed to drag nine million stone blocks into place in a hundred and fifty years. And it makes a plausible case for why this particular sequence of pyramids was built. Mendelssohn has proposed that because of the antagonism between the tribes of the Valley and Delta kingdoms, pyramid building was intended to weld the two kingdoms of Egypt into a unitary state. The goal was to engage the peasants in an all-consuming project that kept them busy and out of trouble every summer while the Nile flooded their farmland, and to impose a single, national allegiance over the many tribal loyalties.

The pharaoh's chief engineer/high priest must have had great organizational skills to mobilize such a gigantic work force of a hundred thousand men, and their families—perhaps half the total population of Egypt. One early move was probably to increase taxes in the form of grain that was taken from the villagers. So much might have been taken that the villagers wouldn't have enough left for themselves for the rest of the season. The priests stored the grain at the pyramid site. So if the villagers wanted to eat, they had to go to the pyramid. Also, there must have been a strong religious motivation. Providing for the pharaoh's afterlife would assist their own.

Six or eight generations of regular service by the peasants building pyramids must have erased their memories of how they used to spend their time during the summer flooding—fighting among the tribes, stealing each others' women and cattle. Egyptian tribal peasants had been turned into citizens of the pyramid nation in spite of themselves—and the nation was stronger and more centralized as a result.

### *Arches and Vaults*

A continuing problem in building construction is the provision of roofing. In order to cover any significant span the builder must use a material that is strong in tension. For most of history, timber beams provided the simplest solution. But in Egypt and Babylon, timber was often not available locally. It had to be transported, sometimes over considerable distances.

An alternative to timber beams was the construction of a vault—an arched roof. In ancient Egypt, arches were built of bundles of tall reeds, bent over at the top. The bent tops of the reeds arched over the space below. Sometimes, the whole reed structure was daubed with mud, which then dried in the sun. Some people think that this original structure explains the appearance of later stone columns with their fluted surfaces and flowered capitals, which seem to be stylized replicas of the reed originals.

Not long after 3000 BCE there is evidence of arches built of brick or stone in the form called **corbeling**. Each successive layer of brick was made to project a short distance beyond the layer below. After a number of such layers were built from opposite walls, the uppermost layers met at the top. Such corbeled arches are necessarily considerably higher than the width they can span.

A semi-circular arch made of radially placed bricks can span a distance double its own height. Known from 2500 BCE or earlier, these true **arches** had to be built over a support (called the *centering*) which was removed after the mud mortar had set. Archaeologists suspect that the Egyptians filled the space to be covered with rubble, built their vault, and then removed the rubble. Over a couple of thousand years, builders in Egygpt and southwest Asia improved their methods of building arched vaults with mud bricks. In a technique that avoided the use of centering, they leaned the bricks of the vault against an end wall, laying them up course by course.

## *Conclusion*

By 2500 BCE, the great ancient civilizations of Sumer and Egypt were in place. Other civilizations were also growing up about the same time on the banks of the Indus and Huang-Ho Rivers in eastern Asia. The civilizations of southwest Asia were exported, with modifications, into Europe, Crete, and Greece, and later Italy. Much later, after about 1000 CE, independent civilizations also arose in Mexico and Peru—among Aztecs, Incas, and others.

Civilization means cities. Cities mean a relatively large population of people who are not primary producers, that is, not farmers, foresters, miners, or fishermen. A city's population depends on a variety of specialized artisans, merchants, and administrators. The city must have sufficient control over its hinterland to be able to be assured of a regular supply of food and other raw materials.

As we have seen, the early civilizations were organized around temples and a priesthood. The priests were the administrators. Devotion to the gods of the temple provided a spiritual "glue" that held people together.

Of course, there were soldiers too. When persuasion doesn't work, force may have to be employed. And the accumulated surpluses of wealthy cities were prize targets for nomadic marauders or the jealous greed of a neighboring city. All the joys of the new world of civilization!

Cities and empires also brought large increases in world population. When we left the world of early farmers at 7000 BCE we'd reached a population of about four million. It reached seven million about 4000 BCE, and then doubled in each of the next three millennia, reaching 50 million at 1000 BCE. The next two doublings each took about 500 years. Shortly after the beginning of the Christian era (with the Roman Empire controlling the Mediterranean basin), world population levelled off at close to 200 million souls. The next doublingtook almost 1500 years. If we assume a close connection between population and agricultural technology, then we might guess that farming techniques didn't change very much over that long stretch of time. We'll see.

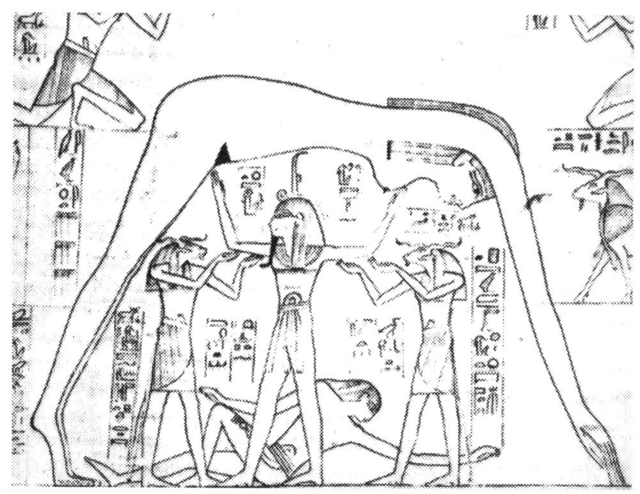

*An Egyptian tomb painting depicting the visible sky.*

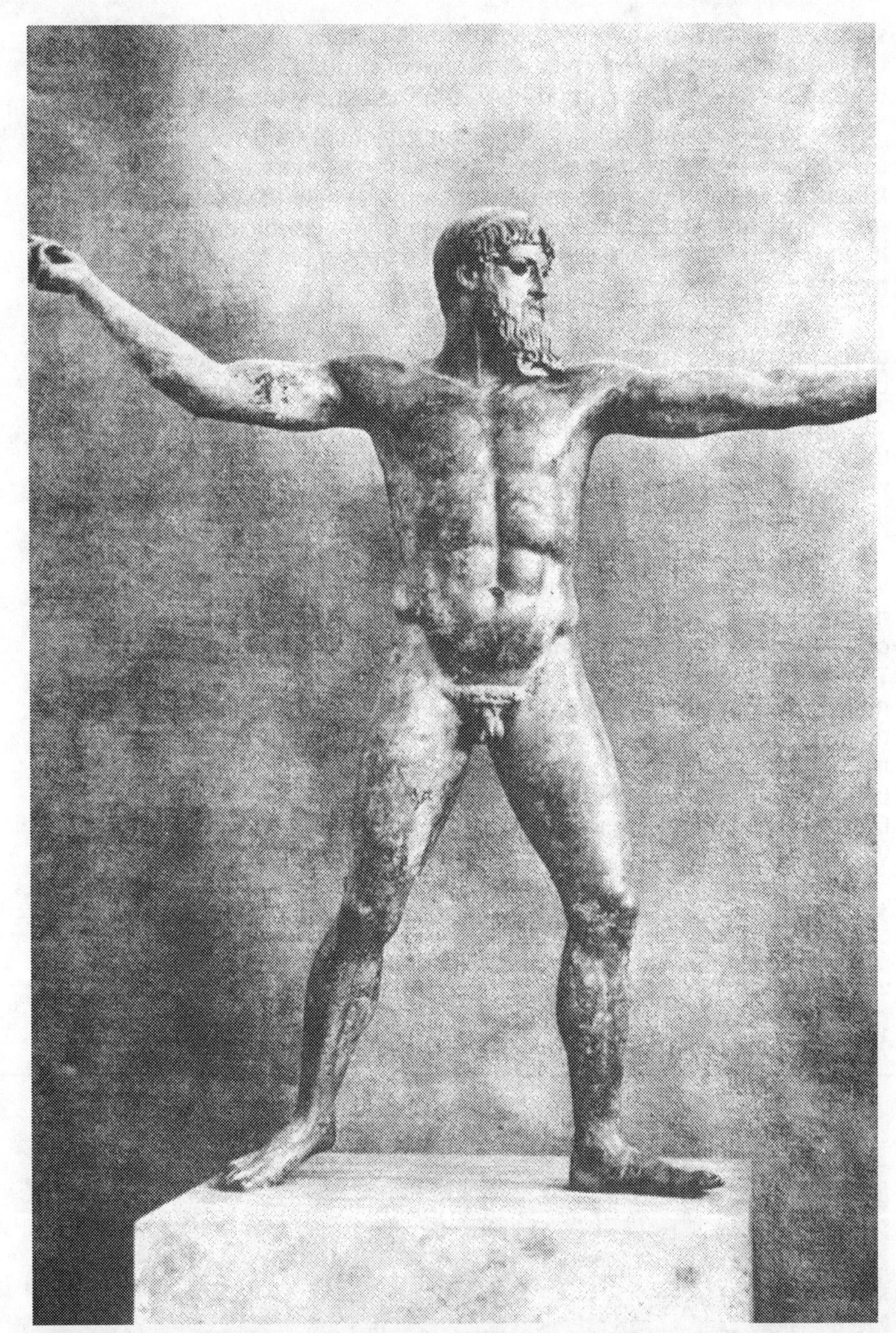

*Poseidon (or Zeus) found in the sea off Cape Artemisium. National Museum of Athens.*

CHAPTER FOUR

# The World of Greece and Rome

In the period from 3000 BCE to 1000 BCE, a number of empires around the eastern end of Mediterranean rose and fell. In the thousand years that followed, dominance over the Mediterranean basin gradually shifted toward the north and west—first to Greece and then to Rome.

The classical worlds of Greece and Rome provided the foundations on which western civilization was built. The ancient classical world depended on two fundamental technologies: the fabrication of iron tools, and writing with an alphabet. These two, very different from each other, started to move power out of the hands of aristocrats and into the hands of the people. Iron and the alphabet were democratic innovations. That is, they made material and intellectual technology available to a larger fraction of the population than bronze implements and pictographic writing ever could.

Iron is more democratic than bronze because its raw materials are more widespread and ultimately cheaper. Soldiers and priests dominated ancient empires and controlled the supply and manufacture of bronze implements. Iron was the first metal to be cheap enough and robust enough to replace the stone and wooden tools of the workaday world.

Alphabetic writing is more democratic than pictographic writing because it is easier to learn. When each word is represented by a separate, complex symbol, you need many years of training to learn enough symbols to write your language. In ancient times, that put the control of record keeping and communications in the hands of the priests who operated the schools. Writing was not something you could pick up in your spare time.

Having an alphabet changed all that. All you need is 20 to 30 symbols to represent the sounds of your spoken language. Alphabetic writing was developed around 1500 BCE, by a Semitic people, the Phoenicians, living on the eastern Mediterranean coast. Their alphabet of 22 consonants was adapted by the Greeks, who added symbols for the vowel sounds. The Phoenician alphabet was also the basis for Hebrew and probably also for the alphabetic writing of India. The Roman alphabet comes at least partly from the Greek.

Alphabetic script is so easy that some bright children nowadays teach themselves to read using television and cereal boxes. Thus, the invention of the alphabet allowed literacy to spread beyond the control of the priests. Merchants could keep their own records without having to hire a scribe trained by the priests. As we'll see, the Hebrew and Greek cultures were able to make great strides because they were not dominated by priests. Writing was used for keeping accounts, for laws, prayers and rituals, and later for literature and science.

After considering the development of the technology of iron, we'll turn to the early stages of science—through the golden age of Greek philosophy to astronomy. We'll conclude our survey of the ancient world with the achievements of Roman technology.

## *The World of the Blacksmith*

Although iron is much more common and widespread than copper, its economic use began later than copper, because of its chemical properties. Being a more reactive element than copper, iron is seldom found in the free state. Its most common ores are oxides and sulfides.

The temperature for smelting iron is about the same as that for copper (about 850° C). At only a couple of hundred degrees more, copper will run out into a puddle at the bottom of the smelter. But the melting point of iron is more than 1500° C, a temperature seldom achieved in antiquity.

When metallurgists mixed finely crushed iron ore with charcoal in their smelters at 850 to 950° C, iron would be left behind as the oxygen combined with the carbon from the charcoal. The iron remained in the form of small nodules scattered through a glassy slag composed of the products of the other minerals mixed with the ore. To retrieve the iron, metallurgists had to keep the mass red hot and hammer away at it to remove the slag and weld the bits of iron together.

You should not wonder that iron did not become an economic product until after 1500 BCE. A lot of accidental discovery and trial and error were needed to get useful quantities of iron. The earliest economic smelters of iron were in the eastern highlands of Turkey. Technical knowledge improved gradually, and iron working spread throughout Europe and Asia, and into Africa. The diffusion of iron making was relatively slow at first, but by 600 BCE iron had largely replaced stone and bronze for tools.

Iron can take a variety of forms, and there are various methods for achieving a desired structure. The iron that came from the smelter after the slag was removed contained a small amount of carbon from the charcoal. Purer iron could be obtained by re-heating with a blast of air to burn off the carbon. But iron with less than 0.3% carbon (wrought iron) is quite soft. It's fine for horseshoes and fences, but not for axes or knives.

With about 1% carbon, iron is quite hard and difficult to pound into shape. Traditionally, iron with from 0.3% to 2% carbon is called steel. The beauty of iron is that the smith can de-carburize it to wrought iron, pound it into shape, and then harden its surface by heating it in a forge in contact with charcoal. That allows the smith to make an ax that is wrought iron inside, with a steel surface. He can then sharpen the edge with a file. Another way to harden iron is to reduce the size of the iron crystals in its surface layer. In the traditional technique for doing that, a smith heated the iron to red heat and then plunged it into cold water. This process is called **tempering**.

Iron was the ideal metal for a wide range of weapons and tools. It even provided the appropriate tools for the blacksmith himself—tongs to hold the hot metal, files to sharpen it, hammer and anvil for pounding it. By 500 BCE, the tools of the blacksmith had achieved the form that survived down to almost yesterday.

From then on, all the buildings, implements, and machinery of civilization were constructed using iron tools. They were used to build the great empires of classical Greece and Rome.

## *The World of Nature*

Our stone age ancestors lived in their minds as part of nature. They did not, as we do, recognize nature as something "out there," apart from themselves. This is shown partly by the totem animals that various tribes used to identify themselves. Members of a tribe that belonged to the hawk thought of themselves as having hawklike features—even if they couldn't fly.

The process of becoming intellectually separated from nature took our ancestors a long time. In brief, the process involved elevating the totem animals to gods, and then somehow separating themselves from their gods, and from nature. This process of separation took two distinct forms—the Hebrew and the Greek.

### *The Hebrews*

The Hebrew process grew out of the religions of southwest Asia. To explain the origin of the world and its inhabitants and other mysterious features of life, the priests of Sumer and Egypt invented gods whom they endowed with supernatural powers. In Egypt, the priests and soldiers used these supernatural powers to control the people, (e.g., by treating the pharaoh as a god on earth). The peasants built the pyramids more as a religious exercise than because they were forced to.

Egyptians had many gods, whom they endowed with family characteristics. And the Egyptian myths of creation were related to their own experience. Using the model of the flooding of the Nile, one myth supposed that the earth had grown up in a hill as the great waters subsided. Using the model of birds, another myth supposed the earth to have been born from a great egg. Yet another myth attributed the origin of the earth to a god masturbating—sowing his own seed.

In contrast to the many gods of their neighbors, the Hebrews conceived of their tribal god, Yahweh, as the one lord of all—the God of gods. Instead of gods of earth, sky, sun, and stars, Yahweh ruled over the whole of creation. Since Yahweh was all-powerful, the Hebrews elevated him far above themselves. Gone were the family relations among the gods; Yahweh became almost totally "other." And, in the minds of the Hebrews, Yahweh took nature with him, so to speak. As lord of *everything* in the world, Yahweh ruled nature, independent of humans. But by separating themselves from Yahweh, the Hebrews also separated themselves from the nature he controlled. No longer were they full participants in the world of nature like the Egyptians and Sumerians.

You can see the Hebrews' mythological treatment of this separation in the story of how Adam and Eve left the garden of Eden. Of course, being the god of the children of Israel, Yahweh had a special concern for them. So, the myth of Adam and Eve also demonstrates the Hebrew belief that humans are superior to the rest of creation because of Yahweh's special interest in them.

The Hebrews thus objectified nature—they no longer associated themselves with totem animals. But, because of Yahweh's control over nature, and the Hebrews' belief that they must obey his laws without question, they did not develop a critical view of nature. As we'll see in a moment, that step was taken by the Greeks.

An important feature of the Hebrews is that they were a pastoral people—semi-nomadic, herding large flocks of animals. Their culture was not centered in great cities, and they were not dominated by priests and soldiers as were the peoples of Sumer and Egypt. The Hebrew people had a considerable degree of cultural and economic independence from the empires around them. They also had a degree of religious and intellectual independence.

This Hebrew distinctiveness was evident as early as the time of Abraham, about 1700 BCE. It was firmly entrenched by the time of Moses, about 1400 BCE. The story of their history, their trials and tribulations, their deeply felt beliefs, were passed down orally through the generations. Only at about 700 BCE were they committed to writing in the Torah—the Law and the Prophets—what Christians call the Old Testament.

### *The Greeks*

The Greek histories and myths were committed to writing at about the same time as those of the Hebrews. The poems of Homer, the *Iliad* and the *Odyssey,* are full of the adventures of the Greek gods, Zeus, Hera, Poseidon, Aphrodite, Athene, Hermes, and the rest. Mixed in with this mythology is the history of the early Greeks, the war with Troy, and the voyages of Odysseus. This history dates back to about 1200 BCE.

Like the Hebrews, the Greeks were not a priest-ridden culture. They occupied territory on peninsulas on the mainland of Greece and Turkey and on the islands in the Aegean Sea. They were a hardy, adventurous people, who lived in city-states devoted to commercial enterprise.

In his tales of the gods, Homer described how the gods had originally been masters of particular territories, like Poseidon ruling over the sea. But Homer endowed the gods with so many human features that they lacked much of the mystery the Hebrews attributed to Yahweh. Indeed, the Greek gods seemed to take on the features of a tribe of their own, occupying Mount Olympus. The Greek gods engaged in so many human-like activities that they seemed to have little time to rule over their territories.

Homer described the Greek spiritual odyssey, which moved the gods out of control of regions of earth and sky, leaving behind a bare nature. Both the Hebrews and the Greeks separated themselves from nature by attaching nature to divinity. But, when the Hebrews elevated Yahweh above themselves and nature, they left him in charge of both—as lawgiver. The Greeks, on the other hand, debased their gods, and (so to say) pulled them out of nature through the bottom. Then, if the gods no longer ruled nature (or humans either for that matter) what did?

The Greeks answered this question by attributing the uniformities of nature to the rule of abstract principles. These principles at first were derived from rules governing human society—principles like fate, justice, love, necessity, and strife. Later, the principles became less personal and more mechanical.

An interesting illustration of this notion can be found in the play of Aeschylus, *Prometheus Bound*, written about 465 BCE. Prometheus, chained to a rock by minions of Zeus, is being comforted by the Chorus, the daughters of the god Oceanus. They think Prometheus ought to be clever enough to get free from his bonds. But he replies:

> *Prom:* Cunning is feebleness beside Necessity.
> *Chor:* And whose hand on the helm controls Necessity?
> *Prom:* The three Fates; and the Furies, who forget nothing.
> *Chor:* Has Zeus less power than they?
> *Prom:* He cannot fly from fate.
> *Chor:* What fate is given to Zeus, but everlasting power.
> *Prom:* That is a thing you may not know; so do not ask.

*Prometheus bound to a rock.*

So Zeus is ruled by fate, apparently as much as men are. And while Aeschylus somewhat personifies the Fates and Furies, they are not so much personal rulers (not God like Yahweh) as abstract principles. You almost get the feeling they've been made persons merely for the sake of the drama. Indeed, in this play, two of Aeschylus' characters are Strength and Violence, the very creatures who bound Prometheus to the rock.

In this intellectual atmosphere, the Greeks invented philosophy and logic. From about 600 BCE to 300 BCE, they created for the first time a system of critical thought about a nature that was separate both from humans and from gods.

### *Greek Natural Philosophy*

The process of creating a system of natural thought didn't happen all at once. We can trace a sequence of developments through about nine generations of Greek philosophers, from Thales in 600 BCE to Aristotle, who died in 322 BCE. In this process, the followers of each man examined his speculations, criticized them, and altered them. Instead of treating his words as if they had the authority of a divine oracle, these men thought for themselves.

Thales and his immediate successors lived on the Aegean coast of Turkey in a region known as Ionia. Thales seems to have wondered about the material basis of life: What are things made of? He made the startling assertion that "all things are made of water." That is, beneath all the variety of materials in our experience there must be some common substrate; and Thales opted for water.

That was too substantial for his successor Anaximander, who chose to call the material substrate the Boundless. His successor, Anaximenes, found that too vague, so he chose air. Now, at the same time as these men wondered about the material basis of things, they also questioned the process of formation of the various materials from the One. They were seeking both for principles that unify nature, and for explanations of the obvious diversity in nature. Anaximander spoke of a material "paying the penalty" for its transformation from the Boundless to its matter and form. The penalty was to return again to the Boundless. Anaximenes used less emotive language, saying that solids and liquids came from air being condensed, and fire from air being rarefied.

About 500 BCE, the Ionian Heraclitus made change the most permanent feature of experience. He wrote:

> The world was not made by men or gods. It has always been and shall be an ever-living flame, being lit and going out in measured amounts.

Heraclitus' image of the flame is meant to convey the idea that all things are like flames in that they can retain their form and substance, even though we know that the material in them is constantly changing.

About 450 BCE, a citizen of a Greek colony in Italy challenged Heraclitus' ideas. Parmenides said that change is only an illusion. We need not suppose that Parmenides really believed that nothing changed. Rather, he was testing our ability to talk logically about the world of nature. He argued as follows: things around us exist. They have being. They are. Things that have no being do not exist. But for a thing to change, it must move from being what it is now to being something it is not now—to not being what it is. But "not being" does not exist. Therefore things can't change. The world is full of being, constant and unchanging. For Parmenides, such Being was all there was—and all there was to understand or reason about. Change, being an illusion, was therefore "unreasonable."

This argument of Parmenides was a direct challenge to our human capacity to understand anything in the world of our experience. About 400 BCE, this challenge was met head-on by the Ionian, Democritus. He proposed an atomic theory—based on no more experience than Parmenides' ideas. But Democritus wanted to explain change as more than just an illusion. Briefly, his atomic doctrine was this: There's nothing in the world but atoms and the void. By void he meant empty space—nothing but atoms and the void. All existence is merely empty space with atoms bouncing around in it.

*Atom* is the Greek word for "uncuttable"—atoms are the tiniest pieces we can imagine matter broken into. Of course, they're far too tiny to be visible. Democritus asserted that all matter is composed of atoms. But he also asserted that the void exists, too. It's as if he was saying, "Nothing *does* exist, Parmenides. It has to if there's to be any change." And that's what Democritus wanted to be able to explain—change. The atoms may differ in size and shape. The way they're arranged or rearranged determines the nature of things. And atoms are able to move and re-arrange to form different things because there's empty space between them—the void.

That's all Democritus felt was needed: some kind of primary substance distributed among the atoms in an infinite number of tiny chunks, spread throughout the void—infinite space. The atoms themselves provided constant Being to satisfy Parmenides' argument, and the void allowed for change in a way that would have pleased Heraclitus.

*Aristotle (384–322 BCE)*

Now the scene shifts to Athens on the Greek mainland. Socrates was about the same age as Democritus. He was immortalized in the dialogues of his disciple Plato. Plato wrote about nature, but didn't add much to the arguments discussed here. However, Plato's pupil, Aristotle, did—and created a synthesis of natural philosophy that would last almost 2000 years.

*Aristotle.*

Aristotle examined and criticized all his predecessors' arguments about nature. And then created a whole system of nature of his own. He wrote about it in a work with the title of the Greek word for Nature, that is, *Physics*.

Aristotle criticized atomism because it was too broad and vague. The atomists hadn't been able to say exactly how the atoms were arranged to compose the wide variety of substances we see around us. Aristotle assumed that Democritus had thought all atoms were composed of one basic material—but he hadn't said what kind of material. And if atoms differed in their shape, how many different shapes were there? Democritus didn't say. So Aristotle considered the atomic theory to be an unsatisfactory solution to the problem of change.

Aristotle also objected to the idea that there was only one basic material. So he opted for four elements—earth, water, air, and fire. They've existed from the beginning. He considered these elements to be pure substances, not the ones we see around us, which are mixtures. For example, flames are fire mixed with earth—that's why they're smoky. And air always has water mixed with it—that's where clouds come from.

Aristotle's four elements were based on ordinary everyday experience. In something like wood, for example, you can find liquids and solids. When wood burns, the flame shows the presence of fire, and the smell may indicate gases. With four elements Aristotle solved part of the problem of change—he didn't have to worry about how elements change into one another. But that was only half the battle. Aristotle still had to find a way to reconcile the dilemma between Heraclitean change and Parmenidean permanence.

Aristotle's solution was a kind of compromise. He agreed with Parmenidea that the universe was full—no void anywhere—just total being, but not pure unchangeable being. Aristotle built change in by saying that materials contained a principle of becoming—change built right into the foundations. Everything that *is* has the potential of *becoming* something else. An acorn is not merely a nut, it's also a potential oak tree.

There are many other important aspects of Aristotle's thought. But we'll save them to be discussed in the context of how they were used later by others.

This Greek view of nature, in the form organized by Aristotle, was the dominant European view among scholars for two thousand years. It provided the framework people used as they thought about the structure and behavior of the world of nature around them. In particular, in ancient times, Aristotle's natural philosophy formed the context within which Greek astronomy developed.

### *Astronomy*

Even before the invention of writing at about 3000 BCE, neolithic villagers must have noticed the regularity of the motions of the sun, moon, and stars. When we get direct evidence of what they noticed in the sky in the early writings of Sumer and Egypt, we find they are treating those lights in the sky as deities. The sun rules the day, the moon the night. And the sun's cycle defines the time unit, the day.

*Astronomy in Babylon and Britain*

What is so striking about sun, moon, and stars is that they move in regular, repetitive patterns. Before long the Sumerians were using the regular reappearance of the new moon to define the time unit, the month—twenty-eight or twenty-nine days in duration. Eventually, they also noticed the annual motion of the sun. At the end of June, the sun is almost overhead in southern Iraq. At the end of December, it is about midway between zenith and horizon.

The Sumerians (whose civilization was centered in Babylon by 2000 BCE) saw the motions of sun and moon as very much more regular and consistent than the behavior of most of the rest of the world around them. The certainties of their sky gods contrasted sharply with the great uncertainties of daily life. On their clay writing tablets they began to keep records of the positions of sun and moon, and later, the planets.

Away off in Britain, writing did not arrive until the Roman conquest, about 50 BCE. Yet there is significant evidence that early Britons were recording particular positions of the sun before 2000 BCE by using alignments of large stones. The most famous of these is the megalithic monument at Stonehenge in southern Britain. Archaeologists have found stone alignments there that point to spots on the horizon where the sun rises or sets farthest north in the summer and farthest south in the winter.

*Stonehenge.*

Another stone alignment on the west coast of Scotland is simpler, probably older, and in a way, more striking than Stonehenge. If you stand on a little stone platform and look toward the southwest over the top of a nearby vertical stone, you can see a notch formed by two mountains on an island 45 km away. On 22 December the sun can be seen to set exactly within that notch along the sightline. On that day the sun rises in the southeast about 9 am, gets about 10° above the horizon at noon, and sets in the southwest about 3 pm. This occurs at a latitude of about 56° north, the same in southern Alaska and northern Ontario as in Scotland—six hours of sunlight.

Six months earlier, they'd had 19 hours of sunlight. We can imagine that neolithic farmers had realized the sun had something to do with warmth and growth. As the days got shorter through the summer and fall, they might well have supposed their sun god was deserting them. (Indeed, above the arctic circle, there are days in winter when the sun does not rise above the horizon at all.) If some medicine man had figured out the sun's cyclical motion, he could propose a ritual that would be effective in bringing the sun back. With his sacred observatory on the coast, he'd be able to know exactly on which day to perform the ritual. It always worked!

There are more than fifty of these ancient astronomical sites scattered across Britain. They indicate the importance of astronomical motions for the neolithic farmers, and show how they could keep records (at least crudely) without being able to write.

*Early Greek Astronomy*

The Babylonians could keep much more elaborate records than the Britons. Eventually they were able to predict future positions of heavenly bodies by arithmetical extrapolation of the time-series in their records. Their predictions were reasonably good, because the Babylonians were careful observers and expert arithmeticians. Over the centuries, they accumulated a large data base of astronomical cycles. They used their data base and their arithmetical rules to make astrological predictions; that is, they tried to match events on earth to particular conjunctions of positions in the heavens.

The cyclical regularity of astronomical positions provided the Babylonians with the factual basis for a science. But, for astronomy to become fully scientific, it also needed a theoretical foundation, an explanatory model. Such models began to be constructed by the Greeks about 400 BCE. The Greek models were built using geometry. From Thales at 600 BCE down to Euclid at 300 BCE, the Greeks created a systematic, logically reasoned structure of thought about shapes and magnitudes.

Starting about the time of Plato, Greek mathematicians imagined a theoretical universe, composed of spheres within spheres to explain the motions of sun, moon, planets, and stars. To a very rough first approximation, those bodies appear to circle about us overhead. But from the Babylonian records, the Greeks knew that there were numerous deviations from any perfect, circular regularity. Nevertheless, they chose to assume that somehow they could find regularities among those deviations. Using their geometry, the Greeks tried to add complications to the simple spherical system in a series of steps, so the positions of the bodies in the model would get closer and closer to the positions observed in the heavens.

About 350 BCE, the mathematician Eudoxus tried to work out a scheme using spheres nested one inside the other. For example, Mars might be located at some point on one sphere. That sphere rotated about an axis attached within a second sphere. This second sphere could rotate at a rate different from the first, on an axis set at an angle to the first one. By the addition of other spheres, Eudoxus might theoretically have been able to approximate the observed motions of Mars. In fact, the task was too daunting.

However, the Eudoxian system of spheres looked promising enough that Aristotle adopted its form as the basis for a set of astronomical principles. Aristotle proposed a philosophical structure for the universe, a **cosmology**: what we must imagine the universe to be like in order for it to be intelligible.

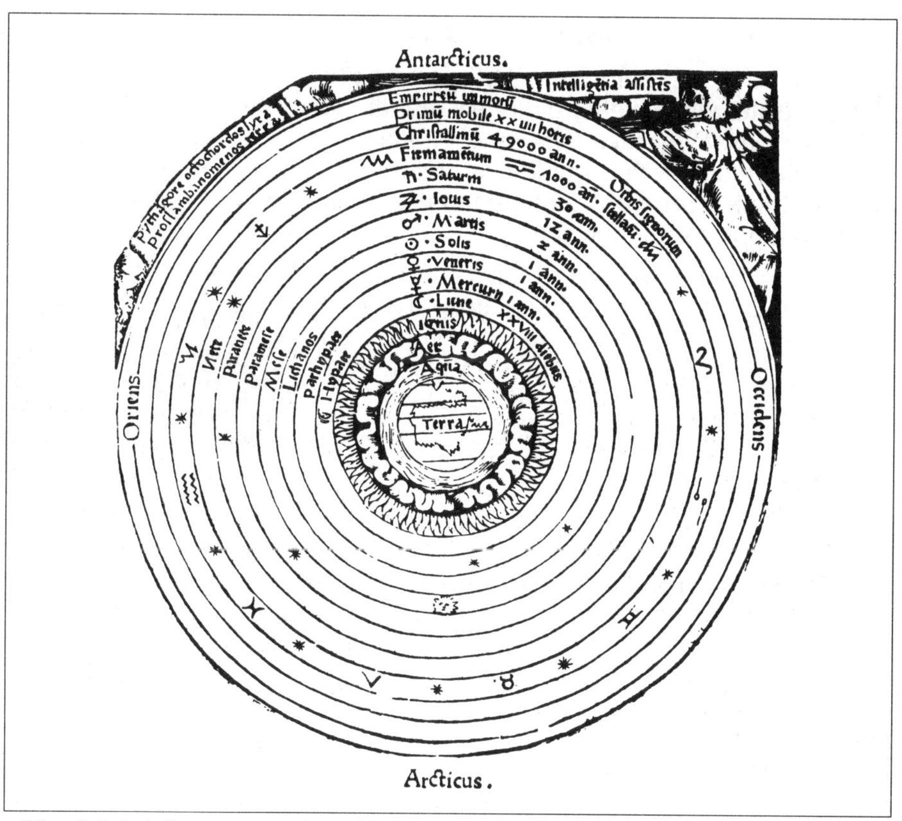


*The Aristotelian conception of the universe, drawn for the edition of Aristotle's* On the Heavens, *edited by Johann Eck (Augsburn, 1519).*

Aristotle's principles were: first, that the earth is at rest in the exact center of the system; second, that the stars and planets move in perfect circles around the earth; and third, that they move at constant speeds.

A perfectly centered, perfectly spherical, perfectly uniform system of motions. For 2000 years, these principles were considered to be the ideals to be achieved by any model of the universe. And that is so, despite the fact that the first practical set of astronomical models did not conform fully to these principles.

Aristotle was trying to understand the universe. As with his natural philosophy, his explanation was based on principles, not on the whims of gods. To fit his cosmology into his matter theory, Aristotle proposed that the heavens were composed of a fifth element (*aither* in Greek, *quintessence* in Latin), a perfect element exactly suited to the circular perfection of the heavens. (For Aristotle, the four earthly elements had natural *straight-line* motions, while *aither*'s natural motion was circular.) Although Aristotle was strong on intelligibility, he was weak on correspondence with actual observations.

### *Ptolemy (AD 100–AD 170) Makes Astronomy Scientific*

After Aristotle, astronomers tried to stick to his principles while creating models that did fit observations. They tried various combinations of circular motions that would give the planets in their models positions that agreed with their observed positions in the sky.

*Claudius Ptolemy.*

Now, a planet like Mars deviates from perfect circular motion in several major ways. First of all, sometimes Mars is closer to the earth than at other times. Second, though Mars moves through the sky (against the background of the "fixed" stars) an average of half a degree a day, it sometimes moves more slowly, and sometimes more quickly. Finally, about once a year, Mars appears to stop in the sky, move backwards for a few months, then it stops again, and resumes its regular motion.

The backwards motion of Mars is called *retrograde*. All the planets are normally seen on successive nights to be to the east of their earlier positions; that is, they move eastward against the background of the stars (which move westward, completing one revolution every 24 hours). However, during a

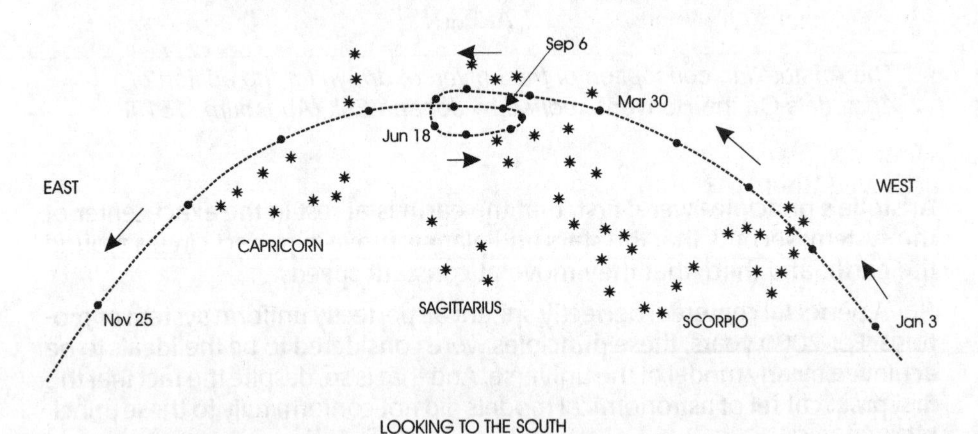


*The apparent path of Mars against the stars. Each dot represents the position of Mars at twenty day intervals, as seen from the Earth. In the year represented here, Mars moved steadily eastward from January to June against the background of the stars. In the middle of June, it appeared to stop, then go backwards (retrograde) until about the beginning of September, when it stopped once again and then resumed its easterly motion. The retrograde motion of Mars occurs about once every two years, lasting about three months.*

planet's retrograde phase, its subsequent positions are to the west of earlier ones.

Convinced of Aristotle's principles—but limited by their mathematical competence, astronomers found a combination of circular motions that suited the motion of Mars pretty well. The full scheme was put into a book by Ptolemy of Alexandria about AD 150 —not only for Mars, but also for the sun and moon, and the four other planets known to the ancients. He provided a separate model for each of the seven bodies.

A model for Mars can be constructed in Ptolemy's way as follows. Draw a circle with center *C* having a radius of 10 cm. Draw a vertical diameter crossing the upper circumference at *O*. Mark *E* 1 cm below *C* on that diameter. *E* is the earth. Mark *Q* 1 cm above *C* on *CO*; *Q* is called the **equant**. With center *O*, draw a circle of 6.6 cm radius, and locate M at the top of that smaller circle, on *CO* extended. The smaller circle is called an **epicycle**. *M* is the location of Mars at a particular time.

*O* moves counter-clockwise along the circle centered on *C* 0.524° each day, at a constant rate measured not about *C* but about *Q*; that is, *O* moves to *O′* in one day, making the angle *OQO′* = 0.524°. (Thus, the motion relative to *C* is not constant.) At the same time, *M* moves uniformly counter-clockwise about *O* at 0.986° per day. If you programmed these parameters into a graph-drawing computer, you'd get a series of positions for *M* that would fit astronomical observations to within a degree or so. Of course, since we're on the earth (at *E*), you'd have to measure an angle *MEM′* to see how far Mars would appear to move in some number of days.

Ptolemy had a similar set of parameters for each of the other six planets. He achieved them by a process of trigonometric analysis of observed positions of the planets, adjusting the various parameters of his model until he got the closest fit he could. In his analysis, Ptolemy followed the procedure that has been the hallmark of science ever since. He started with both data and a model. Not content with either a Babylonian string of observed

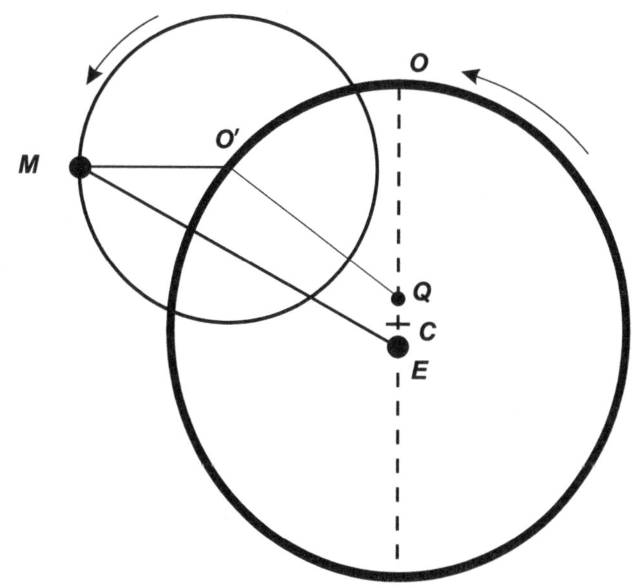

*Ptolemy's Model for Mars.*

locations, or a purely theoretical model like Aristotle's, Ptolemy combined the two. The result was a set of models for the seven planets that predicted future positions of the planets, not by extrapolating the data, but by working the model.

Notice that Ptolemy tried to keep his model for Mars close to the principles of Aristotle. He had the earth motionless in the middle, but not at the exact center of Mars's motion. He had Mars moving along perfectly circular paths—two of them. The motion on the epicycle was uniform about its center *O*. And the motion of *O* along the larger circle was uniform—a constant angular motion each day—but about the equant *Q*, not the center *C*. It is a tribute to Ptolemy's concern for precision that he could modify theoretical principles to make his models conform to observations. Yet he made what seemed to him to be the minimum necessary modifications.

Ptolemy collected his astronomical procedures and results into a magnificent treatise of 13 books. In Greek it was called *Mathematike syntaxis* ("Mathematical composition"). The first book provided the foundations of the spherical trigonometry required in the rest of the work. In the following books, Ptolemy worked systematically through the universe: the earth, time, and seasons; the sun; the moon and eclipses; the stars; and the planets. No other astronomical work in antiquity was so complete. This fact was duly noted by Ptolemy's Arabic translators around 800, who dubbed it "The Greatest" (*megiste* in Greek). From then on the work came to be known as *Almagest* ("al" being Arabic for "the").

Ptolemy created the science of astronomy by using mathematics to combine careful measurements with the simplest model he could use. In its major outlines, this achievement stood for more than 1300 years. As we'll see in a later chapter, Ptolemy's work also provided the methodological foundation for the new science of motion in the seventeenth century.

## *Early Alchemy*

Even in ancient times, craftsmen knew quite a lot about the properties of materials and how they behaved. Cooking, potting, the smelting of metals, all involve chemical processes. The people who first tried to make a consistent theory for these processes were called alchemists.

The word **alchemy** comes from the Arabs. They derived it from Greeks who lived at Alexandria in Egypt. The word may come from the Greek *Khem* meaning "Egypt"—making alchemy simply the Egyptian art; or it may be from the Greek *chyma* meaning "molten things." Eventually, alchemy came to mean the art of transforming substances.

One of the alchemists' main objectives was to find objects, materials, or principles that would transform base metals into gold. Another was to find a liquid that would render human life eternal. In folklore, these two principles became the "philosophers' stone" and the "elixir of life." During their fruitless search for these principles, the alchemists learned a lot about the behavior of materials, and explored a variety of theories about matter.

Aristotle had a principle of development in his theory of matter. He supposed that all things could grow and develop, feeding on their surroundings. Not just plants and animals, but rocks and minerals too. Such development was simply a normal part of nature.

Since no one had actually seen minerals growing, the alchemists supposed that natural growth was very slow. One alchemist wrote that gold was formed in the womb of the earth from shining mercury and clear red sulfur. It was concocted over a hundred years with the help of heat from the sun. The alchemists tried to mimic this natural process going on in the womb of the earth. They considered their apparatus to be an artificial womb, where they'd use high temperatures to speed up the processes. When they subjected various substances to great heat they often drove off vapors which they then condensed and collected.

We call that process **distilling**. When the alchemists distilled wine and other fermented drinks, they got alcohol in the receivers of their stills—the original "moonshine." At 100 proof or so (50% alcohol) the resulting liquor was often used to sustain life in difficult circumstances. The *akvavit* of Scandinavia is a particularly potent brew, with a name derived from the Latin *aqua vitae*, the water of life—the best elixir the alchemists could achieve.

While the alchemists failed in their major objectives, they contributed to social life and to the experiences from which chemistry would be born as a science during the eighteenth century.

## *Technology in Greece and Rome*

With their iron tools, Greek artisans became fine carpenters. They built seagoing vessels as long as ten meters, propelled mostly by oars, twenty or more, operating in single or double banks. With these vessels the Greeks established trade and colonies throughout the Mediterranean. The Greek homeland exported wine, olive oil, and pottery, and imported grain and other essential products. The economy of the Greek city-states expanded greatly after 500 BCE.

*Model of a seagoing Greek Trireme.*

Greek carpenters also became expert in the timber-frame construction of buildings. About 400 BCE, the Greeks began to translate timber construction into stone for public buildings. Because stone beams are weak in tension, Greek stone buildings feature a large number of supporting columns. In this **post and beam** construction, Greek stone masons fitted the joints with great care, and also used iron pieces to secure the joints.

Following 500 BCE, Roman power began to expand in the central Mediterranean. During the next 500 years, the Romans spread out from Italy and eventually controlled the whole Mediterranean basin, as well as much of central and western Europe. Initially, Roman technology followed Greek patterns. But, by about 200 BCE, the Romans had introduced a significant building innovation into Europe—the arch. They likely learned about arches from Egypt.

*Roman Aqueduct at Nîmes in France.*

The Romans built their arches of stone, over a timber centering frame. Once the keystone was cemented in place, the centering was removed. After the Romans had invented a suitable cement for stone, they used arches to span larger areas than is possible with stone post and beam construction. One of the most lasting applications of the arch is the system of aqueducts, which the Romans used to carry water across long stretches of land to supply their cities with fresh water. The Romans also built large buildings with barrel-shaped vaults supported by massive walls.

In addition to water transport, the Romans excelled in road construction. Roman roads stretched across Europe to expand trade and to allow their troops easy access to conquered territories. Instead of the previous paths and tracks of the Greeks and others, the Romans built their roads with great care. They dug deep trenches and filled them with sand, gravel, and stone.

Throughout ancient times, the major source of power was from human and animal muscles. By Greek times, this power was being assisted by various combinations of levers and pulleys. At sea, the power of oars was supplemented by single sails when the wind blew in the right direction. Steering was done with an oar on one side of the vessel near the stern.

Around 100 BCE, the Romans began to use the power of flowing water to turn wheels with blades on them. These water mills were attached to large circular stones used for grinding grain. However, these water mills did not come into widespread use. That was partly because water flows in the Mediterranean basin are seasonal and irregular, and partly because there was usually a sufficient supply of labor that investment in mechanization did not make good economic sense.

About 250 BCE, Archimedes used Euclidean mathematics to analyze certain mechanical systems. He wrote treatises on the lever and on the buoyant forces of water. He is also reputed to have invented various ingenious mechanisms, such as complex sets of pulleys and a screw for raising water.

Soon after, engineers in the Greek city of Alexandria in Egypt invented a variety of mechanisms that used compressed air to control motions. Some authors have said that these men were ingenious enough to have created a technical revolution, that is, to have transformed the methods of producing goods. They attribute their failure to do so to slavery, arguing that the supply of human labor was more than adequate for the level of production that was required.

That may be part of the story, but we should consider the whole context of the times. At least 90% of the population were peasants who were largely self-sufficient. They built their own homes and made their own clothes. From the point of view of their daily lives, the currents of history that impress us today were a very thin, superficial layer of activity. It would take a very different context for ingenious men to transform a whole way of life.

The worlds of Greece and Rome may well have laid the foundations for what we call western civilization. But they did so only because we have a historical memory, not because there was any regular building up from those foundations. As we'll see in Part II, the trip back to our roots takes a very circuitous and precarious route.

At 100 CE the Roman Empire stretched from Britain in the west to the Black Sea in the east, and controlled the whole coastline of the Mediterranean Sea. After 400 CE that empire was in disarray and under siege by marauding tribesmen from the north and east. Roman control collapsed, and Europe sank back to largely local enterprises.

## *Conclusion*

For a thousand years, from 500 BCE to 500 CE, the world of Greece and Rome dominated southern Europe. During that time, those classical civilizations developed ideas in technology and science that would eventually be built upon by their European successors.

The alphabet and languages of Europe are either derived from Greek or Latin, or influenced by them. Though technology declined after 400, the basic knowledge was never completely lost. Ideas about nature and astronomy—science—were preserved in writing by the Arabs. Technical ideas, such as the making of iron tools, were preserved in craft traditions across Europe.

*Notre-Dame d'Amiens. The vault reached to 42 m.*

# PART II

## *The Medieval World*

CONTENTS

While western Europe languished after the decline of Roman supremacy, the focus of power shifted eastward. Although the western Roman Empire had collapsed by 500, the eastern Roman Empire continued to dominate in Greece, Egypt, and southwest Asia. Its capital was Byzantium (modern Istanbul), and it eventually came to be called the Byzantine Empire.

Around 650 a new power arose in southwest Asia—Islam, the religion of Muhammad. Within a hundred years, the Muslims were in control of a territory stretching from Spain, across north Africa, and through southern Asia to India. The Byzantines fell back to the region around the Aegean Sea. Muslim domination continued in the Mediterranean region until after 1300.

Meanwhile, in western Europe, a degree of centralized power began to emerge among the descendants of the invading tribes that had conquered Rome. It arose first in France, and spread gradually to Britain and Germany. In 800, Charlemagne was crowned by the pope as the Holy Roman Emperor. The control of his successors over western Europe was never complete. However, a European culture based on Christianity and the Latin language did emerge. At the same time, western Europe grew in population and power as the result of technological innovations in agriculture and warfare.

Although Latin was the main language of the church, children in medieval Europe learned a different tongue at their mothers' knees. The vernacular European languages we know today developed during the medieval period. In the south, Italian, Spanish, and French grew out of Latin. In the north, German and Anglo-Saxon derived largely from the language of the Teutonic tribes who had occupied southern Scandinavia since about 2000 BCE.

Actually, most of the languages of Europe derived from an ancient common source called Indo-European. A cluster of tribes (sometimes called Aryan or Iranian) occupying Turkestan (east of the Caspian Sea) spread out from there in prehistoric times—perhaps as early as the third millennium BCE. Linguistic scholars today have inferred thousands of Indo-European root words by reading back from Greek, Latin, German, and Sanskrit (the ancient language of India). There is some evidence that Slavic peoples also derived from this cluster of tribes at an early period. Although Slavic languages today have few cognates with German, Latin, or Greek, the Russian language is written in characters derived from the Greeks, through the Christianizing influence of Byzantium. A lasting influence of the western Roman Empire can be seen in the widespread use of the Latin alphabet throughout most of the rest of Europe.

The English language is perhaps the most synthetic in the world because of its origin in the confluence of Latinate French and Teutonic Anglo-Saxon. While the Anglo-Saxon peasants tended their flocks of sheep and herds of cows, their Norman French masters found them at table as mutton (*moton*) and beef (*boef*). (That we don't have a name derived from *cheval* for a table meat, suggests that medievals relished horse-meat as little as we do.)

Since this is not a political history, we'll not be concerned with the shifting boundaries between nations, and will simply identify regions by the common language of their native populations. Although Spain, France, and England were kingdoms in medieval times, the Italian and German nations were not unified until the nineteenth century. Nevertheless, we can use the terms Italy and Germany in the medieval period to mean the geographic areas distinguished by the languages spoken there.

## *In a Mind's Eye*

Frère Jacques awoke with a start. Was that the alarm bell? He struck a flint to light the candle and looked at the clock. Yes, it was time for *prime*. He poured more water into the reservoir, re-set the hammer, slipped on his sandals, and padded along to the church tower. He reached high up on the bell-rope and leaned into his task. Soon the big bell was chiming his brothers to chapel.

After they'd sung their office to greet the new day, the Cistercian brothers set out on their appointed rounds. Jacques returned to his bell-ringer's cell to adjust the alarm to ring again for *tierce*. Just to be on the safe side, he turned the hour glass to provide a check on the waterclock. He was new to this job and he knew the abbot was punctilious.

Jacques decided he liked the job. Having to ring the bell eight times a day meant he had to keep pretty close to the clock. After all, he couldn't be expected to work in the fields or the brewery with an hour glass hanging from his neck. He had a good enough sense of time that he could wander around the monastery quietly and still be within earshot when his alarm struck.

Now he climbed to the top of the tower to inspect the bell. The rope was not very frayed, and the bearings were still well greased. Jacques turned to look out over the fields. The grape harvest was well under way, and already the brothers were directing the peasants in setting up their equipment to plow under the wheat stubble. A plentiful year, Jacques thought, rubbing his belly, even if we do only get one good meal a day.

Then he looked beyond the fields, and swept his gaze around from southwest, through west, to northwest. Bourges, Chartres, Paris, Reims, he chanted under his breath, I've seen them all, those glorious cathedrals, monuments to the power of Almighty God. All within 50 leagues of here. Only recently, he'd been privileged to be the novice who accompanied the archbishop on tour. Now, with his vows completed, he had his first posting. Would the bishops they met remember him in future times? Might he get to be abbot some day—or even higher?

Vain ambition! The Lord will choose the worthy, and brother Jacques is the lowliest of the low. Tempt me not, Satan!

He snapped back to his duty and returned to his cell. The alarm struck for *tierce*; he re-set the alarm for *sext*; and rang the bell. After the brothers had completed their prayers, he decided to survey the monastery buildings.

He walked to the south side where the channel led water diverted from the river Aube to the top of the great wheel. Within the mill geared shafts connected the wheel to the millstones, where even now the threshed grain was being ground to flour for our daily bread.

Outside again, Jacques traced the watercourse into the kitchen and the brewery. On the other side, the fullers had a small wheel driving hammers to pound the woolen cloth obtained from the weavers.

On the water ran, through the tannery and the latrines, finally to flow back into the main river. Jacques marvelled at how much muscle power was saved by having this graceful stream laboring to supply his food and clothing.

Time for *sext*! Prayers, then dinner, and a brief nap before the alarm struck for *none*. After the *none* offices, Jacques returned to his cell and stared at the waterclock. He'd have to find out soon what he was supposed to do in midwinter if the water froze.

> Frère Jacques, Frère Jacques,
> Dormez-vous? dormez-vous?
> Sonnez les matines, sonnez les matines,
> Din, din, don; din, din, don.

## *For Further Reading*

Crombie, A. C. *Augustine to Galileo.* 2 vols. Harmondsworth, Midddlesex: Penguin, 1969.

Gimpel, Jean. *The Medieval Machine: The Industrial Revolution of the Middle Ages*. Baltimore: Penguin, 1976.

Landes, David S. *Revolution in Time: Clocks and the Making of the Modern World.* Cambridge, MA: Harvard University Press, 1983.

Panofsky, Erwin. *Gothic Architecture and Scholasticism*. New American Library, nd.

Stoddard, Whitney S. *Art and Architecture in Medieval France*. Harper & Row, 1972.

White, Lynn. *Medieval Technology and Social Change*. London: Oxford University Press, 1962.

*A peasant uses a heavy to plow to till a field for spring crops outside the Chateau de Lusigan. From the Duc de Berry's* Book of Hours.

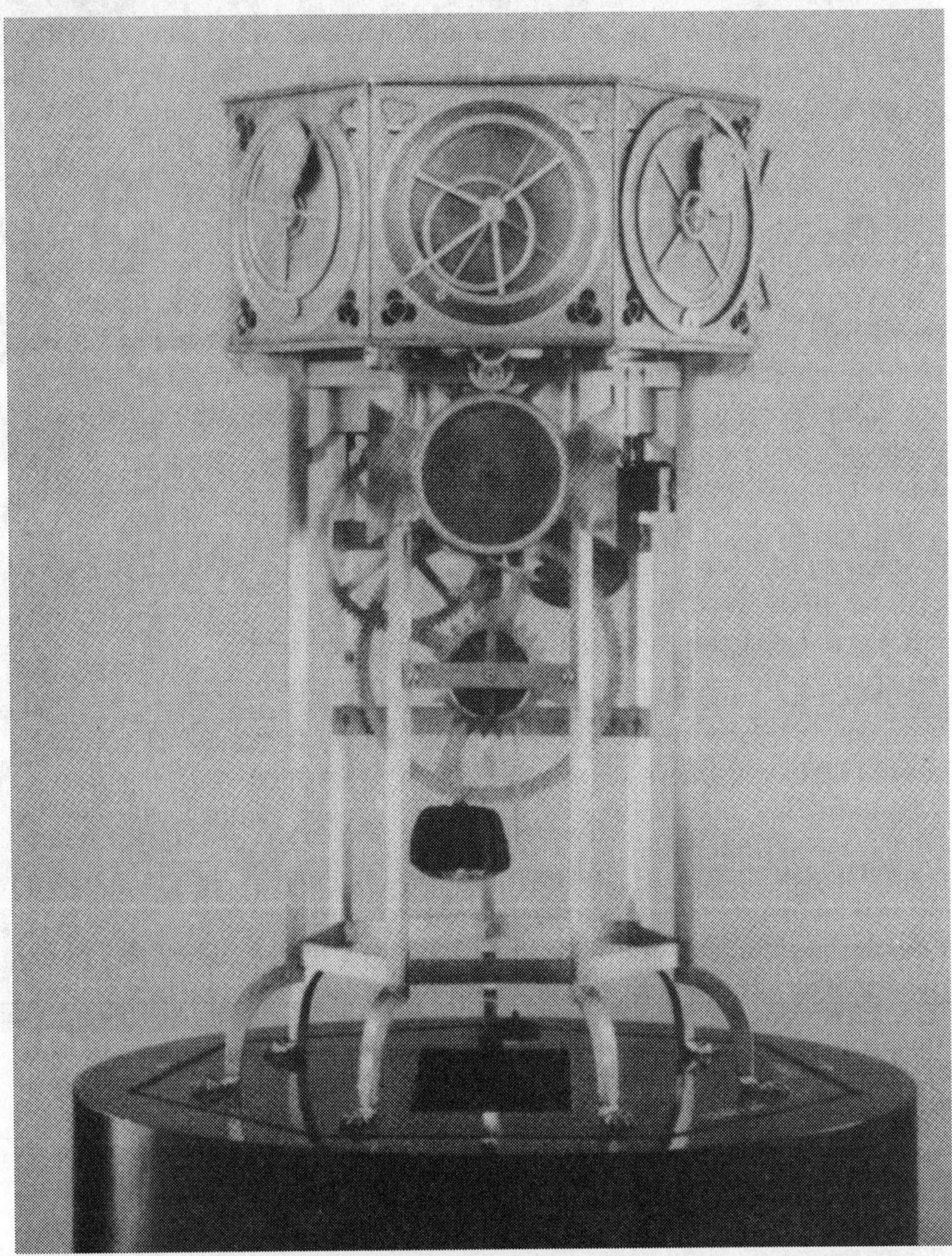

*The Astrarium clock of Giovanni de Dondi, showing the positions of all of the seven known planets, based upon Ptolemaic astronomy.*

## CHAPTER FIVE

# Emergence of European Power

After 400, with the decline of Roman power in western Europe, the local inhabitants soon reverted to the subsistence living typical of pre-Roman times. Indeed, Roman ingenuity had probably never extended very deeply into the countryside. Thus, peasants in the remoter areas of Britain, France, and Germany probably followed a life-style during the Roman period very little different from a thousand years earlier. That is, they had a self-sufficient economy where they used mostly wooden implements, made their own clothes, and eked out a bare existence from crops on small plots of land, with a few animals. In regions closer to the Roman settlements, peasants were required to produce a surplus sufficient to supply the soldiers and administrators. Their agricultural production and living accommodations may have benefitted from Roman iron fabrication and building technology.

After 400, tribes from northern Europe and central Asia drove into southern and western Europe. As classical scholar S.J. Tester (1987) put it:

> This was not the sort of "invasion" which simply crushed and destroyed what was there before. It was, after the battles were done, a movement of whole peoples into lands in large measure uncultivated and underpopulated. There was living space there, so they occupied it, and were assimilated. Though the offical imperial administration disappeared, the new rulers changed less than might be imagined, and many of them, like Theodoric in Italy, were even rebuilders of Roman cities after the ruins of the fifth century. While some of the great Roman landowners fled, many stayed and "collaborated," or became important churchmen; and it probably did not make a great deal of difference to an oppressed and overtaxed peasantry that their masters had changed.

The main remnant of Roman technology was a knowledge of iron smelting.

### *Agricultural Improvement*

In the subsistence economy of Europe from 400 to 900, iron tools were prized possessions. The scarce iron was used to provide sharp edges on implements made mostly of wood.

About 700, a new innovation in agriculture came to western Europeans from the northeast. It was a heavy plow that gradually allowed them to expand the amount of land under cultivation and improve their diets. The significant feature of this plow was that it cut into and under the soil and turned it over. Previously, most soil manipulation had been done by hand hoeing.

There had been simple scratch plows in the Mediterranean regions for many centuries. They were quite satisfactory for light, sandy soils. However, much of the earth of western Europe was heavy, wet clay. When the new plow turned over chunks of clay, exposing them to sun and air, they dried out and could be broken into soil fit for planting. (See illustration on page 71.)

The new heavy plow could not be pulled by a human, or even a single team of oxen. The force required to cut and turn the heavy soil took the effort of as many as eight oxen. The framework of the plow was wooden, with vertical and horizontal iron blades bolted on for cutting through the soil. The angled mould board for efficient turning of the soil was a later innovation.

In the earlier centuries, no single farmer was wealthy enough to own a whole plow. The cooperative spirit of village life is clearly shown in a Welsh law of about 945, described by Lynn White in his book *Medieval Technology and Social Change*:

> ...each plough-team is to plough at least twelve acre-strips before it breaks up, assigning one strip apiece to the ploughman, the driver, the owner of the plough-irons, the owner of the plough-frame, and to the owner of each of the eight oxen.

The new heavy plow (occasionally with wheels attached) improved the efficiency of cultivation. Land was better plowed, and more land could be brought into use. Gradually, over several centuries, as the economy improved, more blacksmiths could make more tools, which improved implements, which improved agriculture, which allowed for more blacksmiths... And soon with more and better iron tools, such as axes, trees could be felled more easily to make more land available. Indeed, trees had to be felled to supply the charcoal needed in the smelting and fabricating of iron.

Another feature of agricultural improvement was crop rotation. In neolithic times, peasants farmed a plot until the soil was depleted and then moved to a new plot. In early medieval times, peasants developed a technique of alternation. Each year, half the land was cultivated, and the other half used to pasture livestock. The animals' manure helped to improve the soil. The next year, the two fields were reversed.

One field was plowed in the fall and planted with wheat or rye. The other field was left fallow. Late the following summer, the first field was harvested and then left fallow for the next year. The previous fallow field was plowed and planted after the harvest.

Then, in the late eighth century, a three-field style of crop rotation began to be used. Over the next five centuries this new style of agriculture spread quite widely across western Europe. Frequently, the new style was more easily instituted by adding a third field, that is, by bringing new land under cultivation.

Added to the previous two fields, the third field was plowed in the spring and planted with peas, beans, oats, or barley. In the fall, this field was left fallow, and the previous fallow field plowed and planted with wheat. Next spring, the previous wheat field was plowed and planted with the new summer crops. Over a three-year cycle, a single field would be used in succession for fallow, fall planting of wheat, and spring planting of peas or beans.

Three-field rotation had three main advantages. First, two-thirds of the land was producing crops each year, instead of only half of it in the earlier scheme—a significant gain in productivity. Second, the new legume crops (peas and beans) were able to extract nitrogen from the air and supply essential nitrates to the soil. This helped to keep the land fertile. Finally, peas and beans improved the medieval diet by adding essential proteins. Grain crops are predominantly carbohydrates—the medievals obtained protein from meat, fish, dairy products, or legumes.

All these various agricultural improvements enlarged agricultural production. With some surpluses above the requirements of the producers, food was available to support more people in occupations other than farming. Artisans and merchants became more numerous, and commercial activities gradually increased. Improved agriculture also led to a significant increase in the population of western Europe. From about nine million in 700, the population doubled to 18 million by 1100 (400 years), and doubled again in 200 years, reaching 36 million by 1300. No rate of population increase exceeded that pace again until after 1750. Indeed, the population of western Europe declined drastically during the fourteenth century and did not surpass 36 million until after 1500.

## *The Horsepower Revolution*

Although horses were introduced in the Middle East around 1500 BCE, for many centuries they were used mostly for riding or carrying packs. Donkeys and camels were also used that way, as well as for turning some early mechanisms designed to grind grain or move water.

Horses became significant "engines" of the medieval economy only after the introduction of some clever harnessing gear.

### *The Technology of Farm Horses*

Horses began to be a factor in European agricultural improvement after about 1100. While these "oat burners" may be more expensive to feed than cows, they have more speed, strength, and endurance. Plows could be pulled by teams of two or four horses, instead of by four or eight oxen (an ox being usually a castrated bull). Horses improved transportation and commerce. They could be used to carry workers to fields farther from home, thus expanding the area that could be cultivated. They could also pull carts to transport produce to market, or carry a rider or pack. Just imagine slapping a saddle on a cow to ride to town.

However, before horses could be adapted to farm work, two major technological innovations were required. The first was iron horseshoes, to protect their hooves. Shoeing of horses became common after about 1000 as the number of blacksmiths increased. The second requirement was an appropriate harness. Oxen were often harnessed with a wooden yoke across their horns or shoulders. Because of their different anatomy, horses cannot use the harness appropriate for oxen. Early harnessing for horses with flexible straps was inefficient because it tended to choke them. The stiff leather horse collar appeared in western Europe in the tenth century, and was gradually improved by padding. The collar rested on the horse's shoulders and left ample room for its neck. With such a harness, horses could put their full strength into pulling loads.

Since horses don't normally provide milk and meat, how is it that horses made their appearance on European farms?

*Horses in Warfare*

The answer is: farmers used horses that were bred for battle, but were surplus from the needs of warfare. Horses had become prominent in warfare earlier as the result of a significant invention that appeared in western Europe during the seventh century—the stirrup.

In Roman times, horses were used to pull light chariots. Later, in western Europe, with many forests and hilly terrain, chariots were not of much use. There, horses were ridden to battle, where the soldiers would dismount and fight on foot. The introduction of the stirrup changed all that.

The stirrup originated in central Asia around 400. It was brought across Asia and into western Europe by tribes of nomadic warriors. During the seventh century, western Europeans adapted the stirrup for a new form of battle—mounted shock combat.

This D-shaped piece of wood or iron was hung by straps from the saddle, which was strapped around a horse's belly. The stirrup provided the rider with such a firm seat that he could rein the horse with one hand and wield a lance or sword with the other. During the course of centuries after the seventh, the armor of the warriors became more elaborate and heavier, and horses had to be bred specifically for carrying them.

*Equestrian statue of Luis d'Orleans.*

The system of shock combat–lance-bearing armored knights firmly mounted on their horses–was associated with the development of a whole new economy in western Europe.

## *The Feudal System*

Prior to 700, European villagers typically farmed the local fields in common. Each family was entitled to a share of the produce. After about 750 this scheme began to be modified. Kings assigned land to barons who agreed to supply knights for battle. The agreement between king and barons entailed a mutual fee (*feudum* in Latin, hence "feudal") or obligation. The barons swore oaths of allegiance to the king in return for his recognition of their ownership of an estate. In like manner, peasants gave their allegiance to the local baron, pledging to supply his labor in return for his protection.

In addition to feeding themselves, the peasants had to supply the baron and his retinue with food. And with horses. In the course of breeding horses for mounted combat, the peasants were able to use the culls for farm work. Thus, an innovation in warfare eventually led to significant agricultural improvements. Of course, in addition, the demands of warfare also increased the number of blacksmiths, who produced weapons and armor, and agricultural implements.

A lot of the battles were fought between neighboring barons and kings, descendants of the marauding tribes from the east. Also, there were still northern nomads, such as the Vikings, against whom the peasants needed protection. But one of the most enduring threats perceived by western Europeans came from the Muslims.

## *The Rise of Islam and the Crusades*

Islam, meaning "submission to God's will," originated with Muhammad in Arabia in the early seventh century. Muslim means "one who has submitted to God's will." As God's prophet, Muhammad was able to convince many Arabians to submit to God. Soon, Islam provided a strong unifying force among the Arabs. The period from about 650 to 750 saw a great expansion of territory controlled by the Arabs: extending from Spain in the west, across northern Africa, and southern Asia as far as India.

The Muslims seemed to be a threat to Europe at both its eastern and western frontiers. In the east, the Byzantine Empire gradually shrank over centuries until it included little more than the city of Byzantium. In the west, after the Muslims conquered Spain, they crossed the Pyrenees into southern France in 720. In 733, they were defeated at the battle of Poitiers and retreated southward.

Farther, during the eleventh century, Turkey was invaded by tribes from central Asia, who soon converted to Islam. In addition to being a threat to the

Byzantines, the Turks also interfered with Christian pilgrimages to the Holy Land.

The Byzantines appealed to their Christian brothers in the west for assistance. By this time, the pope in Rome had established a certain degree of authority over the kingdoms of western Europe. He and his bishops had become deeply involved in the feudal system. They brought the authority of the Church to the giving and taking of feudal oaths, and had even given up land to feudal barons as a means of resisting the Muslims from Spain.

We should be aware that among both Muslims and Christians, religion and politics were closely intertwined. Not that the priests were the state administrators as they had been in Babylon and Egypt, but they promoted loyalty to the state as a religious observance. Perhaps the most striking illustration was the crowning of Charlemagne as Holy Roman Emperor by the pope in 800. However, the authority of pope and emperor over powerful barons was never complete.

Nonetheless, western European barons did respond to the Pope's appeal in 1092 for the mounting of a crusade to free the Holy Land from the Turks. The feudal system had brought some stability to Europe, and agricultural improvement was encouraging the growth of trade. Although the agricultural surpluses were relatively small, they were sufficient to support the kind of venture represented by the Crusades. Over a period of 300 years, eight Crusades were sent to the Holy Land, with varying degrees of success.

## *Gothic Cathedrals*

Meanwhile, during the twelfth century, townsfolk throughout western Europe were benefitting from the economic growth that resulted from increased agricultural production. Old towns that had been established by the Romans were reviving; and new towns grew up beside some of the baronial castles. In medieval Latin the word for castle was *burgus*. The settlement that grew up beside a castle, frequently for commercial reasons, was called *falsus burgus*. That term became *faubourg* in French—the word for "borough" or "suburb." And from the same root, we derive the word for the middle (originally merchant) class, the *bourgeoisie*. These new towns were the sites of the local markets where the peasants traded their produce for goods manufactured by the growing variety of artisans in the towns: blacksmiths, weavers, clothiers, shoemakers, etc.

In addition, as certain towns developed specialties appropriate to local resources and skills, trade among the towns increased. Wine from France, for example, was traded for woolen goods from Flanders.

In the religious climate of medieval Europe, the building of large cathedrals became a way for towns to display their growing wealth. From 950 to 1150, a number of cathedrals were built in France, Germany, and England in what is called the Romanesque style. This style used semi-circular arches and barrel vaults. The vertical load of the vault was supported on massive pillars.

The lateral forces from the vault were balanced by very thick outer walls. Despite the heaviness of these stone structures, the architects were able to achieve significant dimensions.

Romanesque cathedrals ranged in length from 50 to 100 m, and achieved spans as wide as 12 m, with vaults as much as 30 m above the floor. Their plan was sometimes in the shape of the Christian cross, with a transept forming the bar at right angles to the long axis, which had the nave at the western end and the chancel (including the sanctuary) at the eastern end. Later, some cathedrals had towers at the western end, reaching as high as 50 m. Because of the heavy loads that had to be supported by the walls, the windows in these cathedrals were rather narrow.

About 1150, the Romanesque style began to be displaced by the Gothic style, which adapted some of the earlier features and added novel ones. The most striking aspects of Gothic cathedrals are their high pointed arches and relatively light, glass-filled walls. The pointed arch may have been of Muslim origin, but the Gothic architects carried it to great heights. The Gothic style is marked by the coherence of its design.

The pointed arch allowed greater flexibility, especially in the shape of spaces to be vaulted. The vault was made lighter by incorporating stressed ribs joined by lighter materials. The vault was supported by columns much lighter than in the Romanesque style. The lateral forces were carried by struts (called flying buttresses) across to relatively massive piers outside the building. With the internal columns and external piers supporting the roof and wind load, the walls were made relatively light. The walls were then filled with elaborate stained glass windows.

*Notre Dame de Paris from the east, showing flying buttresses.*

With their growing wealth and skill, French architects conducted a century-long campaign of cathedral building and innovation. Between 1150 and 1250, about 80 significant Gothic cathedrals were built in towns and cities across France. The vaults of the earlier cathedrals reached a height of about 25 m. The vault of the cathedral at Chartres, begun in 1194, reached 35 m, with a span of 20 m. Its outer dimensions are 50 m wide by 135 m long. Though the Chartres cathedral today has two towers at its western end, only the southern one was completed during the thirteenth century, to a height of 107 m. The later tower reaches to 115 m.

Cathedrals later than Chartres had even higher vaults; Amiens, started in 1220 reached 42 m (see picture on p. 66). At Beauvais, between 1225 and 1272, the chancel portion of a new cathedral was completed with the vault soaring to 48 m. Twelve years later the upper part collapsed. Over the next 38 years, a rebuilding program inserted new columns between the earlier ones to support additional ribs to strengthen the vault. The transcepts were not built until the early sixteenth century, and the nave of the cathedral was never built. At Beauvais, a 150 m tower was built of stone and wood in the 1560s, but it too collapsed, in 1573, while masons were preparing to reinforce it. They gave up after that. The Beauvais fragment stands today as mute testimony to unrealized ambitions.

During this hundred-year period, architects improved their designs and extended their skills and those of the artisans who did the work. The example of Beauvais suggests that they pushed their abilities and their materials to the very limit. The spire of the cathedral at Strasbourg was completed in 1439 and reached a height of 142 m, not surpassed until the building of the Eiffel Tower (300 m, in steel) near the end of the nineteenth century.

An intriguing association of the Gothic cathedrals to the philosophy current at the time of their building will be related in Chapter 6.

## *Harnessing the Power of Water and Wind*

Technical ability in the medieval period is nowhere better illustrated than in the mills: waterwheels and windmills. Milling is the process of pounding or grinding grain: first, to remove the seed coat; and second to grind the bare grains to flour. (The outer husk is previously removed in the threshing process.) In neolithic times, we imagine that grains were first consumed in porridge, for which the grinding was quite rudimentary. Stone pestles and mortars are known from before 7000 BCE. The presence of baking ovens in Iraqi houses by 6500 BCE suggests that some kind of wheaten cake or loaf was part of the neolithic diet; though there is no clear evidence of when the first flour was milled.

Presumably, after a few millennia, grain was ground more efficiently. After 3000 BCE, there is evidence that Egyptians were grinding grain using a cylindrical stone roller on a slightly concave base. At least from that time, we can consider bread as a staple in the human diet.

Evidence for other than hand milling does not appear until after about 500 BCE. Then, pairs of circular milling stones were arranged so that a donkey or camel could rotate the upper stone while the miller poured the grain into a central hole in the upper stone. The grain was ground to flour between the stones and emerged at the outer edge.

*Water Mills*

The first application of water power to milling dates from about 100 BCE. The simplest form of water power used blades attached to a vertical shaft. The shaft passed through a hole in the lower, fixed millstone and was attached to the upper stone. A stream of water directed against the blades caused the upper stone to rotate. Since the plane containing the blades is horizontal, this type of wheel is called horizontal. (It is the forerunner of the water turbines so widely used in hydroelectric generation from 1890 onward.) Being a direct drive, the stone turned at the same rate as the wheel.

*An undershot waterwheel operating bellows for a furnace. From Boeckler,* Theatrum machinarum, *1662.*

Vertical waterwheels (with the axle horizontal) are more adaptable than horizontal wheels. But they require some device for converting the circular motion in the vertical plane to motion in the horizontal plane to drive the millstone. Greek engineers in Alexandria about the time of Archimedes wrote about devices using gears. Indeed, recent evidence suggests that Greek engineers of that time had developed a variety of very ingenious geared mechanisms. Gearing for waterwheels was much cruder.

The simplest gearing to allow a vertical wheel to drive a horizontal stone used the wheel itself as a gear. Wooden dowels were inserted at regular intervals around the circumference of the wheel, projecting horizontally to the side. As they reached the top of their travel these "teeth" engaged a lantern gear on a vertical shaft connected to the millstone above it. The lantern-wheel looked somewhat like a cylindrical bird-cage; with "top and bottom of circular boards connected by staves inserted at equal intervals along their circumferences." This arrangement not only converted the direction of rotation, but also permitted a speed ratio between the two gears other than 1:1. In later mills, the gearing was transferred to the end of the shaft driven by the wheel.

The Romans used waterwheels to a limited extent. Typically, the wheel was undershot, that is, with a stream striking the blades at the bottom of the wheel. The millstone rotated at about five times the speed of the wheel. For an undershot wheel of 2 m diameter, rotating at 10 rpm, the power output is around 2 kW. This kind of mill could grind 150 kg of grain per hour, compared to about 7 kg for a pair of slaves, or about 30 kg for a donkey-mill.

Water mills seem not to have been widely used in Roman times. Historians still argue about why. Some attribute the low level of mechanization to the abundant use of slaves; others to the desire of administrators to avoid technological unemployment. Evidently, for whatever reason, the need for water-powered mills was not felt strongly.

A stronger need for mills was felt in medieval Europe. The Domesday Book survey of central, southern, and eastern England in 1086 recorded 5600 water mills. This represents about one mill for each 50 households, and about one mill per ten square kilometers. This period is the source of those traditional stories of our childhood about the "jolly miller."

Undershot wheels required a relatively fast moving stream. In hilly areas, the more efficient overshot wheel gradually made its appearance. (Some details of waterwheel efficiencies are given in Chapter 13, where John Smeaton's analysis is discussed.) The water was directed to the top blades (or buckets) of an overshot wheel, providing more than double the force per volume of water available from an undershot wheel. The overshot wheel required the mill to be situated so that there was a fall of water of a height equal to the diameter of the wheel. Later, slower moving streams were dammed to provide the appropriate fall.

Between 1000 and 1300, there was a many-fold increase both in the number of waterwheels in Europe and in the variety of their uses. For a long time, milling grain was the major use. Gradually, water power was adapted for other tasks. The circular motion could be used to drive a continuous chain of leather balls, which rose through a pipe to lift water. Eventually, lathes were operated by water power.

A more dramatic use was the conversion of circular motion to reciprocating motion. The cam-shaft was known to the Greek engineers, but not widely applied until medieval times. Cams projecting from the horizontal axle of a waterwheel alternately engaged, lifted, and released one end of a pivoted lever. A heavy hammer at the other end of the lever was used for many different pounding operations—the fulling of cloth,

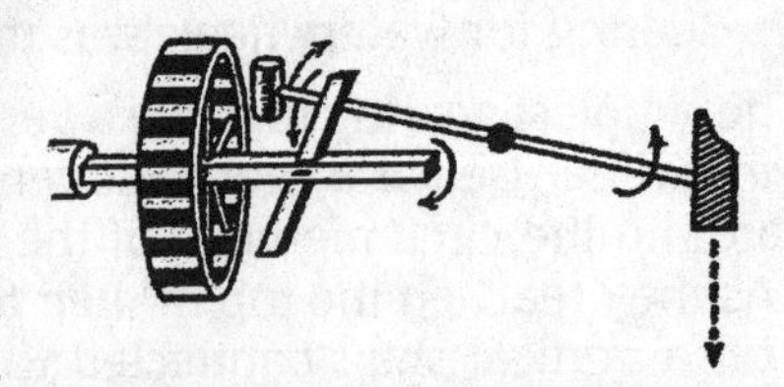

*A cam to convert rotary to reciprocating motion.*

paper making, iron forging, pounding oak bark in water for tanning leather, etc. Mills were also adapted to reciprocating motion to drive saws for cutting wood.

In some districts in France, the number of mills increased ten-fold or more within two centuries. The variety of uses can be seen from the region around Troyes, 130 km east of Paris. In 1493, they had 40 mills: 20 for grinding grain, 14 for paper making, 4 for cloth fulling, and 2 for leather tanning.

*Windmills*

Windmills seem not to have been known in antiquity. The earliest evidence for the mechanical use of wind power (other than for ships) comes from China about 400. Cloth scoops attached to a vertical shaft were used to turn prayer wheels. By 900, windmills of the same basic design were being used in the Middle East to grind grain. Allowing for the switch from water to wind, their basic structure is like the horizontal waterwheels, except that the stones are below the vanes in a windmill.

*A post windmill. The entire structure could be rotated to face the wind.*

In western Europe, the typical windmill had four vertical sails attached to a horizontal shaft. It seems to have appeared quite suddenly in the decades just before 1200 and spread quickly into regions rich in steady winds and poor in flowing streams. Later, various ingenious devices were devised to allow the upper part of the structure to rotate in order to face the vanes into winds from different directions.

## Clocks and Other Mechanisms

Keeping time by the sun goes back far before recorded history. Mostly, people arose at sunrise and went to bed at sunset. By watching the sun, they had no trouble distinguishing morning, noon, afternoon. And that was enough.

By 3000 BCE, Babylonian and Egyptian astronomer-priests were able to be more precise about the time of day. They could keep track of the angles made by the shadow of a vertical stick—a primitive sun dial. At night, they kept time by noting the rising of particular stars.

Later, for keeping track of shorter time intervals, the ancients developed devices like graduated candles and waterclocks. In Greek times, various mechanisms were attached to containers of water from which the water was allowed to flow in a thin stream. With elementary gearing, a waterclock could be used to turn a hand on a dial.

For thousands of years, the majority of people had no need even for rudimentary clocks. Farmers get enough information from the sun. And town dwellers followed much the same kind of daily routine as on the farms. Then, in medieval times, the situation changed because of the requirements for prayer. In the daily round of Christian life, prayers were said six times a day. Priests in the cathedrals and monasteries rang bells at the appropriate times to call the faithful to prayer. One of the tasks of the priests or monks was to be sure that the bells rang at the correct times.

From about 1200 there are indications of mechanisms attached to waterclocks to ring bells at regular intervals. But a waterclock suffers from two main disadvantages. It must be refilled with water at relatively short intervals, and it can freeze in the winter. Of course, freezing had not been a frequent problem in the Middle East or around the Mediterranean. In northern France and England, winters were colder.

Eventually, shortly before 1300, the weight-driven mechanical clock was invented somewhere in western Europe. The actual origin is entirely obscure, and historians today are still guessing about the sequence of development. Yet it seems to be clear that weight-driven clocks did exist before 1300.

In its basic structure, a mechanical clock is driven by a falling weight attached to a cord wrapped around a shaft. As the weight falls the shaft turns. The major design problem was to find a way to keep the weight from falling all at once. A system of reduction gearing attached to the shaft could slow the rate of fall, but crude iron gears could not be given fine adjustments, particularly because of abundant friction. A way had to be found to alternately check and release the fall of the weight.

The device to accomplish that is called an **escapement**. The earliest escapement mechanism is called a **verge and foliot**. It had to allow the stop/start of the falling weight, at regular intervals. This was achieved by having the teeth of a vertically set gear wheel (a crown gear with its teeth perpen-

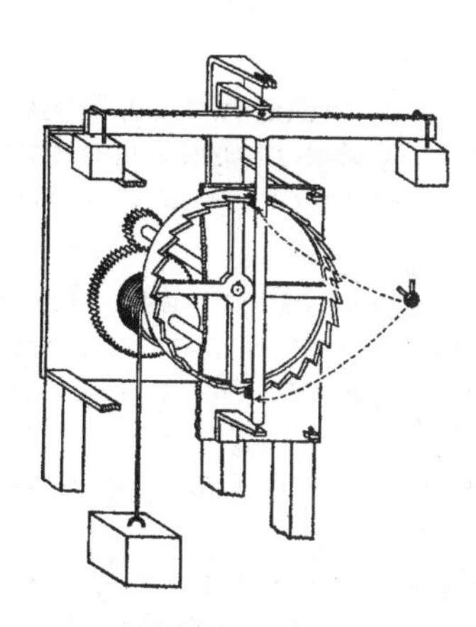

*A verge and foliot escapement mechanism for a weight-driven mechanical clock.*

dicular to the plane of the gear) alternately caught and released. The verge was a metal rod set up alongside the crown gear with two small metal tabs, called **flags**, sticking out. The flags were set at different angles. One flag engaged teeth at the top of the gear, the other at the bottom. At the top end of the verge, there was a horizontal bar (the foliot) with movable weights on it.

To start the clock, you swing the foliot aside. That allows the crown gear to rotate, letting the weight drop a bit. But then, the upper flag on the verge engages with a tooth on the crown gear and stops it (and the falling weight). This dynamic action causes the tooth to kick back on the flag and swing the verge and foliot back in the opposite direction. The weight falls unchecked for a moment. The next moment the other flag on the verge engages a lower tooth, which after checking the rotation of the crown gear is kicked back. As the verge and foliot rotate regularly back and forth, the flags alternately engage and release the crown gear, and the weight descends slowly in jerks.

With proper design and adjustment, the clock can operate for twenty-four hours before the weight must be rewound. Hands and bell-ringing mechanisms are attached to appropriate shaft of the reduction gears.

The earliest clocks had gears a couple of feet in diameter and were probably made by blacksmiths. They were fairly crude, with only an hour hand. Yet, in less than a hundred years, some very fine clocks were being made in England and Italy. One of the most mechanically elaborate clocks was the astrarium made by Giovanni de Dondi of Padua in the 1350s.

Dondi built his astrarium to display the positions of the seven planets—a mechanical representation of the heavens. He cut the gears with exquisite precision, some of which had to be elliptical. His clock was a masterpiece of mechanical ingenuity. (See picture on page 72.)

During the fourteenth and fifteenth centuries, clock watching entered fully into the consciousness of western European town dwellers. The first clocks were in the cathedrals and town halls. Soon, wealthy aristocrats and merchants had table clocks. Shortly after 1400, smaller table clocks were being driven by coiled springs. As the spring drive was perfected, it was also miniaturized. Before 1500, pocket watches were available to the well-to-do, more as status symbols than time-pieces, since they were not very accurate.

Mechanical clocks also had an intellectual influence. From Dondi's time on, we find some philosophers referring to the universe itself as a kind of

*The gear mechanism of a weight-driven clock.*

clockwork mechanism. (This metaphor of universe as mechanical clock just wasn't possible until there were clocks.) So, the skills of mechanics like Dondi influenced philosophers, who gradually changed from thinking of the universe as an organism—the way Aristotle taught—to thinking of it as a mechanism.

While clock watching transformed Europeans' sense of time—with regular hours independent of sun and weather—clockmaking transformed technology. The gears of the first clocks could be made by blacksmiths at their forges. The finer clocks and watches required much more subtle fabrication. Goldsmiths added watchmaking to their skills of detail work on precious metals and gems. It is no wonder that even today, a jeweller is a person who deals in watches as well as in earrings and brooches.

By 1400 then, there were a number of trades skilled in making gears of one form or another: millwrights used wooden or iron gears in water mills and windmills; blacksmiths and clockmakers used finer iron gears for large tower clocks; jewelers and watchmakers used very fine gears, usually of brass, for household clocks and watches.

Gears, shafts, and cams represent one collection of devices for transmitting circular motion and converting it to reciprocal motions. In the medieval period other devices were developed for transmitting and converting motions. Simple cranks seem to have been known from the ninth century and were slowly adapted to use in such devices as grindstones. The compound crank was developed in the fourteenth century and was applied to the carpenter's brace by 1420. Connecting rods were attached to compound cranks soon after.

*Textile Machinery*

Improved mechanisms were also applied in the textile industry during this period. Weaving had traditionally been done on a rectangular wooden frame with warp threads stretched out in one direction, while the weaver interlaced the weft threads over and under the warp. Eventually, the warp threads were attached to strings by a harness so alternate threads could be lifted together, allowing the weft to be passed through the space between them on a shuttle. By 1200, a treadle mechanism was applied to the loom to allow the weaver to lift the harness by foot action.

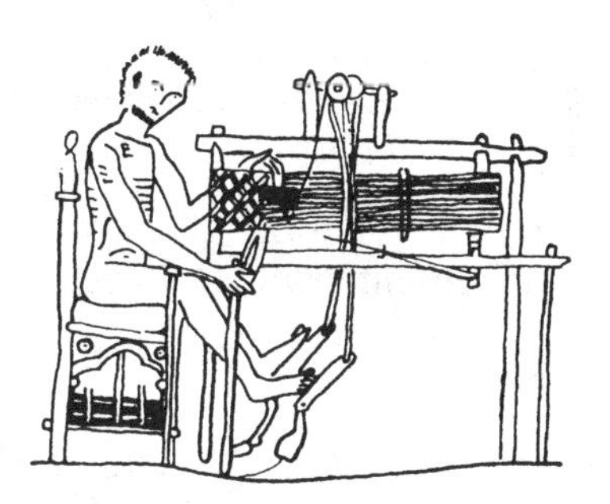

*Weaving loom with treadle. Drawing from the 13th century.*

Traditional spinning had been a hand operation in which the spinster laid in tufts of wool and twisted them into a thread. After that, she wound the thread onto a spindle. Some of the spinning operations were mechanized with the introduction of the belt-driven spinning wheel about 1280. Treadle-operated spinning wheels with crank and connecting rod did not appear until after 1500.

Throughout the medieval period, the manufacture of textiles gradually became more industrialized. That is, instead of spinning and weaving being mainly family occupations on the farm, they moved into the towns where specialized workers congregated to produce thread and fabric for owners of the machinery—the textile merchants. With the introduction of mechanical clocks, hours of work were much more closely regulated than they had been before. In the aftermath of the Black Death, there were considerable labor shortages. This gave the laborers more power, and besides seeking higher wages, they complained about the routines imposed by the clocks. Legal rec-

*The Black Death seen as God's punishment.*

ords of the period give evidence of some of the same kinds of labor-management disputes that are still common today.

## *The Black Death*

The population decline of the mid-fourteenth century resulted from the widespread depredations of the bubonic plague—the Black Death. It struck a European population that was already experiencing a decline in health from the peak reached about 1300.

As we've seen, various technological developments had been improving the European food supply from 1100 onwards. The food supply also benefitted from a slight increase in temperatures during that period. The European climate had become milder. Then, after 1300, there was a small decline in temperatures. Winter blizzards occurred in places that hadn't known them for a couple of centuries. But the population of Europe had already expanded to consume that earlier enlarged food supply so that when the climate became unfavorable, the quality of the diet declined. Between 1310 and 1330 a couple of severe famines swept western Europe.

And then, in 1348, the Black Death struck! The bubonic plague had been endemic in the Far East for many years. During one of its periodic flare-ups, fleas, which carried the bacteria, infested rats and cargoes on ships carrying goods from the east to European ports. The plague spread across Europe like wildfire. By 1351 the population of Europe was only two-thirds of what it had been just three years earlier. This was a terrible epidemic with drastic results. Some people saw the plague as God's punishment and gave all their goods to the church in their panic for salvation. Others decided it was better just to eat, drink, and be merry. At all events, the normal routines of daily life were shattered for many years.

## *Conclusion*

Between about 700 and 1300, western Europe underwent a significant transformation. Its population multiplied four times, to 36 million. Diets improved as agricultural production was expanded and made more efficient. Food production was improved by the use of water mills and windmills. Mills equipped with camshafts and other machinery enlarged the capacity of artisans to produce clothing and other goods.

Politically, western Europe was ruled by barons, some of whom gave allegiance to a king, as in France and England. Much of the rest of Europe was a patchwork of duchies and principalities. At the same time, the pope in Rome exerted considerable spiritual authority and a degree of temporal authority as well. Evidence of the role of the Church in European life can be found in the Crusades and the building of cathedrals.

The six hundred years from 700 to 1300 saw the foundation of styles and attitudes in politics, religion, and technology that were uniquely European. One might almost say that Europe was invented during this period.

As trade increased, towns became larger, and craft skills became more specialized. In the fourteenth century, Nuremburg in Germany had a population of 40 to 50 thousand. From that time, there is a list of 50 different trades and crafts in Nuremburg, comprising more than 1200 separate shops or establishments. More than 500 of those were in the clothing trade: shoemakers, tailors, furriers, dyers, hatters, etc. Artisans concerned with materials and implements such as braziers, tinsmiths, locksmiths, needle makers, and pewterers numbered more than 250. There were more than 160 food producers like bakers, butchers, and fishermen. Artisans involved in transportation, like farriers, coach builders, and stirrup makers, numbered more than 130. More than 50 worked in building trades such as carpenters, glaziers, and stone masons. Just fewer than 50 supplied military gear like mail gloves and shirts and swords. Finally, there were 17 money-changers and 16 goldsmiths.

With the Black Death, the medieval flowering of Europe was brought to a near standstill. Whatever next steps in technology and science awaited Europeans, they were delayed. But only temporarily. By 1450, Europeans were on the brink of conquering the world.

Before we get to that European efflorescence, we need to consider also some of the intellectual components of medieval life.

*A university lecture as portrayed in the fourteenth century.*

## CHAPTER SIX

# Recovery of Classical Knowledge

In Chapter 4 we examined some aspects of Greek natural philosophy. Here we shall consider how that and other work was transmitted to the Latin West in medieval times, and what western Europeans did with it. This was a transmission both in space and time: by a roundabout route from Greece through the Near East to Spain, and thence into France, and spanning a period of well over a thousand years. As we'll see, the very words themselves also followed a circuitous route of translation from language to language.

### *The Preservation of Greek Knowledge*

After the golden age of Athens (to about 300 BCE), the center of Greek philosophical activity shifted to other places, particularly to the Greek city of Alexandria, built on the Mediterranean coast of Egypt. Scholars in Alexandria continued to study philosophy, mathematics, and astronomy into the early period of Roman domination. Their activities were centered in the Museum, to which was attached a magnificent library. In following centuries, conquests and re-conquests took their toll on scholarship and the contents of the library. By 400, with the decline of Roman authority, little of the earlier glory of Alexandria remained.

However, from the start of the Christian era, for some hundreds of years, small pockets of scholarship were established in other places in the Middle East. These largely had the effect of preserving the writings of Athenian and Alexandrian scholars, either in their original Greek or translated into Syriac or Aramaic, re-copied from time to time as their papyrus degraded through slow oxidation.

After 650, Muslim Arabs conquered most of the southwest Asian territory that had earlier been controlled by Rome. Soon they had spread across north Africa, all the way to Spain. In 763, a new Muslim capital was founded at Baghdad, near the site of ancient Babylon, in Iraq. Within a century, Baghdad became the largest, richest, most magnificent city in the world. Along with its pre-eminence in trade, administration, architecture, and art, Baghdad took a lead in scholarship. At first, the Muslim scholars translated works by Aristotle, Ptolemy, and others from Syriac into Arabic. Soon, the Arabs were studying, extending, and commenting on those texts, as well as translating others directly from Greek. The Arabs also did important new work in mathematics and alchemy.

Eventually, many of the ancient Greek writings we know today were available in Arabic, not only in Baghdad, but also at the western end of the Mediterranean in Muslim Spain. From there they were available to medieval Latin Europe.

Although Islamic control of Spain was declining by 1100, Muslim scholarship was still strong there. And in Toledo, in central Spain, there were opportunities for European Latin scholars to meet Arabic scholars. Between 1100 and 1200, many of Aristotle's works were translated from Arabic into Latin. Sometimes this was done by Jewish or Castillian scholars, which meant that Aristotle would be translated from Arabic to Hebrew or Castillian, and thence to Latin. These scholars also translated medical writings and Ptolemy's *Almagest* into Latin, as well as original works and commentaries by the Arab scholars themselves.

### *The Rise of Western Scholarship*

In 1100, there were scholars in western Europe, particularly in France, who were able to profit from the Arabic and Greek learning available in Spain. Three hundred years earlier, there hadn't been.

Even earlier, in Roman times, scholarship had generally been at a much lower level than during the golden age of Greece. Such Roman learning as there was didn't penetrate very deeply into western Europe. Through the influence of Christianity, officially adopted by Rome after 300, Latin religious texts provided the main basis for learning in the west.

Scholarship in western Europe continued at a low ebb until 800. Then, as kings and popes gained more authority, some of them wanted their priests to be better educated. Charlemagne (Charles the Great) had loose control over a wide territory consisting of much of present day France, Germany, and northern Italy. He wanted his priests and administrators to be trained in at least the rudiments of reading, writing, and thinking. So Charlemagne ordered cathedrals in major centers to establish schools.

*Charlemagne.*

These schools prospered fitfully over the next three hundred years. But by 1100 there were a few centers where the tradition of learning was firmly established—places like Paris and Chartres—with a regular curriculum in place. Medieval education had two major components called the trivium and the quadrivium. The trivium had three parts—grammar, logic, and rhetoric—based mostly on Latin authors like Cicero. Grammar describes how to put words together correctly into sentences; logic describes how to put sentences together correctly into arguments; and rhetoric describes how to put arguments together correctly into discourses.

The quadrivium may sound surprising. It consisted of four mathematical subjects—arithmetic, geometry, astronomy, and music. The core of the quadrivium could be traced back through Latin authors to Greek sources, though the European students of 1100 knew no Greek. However, once they'd mastered the trivium and the quadrivium some scholars wondered what came next. They started looking around for writings more advanced than those of the quadrivium. They found them in Muslim Spain.

## *Scholastic Philosophy*

By 1200, Latin scholars in Paris, Oxford, and other centers had access to the intellectual riches of ancient Greece, and of Islam. An embarrassment of riches for medieval Christians. How could they use the wisdom of pagan Greeks and infidel Muslims? What they tried to do was to reconcile two major fields of learning, to incorporate Greek philosophy into Christian theology. That is, the medieval European scholars tried to apply the rules of logic to the central tenets of their faith—to marry faith and reason. During the three hundred years from 1100 to 1400, this intellectual activity followed several courses, which often varied from school to school. And for that reason, this philosophic period is often called scholasticism—the philosophy of the schools.

In the early period of scholasticism, the scholastic philosophers examined the tenets of their faith—such as the nature of God and the Trinity—using the logic they were learning from Aristotle. For example, the philosopher Peter Abelard listed over 150 sets of statements from holy writings that seemed on the surface to be contradictory. With the faith that the holy spirit could not be guilty of contradiction, Abelard tried to find logical ways to account for the differences. His predecessors had been content to teach the doctrines of faith; Abelard analyzed them. And he did it with the conviction that faith and reason could be combined in a wholly consistent way. Others at the same time, felt that the very existence of God could be proved by rational argument, that logic would support faith.

Another feature of the scholastic philosophers was the way they organized their material. They divided books into parts, parts into chapters, and chapters into sections. That's the way we divide technical treatises today, but it was a new style in the 1100s. It put order and relation into their thinking and their writing. Curiously enough, a similar kind of ordering and relating appeared at the same time in the style of the great Gothic cathedrals that were built between about 1150 and 1250.

### *Scholasticism and Gothic Cathedrals*

While the philosophers were dividing and subdividing their texts in an orderly way, the architects were dividing and subdividing the cathedrals in an orderly way. The cathedrals were divided into three parts like a cross: the top

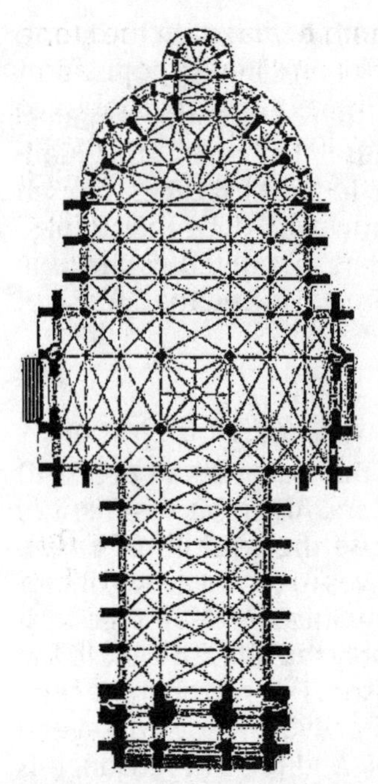
*Floorplan of the cathedral Notre-Dame d'Amiens.*

of the cross is the chancel (or sanctuary), the bar is the transept across the church, and the main support of the cross is the nave or body of the church. The chancel represents the Father, the transept the Son, and the nave the Holy Spirit. Each of these three parts is then divided and subdivided on a plan in which similar forms are repeated again and again. Just as the pattern of scholastic logic is repeated over and over. This conjunction of architecture and philosophy was not accidental. The architects learned the idea of order and relation from the philosophers in the cathedral schools.

One of the important forms the architects used was the pointed arch, which allowed them to build the high vaulted interiors that are the glory of the gothic style. The same pointed arch appeared in the shape of the windows. In later cathedrals it was also the form of various small supports and struts. While the pointed arch was necessary to create the tall vault, it was also used simply as a design pattern in places where other shapes would have done just as well. Everywhere you look in a gothic cathedral you see pointed arches. Just as everywhere you look in scholastic philosophy you see logical analyses.

There's also a logic in the way the architects used the shape of the pointed arch: in windows and other places, the architects set out to inscribe a circular stone tracery within the pointed arch. If you look at the designs of different cathedrals, you'll see that various architects did that differently. And it looks as if they were learning from one another, trying to fit the circle into the arch more perfectly in each succeeding cathedral.

If you think of the arch as pointing upward to heaven, you can imagine it to represent faith. And the circle, so important to Aristotle and Ptolemy, represents reason. So, while the scholastic philosophers were trying to marry faith and reason in their writing, the architects were performing a similar intellectual task in their design of the cathedrals. And in both cases, the men involved were engaged in a developing enterprise—learning, elaborating, and improving as time passed.

These Gothic cathedrals were also a very important part of the lives of medieval people. They were by far the largest buildings in the landscape. Peasants in the fields around a town could see the towers of their cathedral looming skyward in the distance. And they could hear the bells in the towers chiming for prayers or summoning them to worship. The cathedral and the Christianity it represented dominated their lives. When they attended mass on Sunday, they could gaze in wonder at the elaborate images in the stone carvings and the stained glass windows. Designed for an illiterate population those images told all the stories of their faith. The life of Christ, the stories of the Old Testament, the miracles of the saints, the exploits of their heroes—all were portrayed in stone and stained glass. Within the cathedral, the high vaulted nave drew their gaze upward, as the priests encouraged the congregation's minds and thoughts upward too. The cathedrals were the symbols and centers of the medieval age of faith for the common people, even while the philosophers were analyzing that faith with the logic of pagan Greeks.

*The Decline of Scholastic Philosophy*

After 1200, the scholastic philosophers had virtually all of Aristotle's works in Latin. Whereas a couple of generations earlier their work had been to use Aristotle to test and improve their faith, now, they began to use the tenets of faith to Christianize Aristotle. So, men like Albertus Magnus and Thomas Aquinas went through Aristotle with a fine-tooth comb. As well as analyzing his statements by his own logic, they asked again and again, "Is that something a Christian can believe?" If not, they rejected it—as philosophers they had to try to reject it on logical grounds.

For example, Aristotle had written that the universe had existed forever. That contradicted the creation of the earth by God in the early chapters of Genesis, so the scholastics looked for arguments against Aristotle.

For example, if the universe had no beginning, that would mean that its age is infinite. Now we know that Adam has had a large number of descendants in only six thousand years. So, if the world had existed from eternity, then the number of humans now must be infinite. It isn't—so the world can't have existed forever.

A second argument was based on God's infinite power. All agreed that God has existed forever. To say that the universe must have existed forever, would mean that God would have to have created it in the first instant. But that would be a limitation on God's power—by specifying when he had to create the universe. That would destroy God's freedom to decide when to create the universe. Therefore, since God is all powerful, he must have had the freedom to choose when to create the universe. So the existence of the universe is not co-terminal with God, and thus not infinite.

In arguments like these the scholastics were willing to consider and evaluate every possible argument and opinion. Often they engaged in the kind of logical hairsplitting that has come to be called "scholastic" in the derogatory sense of "pedantic."

The scholastics reasoned about the questions they raised. They didn't simply assert that the Bible says "the universe was created at a particular time," they also looked for logical arguments to justify it. That got them into trouble because the very fact that they were arguing implied that they considered that the outcome was arguable. In the 1270s, church authorities in Paris felt that their philosophers were getting out of hand. So they published a long list of statements that philosophers were forbidden to hold or argue. If the philosophers had said, "but our faith is pure—trust us always to arrive at the proper conclusions," the answer would have been: "then if you admit that faith is supreme, what's all the arguing about?"

The fact is that faith and reason are radically different styles, and the scholastics failed in their attempt to marry them. It took them the better part of 200 years to find that out. In the process they learned a lot about logic and a lot about Aristotle. But after 1300, men who were good at logic just stopped talking about faith. And, at least for a while, men who were good at faith stopped talking about logic. Through the course of the 1300s, theologians became more mystical, and philosophers started to concentrate on natural philosophy. They did some interesting work, but over a couple of hundred years they mainly concentrated on the elaboration of issues raised by Aristotle and his Arabic commentators.

## *Astronomy in Medieval Times*

The works of Ptolemy were prominent among the Greek texts that were translated and studied by the Arabs. Muslim scholars studied Ptolemy's *Almagest* in considerable detail and made his work more accessible to later astronomers together with their additions to his work.

Muslim astronomers also preserved and improved astronomical instruments used by the Greeks. One in particular is the astrolabe. The astrolabe is mainly a device for making astronomical calculations. It consists of a metal lattice-work star chart that can be rotated above one or other of several plates engraved with a grid of circles and radii so that a star's position can be read for a given time. A separate plate is needed for each latitude of interest. Usually, the back face of the astrolabe had engravings used for geometrical calculations, as well as sighting pins attached to a bar that can be rotated, for measuring the angles of the stars. Astrolabes were used by astronomers for more than 500 years.

*An astrolabe, from about 1500.*

Ptolemy's great astronomical work was translated into Latin before 1200. During the next couple of centuries, a number of the Muslim astronomical works were also translated. Latin scholars did not make any significant additions to the analysis of planetary motions by Ptolemy and the Muslims. However, they did do two things. First, they wrote simplified summaries, which would make the basic ideas accessible to students.

About 1230, Joannes de Sacrobosco (Latin for John of Holywood, who died about 1250), a mathematics professor at the University of Paris, wrote an elementary treatise *On the Sphere*. Based on Arabic commentaries of Aristotle and Ptolemy, Sacrobosco's book was essentially an introduction to the structure of the universe in simplified terms. He provided arguments for the earth being a sphere in the center of the universe, and described the planetary system briefly. Although Sacrobosco outlined the structure of earth and heavens, his book made no contribution to mathematical astronomy. Rather it provided students of the middle ages with their cosmological setting. Its picture of the universe was the one adopted by Dante when he wrote *The Divine Comedy* in 1314. The *Sphere* was popular for so long that it was one of the early printed books, appearing in 30 editions between 1472 and 1500.

The other medieval work in astronomy involved the updating of Ptolemy's tables for the positions of stars and planets. Because of several long-term shifts in the heavens, the original tables became less accurate over the centuries. Muslims in Baghdad, Toledo, and other places had already made revisions. In the Latin West, the most famous tables were produced by scholars in Spain and France between 1275 and 1325. Based on Arabic sources, these tables were attributed (wrongly) to the patronage of King Alfonso X. The *Alphonsine Tables* appeared in print for the first time in 1483 and were not superseded until 1550.

Other works of Ptolemy were also translated into Latin. One was a geography, that is, a description of the surface of the earth, providing lists of coordinates for the locations of numerous cities and landmarks. (See picture on page 113.) This work, together with another on optics, contained mathematical theorems for projecting a spherical surface onto a plane. Probably the most popular of Ptolemy's works was one on astrology.

## *Astrology and Alchemy*

There are two antique "sciences" that were very influential in their own time—astrology and alchemy. Today, most of that influence has disappeared. However, astrology was, for a couple of thousand years, one of the main motivations for the study of astronomy. And alchemy was the prominent "science" of materials until it gradually gave way to chemistry during the period from 1650 to 1800. Both are worth our notice here because of their earlier importance, and because they did influence aspects of studies we consider to be properly scientific; for, truth comes more readily out of error than confusion.

### *Astrology*

In Ptolemy's time, one main use of astronomical data was for making astrological judgments. That is, to determine from planetary positions the character of a person, or to foretell events. Ptolemy's book on astrology contained much useful information about the influence of the sun on climate and weather, as well as the planetary "influences" that still pervade astrological writings today. It seems clear that Ptolemy was as serious about his astrology

*A caricature of an alchemist's workplace by Pieter Breughel the Elder.*

as his astronomy. Yet, while astrology may have been a major motivation for his astronomical analyses, he made them far more detailed than astrologers could profitably use.

Although the Christian church tended to reject astrology, its influence was felt in Europe for a long time. Many medieval scholastic philosophers took some of the astrological tenets for granted, even as they strove to Christianize pagan learning.

*Alchemy*

Along with astrology, alchemy has been called a "fossil science." Both have their roots in ancient lore, dating back to Egypt and Babylon. And both represent attempts by humans to come to terms with the mysteries around them. One main difference is that astrology tends to be a somewhat passive exercise, while alchemy is more active. An astrologer's main task is to tell you something about yourself and your future; then it's up to you to decide how to prepare for it. An alchemist, on the other hand, has recipes for manipulating materials to produce startling results.

Since alchemy was for a long time a secret art, its adepts often refused to commit it to writing, or else wrote in occult codes. Some Muslim alchemical works were translated into Latin along with the other writings previously described. In addition to alchemy's "magical" goals (see page 62), the Muslim alchemists made significant additions to what we could call genuine chemical knowledge. They were familiar with metals and alloys, soda and potash, and a variety of salts. Before 800, they had a recipe for the preparation of nitric acid.

The Muslim alchemists knew a lot about a number of chemical reactions, including those used in such trades as metallurgy, dyeing, and glass making. Their work is still embedded in our language in words like alcohol and alkali. When the Muslim alchemists distilled wine, they called the condensed liquid, spirit of wine, because they couldn't actually see the original vapor that produced it. Alcoholic liquors are still called spirits.

Sometimes the Muslim alchemists derived the names for substances from the process by which they were obtained, and sometimes from their appearance. Alkali refers to a whole class of materials that were first used to make soap. They were obtained from the ashes of burnt plant materials. The word alkali comes from the Arabic word for frying, because they obtained alkalis by frying plant materials. An artisan poured water on the ashes, which dissolved the alkali out of the ashes. The alkali that comes from wood ashes is potash. The alkali that comes from the ashes of seaweed is washing soda. Soda is another Arabic word, which means splitting. The name soda was given to the alkali obtained from seaweed, because it was used to cure splitting headaches—no kidding!

The alchemists called some shiny minerals by the name vitriol, from the Latin word for glass. They had blue vitriol and green vitriol. In smelting minerals to obtain metals, artisans got copper out of blue vitriol and iron out of green vitriol.

Later, alchemists found that if they strongly heated powdered vitriol and condensed the vapor, they'd get a liquid that would sting the skin. They called it spirit of vitriol. They also identified it as an acid (from the Latin word for "sharp"). And that's how the word "vitriolic" came to mean sharp and biting. The alchemists also made an acid when they burned sulfur and dissolved the resulting vapor in water. They called that, spirit of sulfur. Only after 1600 did European alchemists realize that spirit of sulfur and spirit of vitriol were the same thing. We call it sulfuric acid. But as long as the two things were obtained by different processes, it was quite reasonable to call them by different names.

The translation of the Arabic alchemical texts into Latin made such knowledge available to the Latin West in medieval times. As in astronomy, the medieval Europeans did not make many significant additions to the Muslim achievements in alchemy.

## Conclusion

From a scientific point of view, the medieval Latin period did not add very much positive knowledge to what had previously been done by the Greeks and the Arabs. Nevertheless, it was that Greek and Arab knowledge, in its new Latin dress, that provided the groundwork out of which modern science grew.

Because of the circuitous route of transmission of Greek learning, some of the works may have undergone four translations: Greek to Syriac, Syriac to Arabic, Arabic to Hebrew, Hebrew to Latin. We may be amazed that anything like the original ideas could have been preserved along the way. However, scholars in the Renaissance gained access to manuscripts in the original Greek from Constantinople, which sometimes allowed them to confirm or correct their earlier manuscripts. In this process, we should be aware that we are talking about ideas that originated as early as 300 BC, although none of the manuscripts available today (or likely available in 1500) dates back farther than about 1000. Not only has our knowledge of ancient Greece gone through several translations, it has also gone through numerous recopyings.

*Medieval scribes. An engraving from the early 1500s.*

*The Arabic astrologer Albumasar.*

*A portrait of Andreas Vesalius demonstrating the muscles of the arm, from Andreas Vesalius, On the Construction of the Human Body (1543).*

# PART 3

# *The Renaissance World*

CONTENTS

Renaissance (literally "re-birth"), as a time period, has rather elastic boundaries. It was used by cultural historians in the nineteenth century to mark what they saw as the revival of the classical forms of ancient times—Greek and Roman. In literature, Dante Alighieri (1265–1351) wrote his *Divine Comedy* with the heavily medieval theme of Heaven and Hell; but he wrote it in Italian. Does that make him a kind of transitional figure?

Near the end of Dante's life, Francesco Petrarca (1304–1374), a great Italian lyric poet and scholar, sat on Julius Caesar's tomb and wrote a letter to his ancestor, dead lo these 1400 years. Although Italians had tended to see themselves as continuous heirs of Roman traditions and institutions, Petrarca signalled his recognition of the historical gulf between his times and those of Imperial Rome. *He* was living in a new age; whether his fellow citizens were, may be a moot point.

Around 1300, Giotto di Bordone (1267–1337) was painting frescoes in St. Francis's (1181–1226) church in Assisi. Giotto painted biblical scenes set in an architectural framework, portrayed with a small illusion of depth (sometimes called "soft" perspective). Another transitional figure!

Some aspects of the Renaissance grew out of their contemporary historical context (and thus were not a *re*-birth); others *were* a revival of classical forms; still others (like printing with movable type) were real innovations. Some cultural forms just stayed medieval, for varying time-spans in various regions.

All in all, the period of the European Renaissance, from 1300 or 1400 to 1600 or 1700 (depending on how you stretch the elastic), was a time of vigorous activity in art, religion, and science—a restless and expansive time.

Martin Luther (1483–1546) was not the first to question the authority of Papal Councils, but, with the connivance of German princes, he's the one who provoked the split in the Catholic church—the so-called Reformation. John Calvin (1509–1564) and Henry VIII (1491–1547) and many others contributed their share to religious dissention. Roman Catholic reaction spawned the Council of Trent and the Counter-Reformation, which contributed to the resistance Galileo would encounter, even though he was a good enough Catholic.

In science, generally, we can apply the term "renaissance" with considerable accuracy. In the single year of 1543, four seminal works were published, all based on Greek models: by Copernicus, a reworking of the Greek astronomy of Ptolemy; by Vesalius, an elaborate improvement on the Greek anatomy of Galen; by Tartaglia, an Italian edition of Euclid's *Geometry*, and a Latin translation of Archimedes' treatises on applied mathematics. But these men and their successors did more than merely imitate their Greek forebears. Copernicus introduced the novelty of the moving earth; Vesalius made anatomy vividly realistic; and Tartaglia (in works other than his translations) contributed to the invention of algebra.

Throughout the seventeenth century, investigators combined models of Greek thought with a devotion to the details of experience Aristotle had only occasionally glimpsed. They "tortured" nature to win her secrets; then they fitted the results of trial and measurement into satisfying pictures of the world. Sometimes they depicted little corners of the world (as Harvey did with the circulatory system); sometimes they painted on a broad expanse of canvas (as Newton did with his principle of universal gravitation). But always they aimed at matching their pictures as closely to experience as they knew how.

While European scholars were transforming their pictures of the world, European seamen, soldiers, merchants, and priests were transforming the globe by direct action. With the pretext of winning heathen souls for Christ, they sailed their gun-laden ships to the far reaches of the earth in an insatiable lust for spices and gold. Cultures and civilizations fell before the blasts of their cannon and the lashes of their tongues.

Rather than being a re-birth, the robust renaissance technology grew quite continuously out of medieval achievements in building: from skilled carpenters centering arches at *Notre Dame* to skilled shipwrights laying the keel of *Santa Maria*; from Venetian cannon founders to London gunsmiths; from Cistercian grain mills to German mine pumps; and ingenious mechanisms from medieval architects like Villard de Honnecourt to Renaissance engineers like Leonardo da Vinci.

Until the Industrial Revolution in the eighteenth century, technological progress was gradual—evolutionary, not revolutionary. The material world of Newton was not very different from Leonardo's. Just think of the time periods. Leonardo was born in 1452; Newton, 190 years later. If you go on another 190 years you get to 1832, and James Clerk Maxwell is one year old. Steam railways are running in England. The telegraph is being invented. Isaac Newton would have been much more at home in Leonardo's world than in Maxwell's.

## *In a Mind's Eye*

Mevrouw Smit took off her apron, dusted the flour off her hands, and stepped outside. The sun was nearing the horizon as she walked the few short blocks to watch it sink into the North Sea. After a long dark winter she couldn't bear to waste a moment of its warming rays.

Not that she needed warmth all that much; the baking ovens would provide far too much of that before summer ended. It was more a need for clarity she felt—a clear view across the already shimmering sea and clear air with white gulls etched against the blue sky. Yes, clarity was very much what she needed.

Old Willem had been dead for fifteen years now, and she'd kept the bakery going, even prospering. Their children had all completed their apprenticeships; here in Vlissingen and in Rotterdam they were all well placed—all except Hans of course, wherever he was. Could she believe the tale told by that seaman? Could Hans really be a successful merchant off in New Amsterdam, an ocean away? Her irresponsible little Hans? She doubted it very much.

But—to more immediate problems. With a steady stream of apprentices she ran the bakery with loving efficiency. Willem would have expected no less. Two eight-year-olds had arrived only last week; she was forever mothering other women's children after her own had departed. Mother and master she was to them all. It wasn't easy. And she wasn't getting any younger.

Past her child-bearing years—she was sure of that—still the old urges sometimes stirred within her. What was Jan doing now, she wondered. She'd left him to finish up while Marietje fed the younger ones and put them to bed.

Jan! She'd raised him like her own, and now he was the man of the business. He knew the trade, got along well with their customers and the other shopkeepers on the street. That was important—poor old Willem had argued too much.

But he was half her age! She'd been paying more attention to her appearance lately, even bought a new Sunday frock. Had Jan noticed? Did he care? Had he yet gotten over the marriage of the jeweller's daughter to another?

Silly old vrouw—act your age! So, it wouldn't be a flaming romance; Jan might still see the wisdom of it. A warm companion in his bed, and master of his own shop. She knew he was ambitious—why should he seek elsewhere, with all the trouble of starting somewhere new?

Mevrouw Smit thought about some of the other shops; how old masters had taken young brides when their wives had died in childbirth, and before long bride became widow. Then, widow married journeyman, and before long journeyman became widower, and on the cycle went. Jan could do worse. He might even become wealthy enough to sit for Rembrandt.

Enough of that—she should just overcome her matronly shyness and.... Well, she'd think of something.

The last sliver of sun slid into the sea. Beyond was England. Where her brother had gone to sell Dutch engineering to English lords. Who knew better how to drain the fens?

England, whose Admiral Blake had just been defeated by our own van Tromp. When would they end this foolish war?

She shivered—darkening thoughts of war joined the greying sky.

Then, a strong, gentle arm across her shoulder. You must be getting chilly, Mevrouw Smit, let's go home.

### *For Further Reading*

Caspar, Max. *Kepler*. Trans. and ed. by C. Doris Hellman. London & New York: Abelard-Schuman, 1959.

Cipolla, Carlo. *European Culture and Overseas Expansion*. Harmondsworth: Penguin, 1970.

Cohen, I. Bernard. *The Birth of a New Physics*, rev. & updated ed. New York: Norton, 1985.

———. *The Newtonian Revolution*. New Yorfk: Cambridge University Press, 1980.

Dobbs, Betty Jo Teeter, and Margaret C. Jacob. *Newton and the Culture of Newtonianism.* Atlantic Highlands, NJ: Humanities Press, 1995.

Drake, Stillman. *Galileo at Work: His scientific biography.* New York, Dover, 1995.

Eisenstein, Elizabeth L. *The Printing Revolution in Early Modern Europe*. New York: Cambridge University Press, 1993.

Fauvel, John et al. *Let Newton Be!: A New Perspective on His Life and Works*. Oxford: Oxford University Press, 1988.

Galileo. *Dialogue on the Two Chief World Systems*. Trans. Stillman Drake. Berkeley: University of California Press, 1953.

________. *Discoveries and Opinions of Galileo*. Ed. & trans. by Stillman Drake. Garden City, NY: Doubleday Anchor, 1957.

Guilmartin, J.F. *Gunpowder and Galleys: Changing Technology and Mediterranean Warfare in the Sixteenth Century.* Cambridge: Cambridge University Press, 1975.

Hall, A. Rupert. *The Revolution in Science: 1500-1750*. London: Longman, 1983.

Laslett, Peter. *The World We have Lost: Further Explored,* 3rd ed. New York: Scribner, 1984.

Manuel, Frank. *A Portrait of Isaac Newton.* New York: Da Capo Press, 1990.

Westfall, Richard S. *The Construction of Modern Science: Mechnisms and Mechanics.* Cambridge: Cambridge University Press, 1977.

________. *Never at Rest: A Biography of Isaac Newton.* New York: Cambridge University Press, 1980.

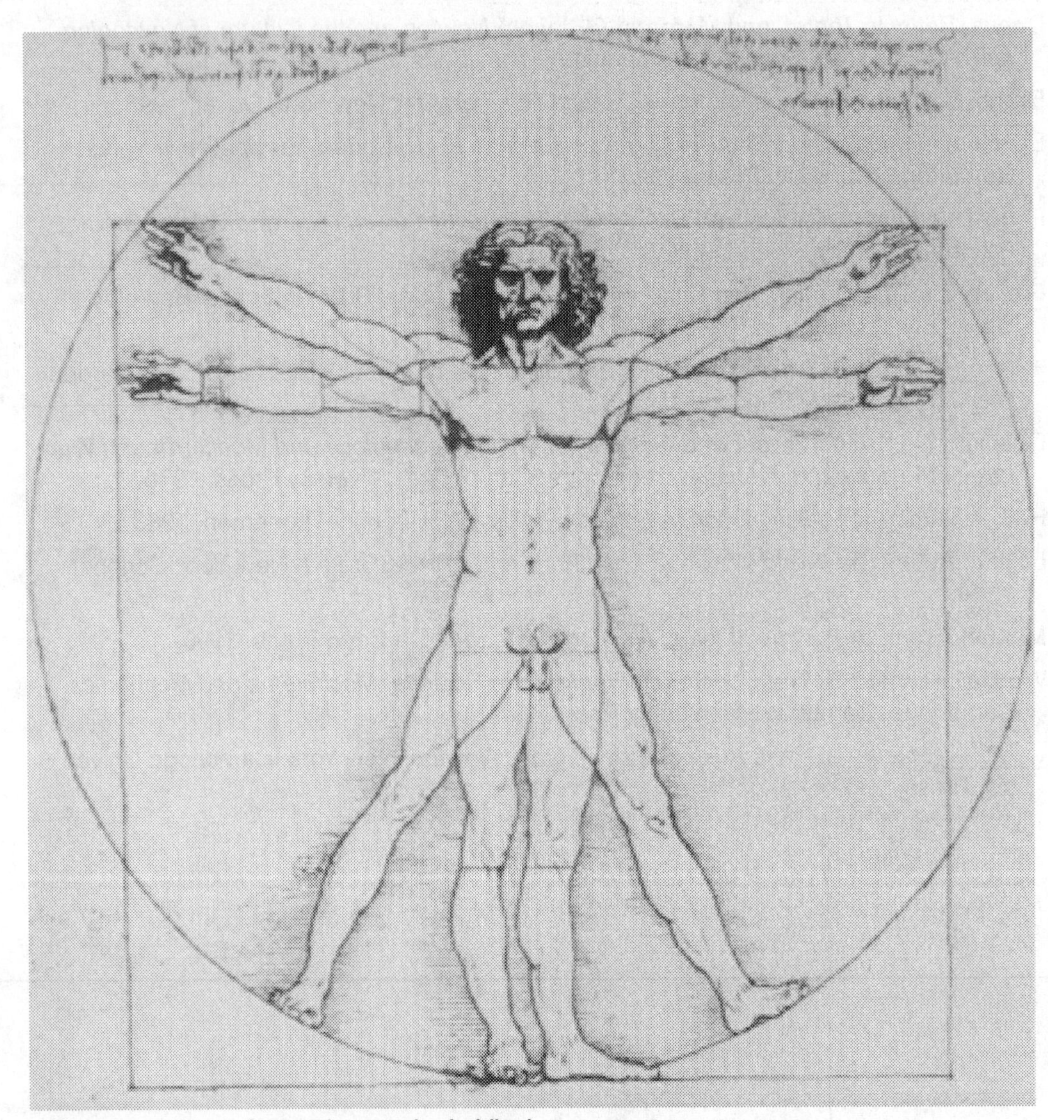

*The proportions of man. Leonardo da Vinci.*

CHAPTER SEVEN

# Engineers and Technology

In technology, there is no obvious point at which we can say, "here ends the medieval world; let the renaissance begin!" Although the Black Death at about 1350 is a convenient boundary, some technological developments seemed to cross it quite readily.

In this chapter we'll look at several topics in technology that really had their start in medieval times, even though they didn't make their full impact on society until later. That's true of mechanical clocks (Chapter 5), as well as of engineering and cannons. While the use of perspective in painting had some fourteenth century precursors, it was not fully formed until after 1400. And printing by movable type is almost entirely a fifteenth century invention.

## *The Engineer's Role*

The term "engineer" appeared first in manuscripts written toward the end of the twelfth century. In Latin and French, the term identified engineers as architects or those responsible for engines. For a thousand years, "engines" were war machines—particularly catapults of various sorts for throwing heavy projectiles against the enemy. Such devices were described in a book written about 25 BCE, *On Architecture,* by Vitruvius Pollo. We can assume from what Vitruvius wrote that in Roman times architects were responsible for the design of such engines, as well as of buildings, roads, aqueducts, etc.

After the introduction of the term engineer, architects continued to be responsible for the whole range of mechanical design activities. We can now think of them as architect-engineers. In the fifteenth century, they broadened their interests to include painting and sculpture. Leonardo da Vinci's activities, for example, spanned the fine arts, architecture, town planning, and military and mechanical engineering. Leonardo was architect, artist, and engineer. Only gradually did particular men begin to specialize more narrowly.

The earliest engineer whose writings are known to us is Villard de Honnecourt. From about 1235, we have a manuscript of what looks like an architect-engineer's notebook. Its sketches and text describe designs for cathedrals and several mechanical devices. Villard seems to have been an itinerant consultant who travelled around Europe making his services available to bishops and barons. Along the way he noted ingenious designs that caught his fancy.

## *Gunpowder and Cannon*

The profession of engineering in its military guise was given a boost by the development in Europe of cannon—tubes of bronze or iron that propelled a projectile by the force of exploding gunpowder. The origin of the recipe for gunpowder—five parts saltpeter, two parts sulphur, one part charcoal—is entirely obscure. Various pyrotechnic mixtures had been known for a thousand years, both in the Middle East and in China. But the application to hurling projectiles was made first in western Europe.

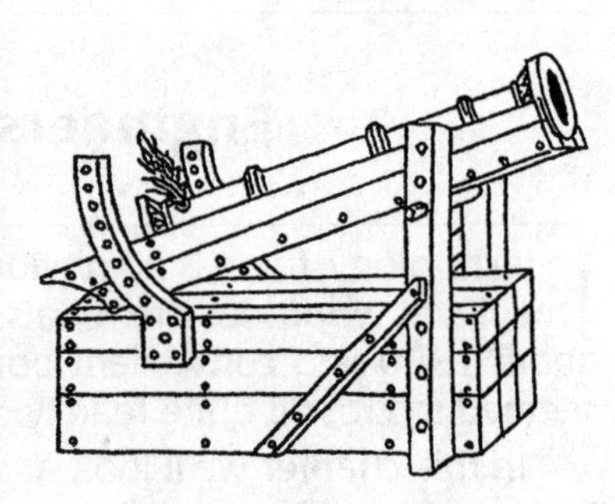

The earliest references to cannon date from the 1320s in France, Italy, and England. The new technology spread rapidly across Europe. But for more than a century after, cannon were used mainly as siege weapons, and were often cast at the site where they were to be used. Although cannon were widely used, they do not seem to have been decisive in battles on land much before 1500. By that time the casting of cannon had become more refined, and wheeled artillery was used with great effect by the French in northern Italy in 1494. Cannon were used more effectively earlier on shipboard, as we'll see in the next chapter.

Until late in the sixteenth century, bronze cannon were superior to iron ones, although always more expensive. Techniques for casting in bronze had been known since ancient times. The skills of bronze founders were developed in medieval times with the great demand for church bells. Such skills were transferred to the casting of cannon with little difficulty. In fact, during wartime, conquering forces sometimes melted bells down for cannon. If the Biblical mark of peace was the beating of swords into plowshares, the medieval mark of war was the recasting of bells into cannon.

The earliest iron guns were built up from wrought iron bars welded together to form a hollow cylinder, with an iron plug inserted at the breech end. Such a technique inevitably created weaknesses, and these early iron guns often blew up. Early cast iron guns were little better, because cast iron was much more brittle than bronze, and subject to cracking.

The casting of iron was hardly known in ancient times. In medieval times, blast furnaces developed gradually in northwestern Germany. Although the melting point of iron is 1500° C, its alloy with carbon (2% to 4%) melts at 1200° C. Strong drafts of air were needed to get smelting fires hot enough to get the molten alloy. The crushed ore was mixed with charcoal in a vertical shaft 3 or 4 m high. Bellows driven by waterwheels provided a blast of air near the bottom of the shaft. The molten alloy was drawn off at the bottom into a rough mold, producing a lump of pig iron.

The metallurgists then took the pig iron to a forge where they could reduce its carbon content by repeated heating and hammering. After 1500, techniques improved sufficiently that iron cannon could be cast. However, the quality of iron depends so much on trace impurities that it took a long time for metallurgists to achieve satisfactory results. For safety, cast iron cannon had to be made heavier than bronze cannon; even so, they could be made for about a third of the cost.

As techniques were refined, the size of firearms was reduced. After 1400 there was a continuous evolution in the design of muskets and pistols. However, the earliest firearms were not as rapid, strong, or accurate as longbows and crossbows in the hands of trained soldiers.

A variety of ingenious applications of gunpowder weapons began to be illustrated in engineers' sketchbooks after about 1400. A number of these have been found in manuscript form in Germany and Italy, in the first half of the fifteenth century. These sketchbooks also illustrate various mechanical devices, such as hoists. They give evidence of a growing sophistication among the fraternity of engineers.

## *Perspective Drawing*

One of the intriguing features of early engineering sketchbooks is their total lack of perspective. Looking at them today, we have trouble sometimes in telling front from back. Now, some of the early Italian engineers were also artists. And around 1425 Italian artists began to portray linear perspective according to rules we still use today.

To consider this revolution in illustration, we must start back in medieval times. The cathedrals of western Europe were filled with images: statues of saints, Biblical scenes in stained glass and on walls. Most of the two dimensional scenes and portraits were really "flat." Artists paid scant attention to the "depth" of a scene.

In linear perspective, depth is conveyed by making background figures smaller than those in the foreground. The size of an object in a scene is determined by the rules of projective geometry. Such rules were not known in medieval times. The size of an object in a painting was determined by its importance; God, Christ, and angels were large, people and animals were small. When painters tried to portray a scene from the Bible, they could have a large Christ apparently in the background, with small people in the foreground.

In fact, painters in medieval times were essentially wall decorators. The same workers were expected to apply a coat of white wash or produce a scene. Of course, as with the stone masons, the more talented workers performed the more demanding tasks. Art in the middle ages was an artisan's occupation: painters, masons, bricklayers, carpenters, cabinet makers. Yet, by 1500, artists like Botticelli and Michelangelo were favored members of aristocratic retinues.

What had happened?

The short answer is "mathematics," but that's too simple. Around 1300, Italian painters came to be more in demand as the commercial wealth of the towns increased. Bishops and princes sought skilled workers to decorate chapels and halls. Giotto di Bordone (1267–1337) was among the leading decorators of his time. In his chapel frescoes he portrayed scenes with more attempt at realism than had been traditional. Giotto often used architectural backgrounds in his paintings. Although he was not following any strict geometric rules, the buildings and walls he portrayed did have a small illusion of depth. We could say that Giotto used "soft" perspective, that is, slightly tilting lines going back into the scene.

Through the course of the fourteenth century, Giotto's successors gradually improved on his techniques. But their skill improved mostly by a process of trial and error. True perspective demanded the rigorous application of geometrical rules. Those rules came into painting from two sources. One of them was the growing study by scholars of the geometry of vision.

Light had a philosophical attraction in medieval times as an emanation from God. As scholars translated more Greek and Arabic texts, they studied light both philosophically and mathematically. They used texts on optics by Euclid and Ptolemy, as well as commentaries and elaborations by Arabic scholars.

The second source of perspective was the *Geography* of Ptolemy. In that work he had considered several rules for transferring points from a spherical surface onto a plane surface. That was necessary to be able to draw maps of large sections of the earth's surface.

Now, these two traditions seem to have combined in the person of Filippo Brunelleschi (1377–1446), an engineer in Florence around 1420. Brunelleschi

*Albrecht Dürer on Perspective.*

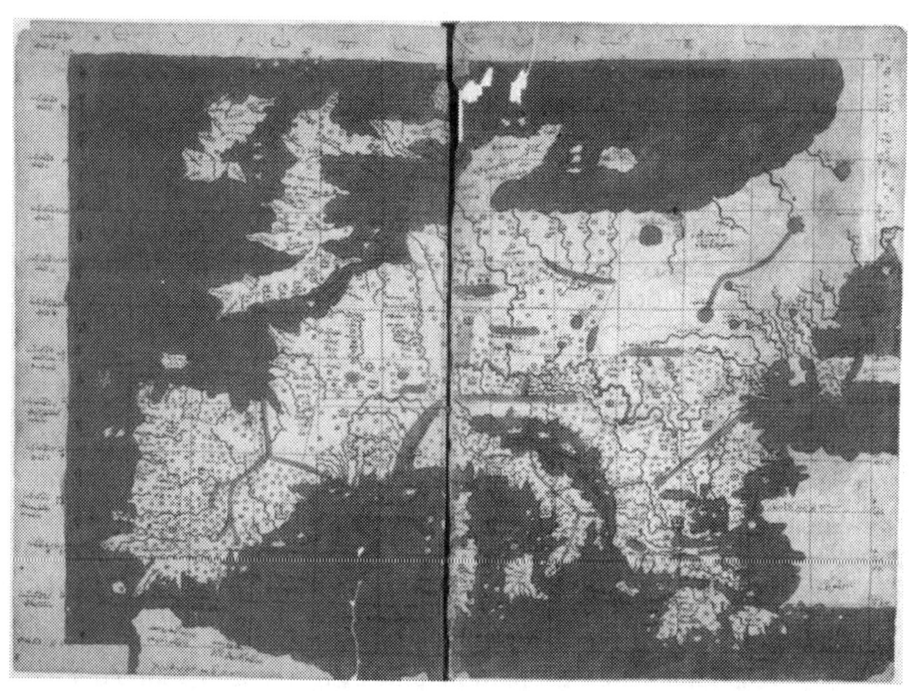

*A map of Europe from Ptolemy's Geography.*

was an ingenious mechanic, architect, and artist. He is said to have learned about Ptolemy's *Geography* from a scholar in Florence, twenty years his junior, Paolo Toscanelli (1396–1482). Toscanelli was part of a study group who hired a Greek scholar to teach them his language. On one of his trips back to Florence, this Greek carried with him a manuscript of Ptolemy's *Geography*. While Toscanelli learned Greek, he also learned projective geometry. (Curiously, there is another important conjunction between Toscanelli and Ptolemy, for which see Columbus in Chapter 8.)

About 1420, Toscanelli realized that projective geometry could assist painters in making their scenes more realistic. By then, some painters were setting their scenes within borders, trying to make them look as though they were seen through a window. Toscanelli told Brunelleschi that artists could really achieve their objective if they used Ptolemy's rules.

Brunelleschi performed a couple of experiments to test what Toscanelli had told him. He painted two panels of architectural scenes in Florence according to the rules of perspective vanishing points.

Success!

Soon, Florentine painters learned the rules from Brunelleschi and applied them in their church frescoes. Among the earliest of these painters were Masaccio and Masolino. In 1435, a young Florentine aristocrat, Leon Battista

Alberti (1404–1472), wrote instructions for following the rules of perspective in his book, *On Painting*. By 1450, numerous artists were filling their sketchbooks with perspective exercises.

From then on, to be a painter required one to have at least some competence in geometry—something to be learned in school. Since schooling was more accessible to families with money, a tendency began for painters to be drawn from a different layer of society than produced your run-of-the-mill artisans. Painting and sculpture moved out of the craft tradition into Fine Art. Of course, then as now, occasionally a talented person from a poor family might still succeed in art.

*Leonardo da Vinci (1452–1519)*

Leonardo was the illegitimate son of a man of good family. He received a fair elementary education and was apprenticed to an artist at age 14. Leonardo first demonstrated his talent more in technical than in artistic realms.

We should recognize that Leonardo was only the most famous of a whole group of artist-engineers, a group that essentially started with Brunelleschi. Of course, Leonardo's fame rests mostly on his painting. When scholars in recent times began to unearth evidence of his mechanical ingenuity, popular opinion began to identify Leonardo as the giant of Renaissance skill and learning. We'll not diminish Leonardo's fame by recognizing that other men may have been better engineers than he, while still others matched or surpassed his artistic talents.

In addition to his art and engineering, Leonardo was a passionate observer of nature. He sketched birds in flight and designed machines that were supposed to fly. He performed dissections on corpses in order to improve the accuracy of his portrayal of human anatomy. And he sketched engineering mechanisms endlessly, whether observing those designed by others or ones originating in his own fertile imagination.

*Leonardo's drawings of devices for automatically unhooking a weight once lowered in place.*

Throughout the numerous sketchbooks of Leonardo that have survived, we can see the great boon that linear perspective brought to technical illustration. Even when we find sketches of gadgets that will not work, we can clearly see the interconnections of parts that Leonardo intended.

Leonardo was a technical genius of a high order. However, if your notion of a "renaissance man" includes a large component of humane letters (well versed in the Latin and Greek classics), then Leonardo was not such a one. He was bright and observant in natural and mechanical things, an outstanding engineer, but he could not read Latin easily.

A better choice for a "universal man of the Renaissance" might be Leon Battista Alberti. Well educated, Alberti had a thorough knowledge of the classics. But he also knew the Florentine artists and worked with them. And he was a fine architect. Alberti shone brightest in his theoretical studies of painting and architecture. Though he had some mechanical knowledge, it was never as deeply technical as Leonardo's.

Perhaps already in the fifteenth century we see the split among educated persons so well delineated by C.P. Snow in his *Two Cultures and the Scientific Revolution*. Of course, the split has much more ancient origins, in the sense that artisans were largely illiterate for many generations. We've suggested that artisans turned into artists as their work required more mathematical knowledge. Such knowledge soon became important to engineers as well. Yet this was not the classical learning so much prized by the humanists. It may well be that the split is between persons who are literate and those who are numerate. Later, we'll see that such men as Galileo and Newton were both, but the conjunction has always been rather rare.

However, there *was* a development in the fifteenth century that gave scholars and artisans a reason for closer contact than had been traditional. That was in the invention of mechanical printing.

## *Mechanical Printing*

The invention of mechanical printing required the conjunction of several technologies: paper to take the impression; ink to make the marks; a press to apply even pressure; and the movable type to form the letters.

Of these, the casting of multiple copies of the letters of the alphabet was the most crucial, and indeed, the most inventive. The other technologies already existed in the 1440s when one or more ingenious goldsmiths worked out the scheme for forming the letters.

To understand the revolutionary character of mechanical printing, we should consider first the situation that had existed up till the invention of movable type.

### *The World of Manuscripts*

In medieval times (say from the tenth century onward) lettering was performed by hand with quill pens on parchment or vellum, using an ink of lampblack suspended in water.

The writing surfaces, the parchment and vellum, were made from the skins of sheep or calves. A considerable amount of scraping and washing of the raw skins was required to prepare a sheet of parchment. Each skin had an area of about half a square meter so that a typical book might take about a dozen skins.

In monasteries, and later in stationers' shops, scribes laboriously copied out sections of manuscripts, letter by letter. That was the process used in the translations of Greek and Arabic texts into Latin, and for the production of successive copies.

In the medieval universities the lecturer read from his text and prepared commentary while the students wrote down his words. Wealthy students could cut classes and buy a copy of the material. As scholarship and learning increased in the twelfth and thirteenth centuries, the demand for scribal output grew for that and for more legitimate reasons. Stationers leased copy texts from the university and returned them when their scribes had made sufficient copies.

Eventually, the demand for books put some pressure on the supply of parchment. By 1300, paper making was becoming widespread in Europe, first in Italy, and soon in France and Holland. The art of papermaking came from the Arabs who had learned it from the Chinese. It consisted essentially of forming a cast sheet from a slurry of pulverized rags. The rags had to be chopped finely and pounded in water. Soon, water mills in Europe were being used to perform the pounding mechanically.

The slurry was poured into a wooden frame with wires stretched across the bottom. A worker shook the frame to mat the fibres together. When enough water had drained off, the sheets were piled alternately with felt and pressed to squeeze out more water. The presses used a lever or screw to apply the required pressure. Before the sheets were completely dry, they were sized with glue and gelatin to reduce their absorbency and give the paper a smooth finish.

Despite these various processes, and although dependent on an adequate supply of linen or hemp rags (from cast-off clothing and rope), paper was about one-fifth the price of parchment. While paper did not by any means immediately replace parchment, it did improve the supply of material to write on. And it made possible the lettering of books by mechanical means, because its surface could be prepared suitably to take the ink. Parchment's surface was not as absorbent as paper's.

### *Inventing Movable Type*

Movable type consists of a number of copies of each individual letter of the alphabet, formed on the end of a small metal slug, called a type. Each piece of type is formed by casting a tin-lead alloy into a mold that has a depression at the bottom in the shape of the letter. This matrix at the bottom of the mold is made of a metal like copper with a melting point well above that of the type alloy.

The matrix is formed by being struck by a punch with the raised letter carved on its end. The punch is made of a metal harder than copper, such as steel or bronze. The skill of the goldsmith came into play in the careful carving of the letters onto the ends of the punches.

Despite valiant searches by historians over many generations, we know few details of exactly when movable type originated, or by whom. Although

*Johann Gutenberg.*

there have been numerous other claimants to the title of inventor, the name of Johann Gutenberg (1400–1468) remains the most prominent. Yet, he is a rather shady character in the annals of the fifteenth century.

Nonetheless, we can say with assurance that a few significant books were printed in Mainz in Germany before 1460. Instead of limiting ourselves to the meager facts, let's imagine a sequence of actions for the printing of one of those early books. Nothing will be entirely inconsistent with what is known about that early period.

About 1450, goldsmith Gutenberg completed his research and development work, and set out to prepare to produce an edition of the Bible. His first task was to prepare the punches for the letters. Assuming two type fonts, one for text and the other for headings, he needed 300 brass blanks about one square centimeter on their ends and 10 cm long. On the end of each punch he carved the reverse shape of a letter or symbol. This meticulous, laborious task must have taken a couple of years.

Then Gutenberg had to prepare the matrices for the molds. Each punch was used to make an impression of its letter onto a small block of copper. The block formed the matrix for the mold in which the type would be cast.

Instead of making 300 separate molds, Gutenberg arranged to use a pair of interlocking L-shaped pieces of copper to form the sides of the mold. Clamped together with a matrix in the bottom, they formed a hollow box into which the molten type metal was poured. After the metal hardened, he removed the mold, and dropped the hardened piece of type into a box. After enough replicas of one letter had been made, he inserted the new matrix for the next letter.

He made the molten type metal from an exact recipe for an alloy of lead, tin, and antimony. Gutenberg cast enough copies of each letter to be able to start setting the pages in type. To set the first four pages of the Bible required 12,000 pieces of type, distributed among the letters according to their frequency of occurrence. The type founder took a couple of weeks to produce that many pieces.

The printer's devil, or apprentice, distributed the type into compartments in two boxes, the upper and lower cases. Next, the typesetter began to compose the first page, line by line. He picked the types out of the cases and laid them onto a composing stick. When he completed a line, he transferred it to a form (called a galley) that would hold the 42 lines of the page. When the first four pages were complete, the type was wedged tightly with slips of wood.

Then, the printers laid the forms for pages one and four on the press, ready for inking and printing.

Manuscript writing had used a water-based ink that was fine for parchment, but would diffuse into paper, which is more porous. Printing ink, which was oil based, worked well on paper, but would not dry readily on parchment. The oil-based ink was developed by Flemish painters in the fifteenth century. In the print shop, the printers smeared the ink over the type, laid on a sheet of paper, and pressed it down to make the impression. The mechanical device (printing press) for ensuring an even pressure over the whole page probably derived from the presses that had been developed for paper making.

In less than a day they could impress a hundred sheets on one side, containing pages one and four. (Within a century, printers were pulling more than 2000 sheets off a press each day.) After each impression, they hung the sheet to dry. Later, they'd repeat the process to put pages two and three on the reverse side of each sheet.

Meanwhile, with another 12,000 types, the typesetter continued on with the next four pages. When the printers had completed the impressions on the hundred sheets of the edition, another worker would remove the type from the forms and distribute them back to the cases. This donkey task was assigned to the printer's devil, who knew the alphabet tolerably well. From time to time he'd be cautioned to "mind his **p**'s and **q**'s," though not usually his **b**'s and **d**'s.

Gutenberg's 42-line Bible consisted of three "folio" volumes, totalling 1700 pages. With each pair of pages taking a day to print, a single press would have taken almost three years to complete the edition of 100 copies.

Three years! Then, what's the big deal about printing with movable type?

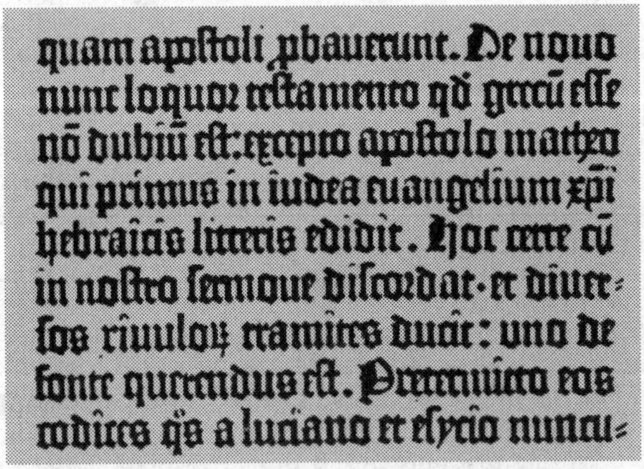
quam apostoli pbauerunt. De nouo
nunc loquor testamento qd' grecū esse
nō dubiū est: excepto apostolo matheo
qui primus in iudea euangelium xpi
hebraicis litteris edidit. Hoc certe cū
in nostro sermone discordat. et diuer-
sos riuulorum tramites ducit: uno de
fonte querendus est. Pretermitto eos
codices quos a luciano et esycio nuncu-

*The typeface used in Gutenberg's 42-line Bible of 1455.*

If the Gutenberg shop had three presses going, the job would take less than a year. But let's suppose only one press, and compare its productivity with scribes performing the same task.

In the print shop, once the punches had been cut, you'd have a crew of six: the type founder, the typesetter, two printers to run the press, and a couple of helpers to fetch and carry.

*A printing shop of the sixteenth century shows the variety of jobs in a printing house. Engraving from a work by Stradanus, about 1590.*

In three years, these six could produce an edition of 100 copies. In a manuscriptory, how many scribes would it take to do the same job by hand?

How fast can a scribe form letters by hand? Try it! Sharpen and split the quill with your pen knife. Dip the pen into the ink pot and start. Form all those vertical strokes of Gothic lettering with their fine detail. After each letter, dip your pen into the ink again. After a couple of pages, refill the ink pot and resharpen your pen. I estimate that a page'll take you two hours; that's five pages a day. So 1700 pages will take a year. For *one* copy. In three years, one scribe can do three copies, while the print shop does 100. You'll need more than 30 scribes, or at least five times the personnel of the print shop. In economic terms, the productivity of the print shop is actually even better than five times that of the manuscriptory.

First of all, we should realize that all 30 scribes need to know how to read and write, whereas in the print shop, only the typesetter needs to. Scribes should be able to demand a higher wage than the laborers who work the presses. Let's make a comparison on that basis. Let's suppose that 30 literate men can work either in print shops or manuscriptories. Suppose their wages are the same—six soldi per day (1 soldo = 1/20 of a gold franc of the period), which we'll write $6. A skilled typefounder gets $4 a day, and two laborers on each press get $2 a day each. So, if our 30 typesetters each go to a small print shop, the daily wage bill in each shop will be $14. (The printer's devils are apprentices who are unpaid—they get only room and board; which we'll assume the others get on top of their wages.)

Thirty print shops would have a total daily wage bill of $420, while a manuscriptory with 30 scribes would have a daily wage bill of $180. But, from the numbers above, the print shops would turn out 33 times as many pages as the manuscriptory, for a wage bill that's only 2.33 times greater; which gives an improvement in economic productivity of 14 times.

Some qualification of that number must be made because of the greater capital outlay for the printing equipment. A fair average value for the period around 1500 is about $1800 per press, which includes the type fonts and all the ancillary equipment. If you amortise that over three years of printing, it adds $2 a day to the wage bill, increasing the bill for 30 shops to $480 a day. That reduces the productivity improvement to about 12.

With your sharp pencil, you can figure on selling printed books for one-tenth the cost of hand-lettered ones, and still make a tidy profit. At least, if you can develop a market that will absorb your massive output. We should not be surprised, therefore, that early printers were shrewd and energetic entrepreneurs.

The actual costs of printed books also depended on the cost of the paper, which ranged from half to double the printing costs, depending on the quality of the paper. Around 1590, two printing firms in France produced an edition of 2500 prayer books at a cost of about $28 per book—with a page size about the same as this book (with four pages printed on each side of a standard large sheet, it's called "quarto" size), and about twice as many pages. If such a book had been produced in a manuscriptory, the cost would have been about $230 each, using paper. If parchment had been used, the cost would have risen to $270. This difference in price by a factor of about ten represents a fair estimate of the change in consumer costs brought about by the transition from hand lettering to printing by movable type.

You might note that scribes or typesetters would have had to work 5 days to purchase a typeset book, and almost 40 days to purchase a hand lettered one. If you think books are expensive today, calculate how long you or your friends have to work to earn the price of a quality paperback book—not many hours, likely.

## *Consequences of Printing by Movable Type*

### *(a) In Technology*

The transformation of book production from hand lettering to movable type was revolutionary. The way books were produced was changed as dramatically as computers today are changing productivity in office environments. (For example, word processors reduce retyping and, therefore, the need for personnel in the typing pool.)

Like computers today, fifteenth century print shops represented an increase in capital requirements at the expense of labor requirements. Despite all the handling of types and sheets, the amount of human handling of a book in a print shop was much less than in a manuscriptory.

As a result, the book business changed. To get into business as a scribe, a person only needed to be literate and neat. Pen knifes and ink pots were about the only equipment a scribe needed. By contrast, a man would need considerable capital to set up a print shop. Even an itinerant printer would need type cases, a press, and other oddments of equipment.

Whereas the scribe needed education, the printer needed capital. He could even be almost illiterate if he could hire a literate typesetter and proofreader. Books and printing became a business or industry, instead of being a skill or profession. In fact, many of the early printers were perfectly able to read, and became quite scholarly, despite coming more usually from the trades than from the professions.

Of course, with more books being produced, bookselling became a much more important trade. For the first century after 1450, publisher, printer, and bookseller were normally combined in a single firm. Only gradually did these functions become as separate as they were until quite recently. Now, word processing and desk-top publishing are trending the other way.

The materials of books also changed significantly. Medieval manuscripts were largely animal products, from the geese supplying the quills, and the sheep supplying the parchment to the scribe himself. The ink, made of lampblack suspended in water, was a product of the animal tallow used for candles.

By contrast, the print shop depended much more on plant and mineral materials. The paper was made from linen rags. The presses and cases were made of wood held together by iron bolts. And the type founder used copper, lead, and tin. Thus, the materials of the book industry shifted from being largely animal products to being vegetable and mineral products.

In the movable type itself, we have the first large-scale example of the production of interchangeable parts. When the typesetter reached for an "e," it didn't matter which one his fingers grasped. Ideally, every "e" coming from the type founder was identical to every other one. It would be more than 300 years before the same principle would begin to be applied in manufacturing other goods. The casting of some parts of clocks and guns began only after 1800. Even then, for several decades, cast parts had to be finished by hand so they'd fit properly with other parts.

*(b) In Society at Large*

Between 1450 and 1500 scholars have counted 40,000 editions of printed books: more than ten million volumes in all. By 1500, there were print shops in 236 towns of western Europe, stretching from Lisbon to Stockholm, Budapest to London. Printing was most densely concentrated in northern

Italy and in the Netherlands. By the mid sixteenth century, a number of prominent master printers were setting the tone of this new business enterprise: Aldus Manutius (1450–1515) in Venice, Johann Amerbach (1443–1513) and Johann Froben (1460–1527) in Basel, the Estienne family in Paris (Henri, 1460–1520; Robert c1495–1559; and their descendents), Christopher Plantin (c1525–1589) in Antwerp, and the Elsevir house in Leyden, founded by Louis Elsevir in 1593 after he'd been trained in Plantin's shop.

With as many as 2000 presses by 1500—some shops might have only one, others, up to ten—you have a work force in the neighborhood of 10,000. It had been zero only 50 years before. If those ten million books had been made entirely by scribes, an army of them approaching 100,000 would have been needed.

Mechanical printing immensely expanded the reading materials available to literate Europeans. At first, the printers concentrated on the classics of Greek antiquity (like Aristotle) and on prayer books and Bibles. Later, living authors began to be published in greater numbers. As adventurers like Columbus and da Gama enlarged the frontiers of European influence, stay-at-homes could read about their exploits in printed books. European horizons expanded both geographically and intellectually.

Gradually through the two hundred years after 1450, printers also began to produce books on subjects that had never before been recorded in any form. Books about handicrafts, pattern books for dressmakers and tailors, and how-to books in many fields, increased the need to expand literacy to artisans and shopkeepers. Reading for recreation also grew gradually, until the novel form appeared in the eighteenth century—clearly a development unthinkable without the multiple copies provided by movable type.

Printers produced handbills and broadsheets as well as books. People's knowledge of current affairs increased rapidly. Word-of-mouth communication was supplanted by the printed word. When Martin Luther questioned the authority of the church in Rome, the issues he raised were widely disseminated via printing. This new knowledge gave literate people more power and helped them to challenge traditional authority.

The ready availability of printed materials also changed learning. Instead of having to go out to talk with the experts, now you could sit at home and read their books. You could plan your own course of study and be "self-taught" in the subjects you decided were important.

Printing gave a strong impetus to the use of national languages instead of Latin and thereby encouraged national feelings. Luther translated the New Testament into German and had it printed. Throughout Europe, the Bible was translated into the common languages of the people. This assisted in the decline of the authority of the Roman pope and also set the stage for the nationalist divisions and wars of the following centuries—speaking and reading different languages tends more to divide than to unite people. Reading the Bible in the common language also seems to have elevated the authority of the "infallible word of God" for some Protestants to heights that few Catholics had ever imagined for the pope.

The challenge of printed books to traditional authority led church authorities in Rome to prepare their *Index of Prohibited Books*, starting in the 1550s. Catholics were supposed to refrain from reading books that had not been properly licenced by the authorities—especially books that were heretical or injurious to faith and morals. The *Index* provided free publicity for Protestant printers as they rushed to print and distribute the forbidden volumes.

### *(c) In Scholarship*

Mechanical printing produced several significant effects on scholarly activities. First of all, it brought scholars into closer contact with commercial affairs, and with craftsmen. The Dutch renaissance humanist, Erasmus, spent three years in Basel at the home of publisher Johann Froben. Printing brought scholars out of the ivory tower.

Second, printed books greatly improved the quality of knowledge. One of the difficulties of the manuscript era had been the multiplication of copy errors. Each successive re-copying increased the number of errors. In printing, a first "proof" could be taken from the galley and checked by the proof reader. Errors could be corrected before the full edition was printed. Errors found later could be signalled in an "errata" sheet, and incorporated into subsequent printings. While manuscripts tended to be degraded through time, printed books got better from edition to edition.

Third, printing encouraged scholars to establish the correct renderings of ancient texts. From 1500 onward, they diligently collated and compared manuscripts of Aristotle and the Bible, seeking to recapture the definitive versions.

By encouraging scholars to establish definitive texts, printing provided them with an endless occupation as they searched for that elusive ideal. They searched every newly found manuscript for evidence that might require some slight emendation here or there. If by "renaissance" we mean the rebirth of ancient wisdom, then the shift to print culture has made it possible for philologists to create an ongoing permanent renaissance. As they pore over dusty manuscripts, they have retreated back to the ivory tower.

### *(d) In Science*

It was otherwise in science. In the age of manuscripts, scholars interested in science had as much concern for defining texts as the philologists. But, their texts were about nature, the world, and the heavens. Once a manuscript of Ptolemy (for example) had been printed, the thing to do was to compare the information in the book with nature, not with other manuscripts. Having the books readily available was a great spur to scientific investigation.

Scientists could now afford to buy books of their own. Instead of having to travel to the libraries containing the manuscripts they needed, they could stay home and collect books. And they could compare information in several books. Tycho Brahe used different printed tables of astronomical data to discover large discrepancies among them and his own observations. That encouraged him to devote his life to the compiling of better data.

Also, botanical and other illustrations were improved. Herbals in medieval times were used to identify plants for medicinal use. As they were copied and recopied by scribes with no botanical knowledge, their accuracy deteriorated. In printed books, wood cuts and, later, copper engravings, were used to produce illustrations. A single correct drawing could be reproduced many times. That meant the publisher could afford to hire an artist to make an accurate sketch. Froben, for example, employed the talents of Hans Holbein.

*An illustration from Otto Brunfel's Living Portraits of Plants, 1530.*

The possibility of maintaining accuracy encouraged scholars to make direct observations in order to get their illustrations right. When Pierre Belon was translating the works of Greek naturalists into French around 1550, he wished to illustrate their works reliably. He obtained support from the king to travel to the eastern Mediterranean to see for himself what the Greeks had described. The same concern for accurate illustrations in printed works can be seen in the geography book (*Theatre of the Whole World*) published in numerous editions by Abraham Ortelius before 1600; as also in the anatomy book of Vesalius (Chapter 11) and the book on mining by Agricola (Chapter 16).

After 1500, science began to grow at rates justly considered revolutionary, as we'll see in later chapters. We have every reason to suppose that such growth would have been impossible without the revolution in communications brought about by the introduction of printing by movable type.

### *Conclusion*

Clearly, the early Renaissance saw great changes in Europe. Clocks, which had started out with an entirely religious function, were now serving the needs of business and industry.

Battles before the fourteenth century were fought with swords and lances, bows and arrows, and catapults. By the end of the fifteenth century, those weapons were being rapidly displaced by cannon, muskets, and pistols. You could easily kill your enemy without getting his blood all over you.

While medieval painting had been decorative and symbolic, the introduction of perspective in the fifteenth century made it much more naturalistic. Artists now strove to make their paintings look like scenes in nature. Leonardo da Vinci studied anatomy with great care in order to make the people he sketched look as real as possible. As we'll see, this tendency in art probably had some influence in making scholars pay more attention to the details of the world around them. That is, art encouraged science.

The introduction of printing by movable type was the most revolutionary development of the fifteenth century. It expanded the worlds of learning and scholarship far beyond anything that could have been imagined during the age of manuscripts.

A scholar who devoted twenty years of her career to investigating the impact of printing on culture, provides a neat summary:

> Intellectual and spiritual life...were profoundly transformed by the multiplication of new tools for duplicating books in fifteenth-century Europe. The communications shift altered the way Western Christians viewed their sacred book and the natural world. It made the words of God appear more multiform and his handiwork more uniform. The printing press laid the basis for both literal fundamentalism and for modern science. It remains indispensable for humanistic scholarship. Eliabeth Eisenstein, *The Printing Revolution in Early Modern Europe* (1983), p. 273f.

A significant trend during this early renaissance period was the increase in the use of metals in technology—a shift from renewable to non-renewable resources. In weapons of war, as in printing, metals became more prominent. Medieval weapons had used a lot of wood and leather. Although iron had been used for swords and armor, it became even more prominent in gunpowder weapons. The gunpowder itself was largely derived from mineral sources. And the chemical energy of gunpowder replaced the muscular energy of the warriors. Our dependence on non-renewable resources for building and powering technology began to increase in this period.

*A three-masted carrack of the late fifteenth century, with square sails on the two fore masts and a lateen sail on the aft mast, illustrates the ships Europeans used to conquer the world. From a contemporary Flemish engraving.*

CHAPTER EIGHT

# Enlarging the European World

While Europeans of the renaissance period were altering their world at home, they were also transforming the world overseas. By 1500 they were more advanced in technology than anywhere else in the world. By combining their technology with a sense of adventurous superiority, they conquered much of the rest of the world.

From the fifteenth century to the present, Europeans have dominated life on five continents. After inventing the modern spirit of nationalism, they carried their national rivalries overseas and imposed them on empires and tribes alike. To support a growing economy at home, they pillaged the world for resources both material and human.

In the fifteenth century, the Europeans established their domination overseas with newly designed ships carrying death-dealing cannon. They literally conquered the world.

## *European Powers, 1100–1400*

In medieval Europe, the most powerful leader was the pope in Rome. For 300 years western Europe readily acknowledged his authority, not only in religion, but also in politics. The Christian bishops and priests administered the feudal oaths that bound society together. They also operated the schools and universities, and they collected the tithes—the church taxes needed to run their large bureaucracy.

The popes in Rome also exercised their power through the Crusades. During this 300-year period, they organized or supported as many as eight expeditions to the Holy Land. The purpose of the Crusades was to capture Jerusalem from the Muslims, who were harassing Christian pilgrims. At the same time, the popes supported the building of a hundred or more Gothic cathedrals, described in Chapter 5.

All this activity was made possible by the growing wealth of western Europe, with its agricultural surpluses, enlarged towns, and increased trade. With greater wealth, people became more aware of their own powers and less dependent on the church. In fact, they began to resent having to send funds to Rome to support the extravagance of the pope and his bureaucracy. The pope's power declined.

As the pope became less powerful, monarchs in England, France, and Spain became more powerful. Kings now imposed more control over the feudal barons and got support from increasingly wealthy merchants. The cement of faith that had unified Europe began to dissolve.

Shortly after 1200, King John in England and King Philip II in France had disputes with Pope Innocent III. The pope claimed sole sovereignty over the clergy and did not allow them to be tried for crimes in national courts. When King John objected, the pope released the barons and the people from their feudal oaths of obligation to the king. So King John withheld the payments to Rome that were needed for administering the church. In France, Philip reserved to himself the right to appoint bishops. Although these conflicts were resolved, they were straws in the wind for the future. Challenges to papal authority would increase as national powers expanded.

Around 1300, Pope Boniface VIII vigorously asserted papal supremacy and refused to allow the clergy in France to be taxed. The king, Philip IV, sent an armed force to Rome to "reason" with Boniface. In 1305, a French bishop was elected pope (Clement V), and he moved the site of church power to southern France. Clement named nine French cardinals to give them a majority in the sacred congregation. The papacy thus came under the domination of the French monarchy.

The papal court remained in France until 1378. Then, a new pope was elected in Rome. However, the French pope did not relinquish his authority, and until 1417 Europe was split in its allegiance to two papal courts. These events did nothing to enhance the pope's reputation or power.

As the authority of the church declined, national monarchies got stronger. The kings subdued their feudal barons and then began to fight with each other to extend national territories. One famous struggle was the Hundred Years' War between England and France, fought from 1337 to 1453. As they fought over territory, the French and English developed new techniques of warfare. With gunpowder weapons—cannons and muskets—they transformed the art of war.

By the mid 1400s, Spain and Portugal too were becoming stable monarchies. And they were the nations that began the first wave of Europe's maritime expansion. What drove the Portuguese and Spaniards overseas?

## *Changing Patterns of Trade*

As the European economy expanded during the middle ages, commerce increased. Regions rich in resources of one kind traded them with others for goods they lacked.

During the Crusades, Europeans discovered the riches of the east, such things as spices and silks. Spices for preserving meat. Silk for underwear. With their growing wealth, Europeans eagerly added these items to their regular consumption patterns. Trade with the Middle East was conducted mainly by the large maritime fleet of Venice. After 1300, the crusading spirit was spent. And a new power was rising in the east. The Ottoman Turks.

The Ottomans were originally a nomadic tribe from Asia, who settled in Turkey during the thirteenth century and became Muslims. After 1288, under their leader, Osman I, they took over control of the Muslim Middle East, and began to expand westward into Europe—into the Byzantine "empire" with its capital at Constantinople, the site of the ancient Greek city of Byzantium.

The Byzantine Empire was the eastern remnant of the Roman Empire. Until the rise of Islam, it controlled a large area in the eastern Mediterranean. Repeated Muslim incursions had left it an empire in name only. By 1400, the Ottomans had penetrated to the Adriatic, leaving only Constantinople in Christian hands.

After 1400, the Ottomans challenged the Venetian fleets for control of the sea lanes and made trade between the Middle East and Europe continually more difficult.

With the Europeans now dependent on eastern goods, traders started to consider alternative routes to Asia. Portuguese and Spaniards, in the southwest corner of Europe, began to have ambitions to replace the Venetians as traders to the east—by roundabout routes.

In the early 1400s, the Portuguese began to explore down the Atlantic coast of Africa. Cautiously at first. They landed at ports along the coast of Morocco and fought their perennial enemies, the Muslims. Soon the Portuguese ventured farther afield. By 1444 they'd reached Senegal and Cape Verde. Along the northwest African coast they found pepper to replace more exotic spices from the far east. By the 1480s they'd crossed the equator.

## *Evolution of Ship Design*

To make any longer voyages, the Portuguese needed better ships. And in Lisbon, they created a revolutionary new type of ship. They did it by combining two earlier shipbuilding traditions. The traditions of the North Sea and the Mediterranean. Traditions which differed because of the conditions of the seas they sailed. The northern ships were short, broad, and deep. They had to stand up to the rough weather and high seas of the Atlantic. Mediterranean ships on the other hand, were long, narrow, and shallow—fine for protected waters. The Mediterranean ships had two masts, carrying triangular sails, much like those on modern yachts. These triangular sails are called **lateen** sails. Lateen sails are set more or less parallel to the axis of the ship. Northern ships, however, generally had a

*Mediterranean ship with lateen sails.*

square sail on a single mast, with the sail set roughly at right angles to the axis of the ship.

The Portuguese built ships that combined both traditions. They put a large square sail on the main, center mast. Ahead of that, they put a second, somewhat shorter, mast, also with a square sail. And behind the main mast they put a third mast carrying a lateen sail.

*Portuguese carrack with lateen sail on third mast.*

These cross-breed Portuguese ships, carracks and caravels, were the ancestors of the great sailing ships that dominated the oceans of the world for the next four centuries. They were larger than their northern predecessors and sailed better. They could also withstand Atlantic gales better than the Mediterranean ships. Their design was established between 1440 and 1460, and spread rapidly to other seafaring nations.

Such ships carried European adventurers across the oceans. Christopher Columbus's three ships (*Nina, Pinta,* and *Santa Maria*) were designed like that, and were about 23 m long. They were large enough to carry supplies for their five-week crossing of the Atlantic in 1492. And they were small enough to do some exploring round the islands of the West Indies. The great age of sail was under way.

## *Basic Sail Handling*

What were the advantages of these new hybrid ships? Let's consider their sailing characteristics compared to the older ships they replaced. Ordinarily, you expect a sailing ship to travel in the same direction as the wind, that is, with the wind behind you. But what if you want to go in a different direction? One thing *no* sailboat can do is travel directly into the wind. With lateen sails, you can easily travel at right angles to the wind. That's because while the wind exerts a sideways force on the sails in one direction, there's a reaction force from the water pushing the ship in the other direction. The force of the water on the keel keeps the ship from simply drifting with the wind. These two forces combine to propel the ship forward. It's like what happens when you squeeze a watermelon seed between your thumb and fore finger—the seed squirts out straight ahead.

Suppose you want to go southward, on a ship with a square sail. If the wind comes straight out of the north directly behind you, everything's fine. But generally winds blow *around* the earth westward or eastward. In temperate latitudes, the prevailing winds are from the west. Nearer the equator, the prevailing winds blow from the east—from the northeast above the equator, and from the southeast below the equator. (These are called the trade winds because of their regularity. The word "trade" entered English from a German word meaning "track" or "path." These winds were notable for their consistency—they "blew trade," that is, along the same track. "Trade" soon came also to mean the course a person follows in his life, that is, his occupation or calling. In French, the trade winds are called *vents alizés*, derived from the Spanish *alisios*, meaning "smooth, regular.")

Suppose your square rigger on its southward course has the wind coming from the northeast at a 45° angle to the direction you're going. You have to set the angle of your sail so it's diagonal across the ship. To do that you pull in on the sheet at the bottom corner of the sail on the side away from the wind—the western side. Then, if the wind shifts so it's coming straight from the east you pull harder on the sheet. You'll keep going, but not so fast. But if the wind shifts to blow from the southeast, you're in trouble. Lower your sail—or change your direction of travel.

On the other hand, the triangular lateen sail is designed to go best when the wind's blowing across the beam of the ship. That's because the lateen sail is attached to the mast so it's normally parallel to the axis of the ship. If the wind blows from the east, that pushes the sail out toward the west and you let it swing out part way. And then hold tight on the sheet. A lateen-rigged ship goes southward easily when the wind's blowing from the east. Then if the wind comes from the southeast, you pull hard on the sheet to make the sail almost parallel to the ship's axis.

In summary, for our southbound ship, the square sail can handle winds from east around through north to west. But not from the southeast, south, or southwest. The lateen sail can handle winds from southeast around past east, north, and west, all the way to south west. Or, to put it another way, square sails can use 180° of wind direction, and lateen sails can handle 270° of wind direction. But neither of them can sail directly into a wind. And lateen sails don't do quite so well with the wind directly behind them.

The new Portuguese ships couldn't sail quite as close to the wind as a full lateen rig. But they were larger and more stable than the traditional Mediterranean ships.

## *Christopher Columbus (1451–1506)*

With their carracks and caravels, European mariners ventured forth on the great oceans of the world. Until the 1490s, their experience was largely limited to sailing within sight of a coast. They used the appearance of the land to tell where they were. But, out in the Atlantic, their only "landmarks" were the stars. To use the stars to navigate required more knowledge and mathematical ability than most mariners had. In fact, Columbus hardly used the stars at all. Instead, he used what is called **dead reckoning**. Dead reckoning means keeping track of the distance and direction you're sailing. For direction, Columbus used a magnetic compass, which had been imported from China in the thirteenth century.

*An early compass with hooks for suspending on wires.*

A magnetized needle was attached to a circular card that had directions marked on it. The card was then suspended by pivoted rings so it could rotate freely however the ship pitched and tossed. To keep a steady course, the helmsman simply had to hold the rudder so that the compass card remained steady, sometimes pulling harder on the tiller to adjust to slight wind shifts. By the way, the stern-post rudder controlled by the tiller was another import from China.

Now, while magnetic compasses point generally northward, the exact direction varies over the surface of the earth and has to be adjusted for. On Columbus's voyage, the variation of the compass was small enough to require adjustment only once in a while.

The way Columbus measured distance was much cruder. He watched some floating object—maybe a bubble—and estimated how far the ship sailed by it during a brief measured time. To measure time, he'd recite a standard prayer—like the Ave Maria,

> *Ave Maria, gratia plena, Dominus tecum, Benedicta tu in mulieribus, et Benedictus fructus ventris tui,   esu. Sancta Maria, mater Dei, ora pro nobis pecatoribus nunc et in hora mortis nostri.*

This might take ten seconds. If the whole ship passed the object in that time, the speed was 4.5 knots—about 8 km/h. Then Columbus would use an hour glass to keep track of how long he travelled at that speed. From that he'd mark on his chart how far he'd gone along his course. For example—in 10 hours at 8 km/h—80 km westward.

In August of 1492, Columbus made the short trip from southern Spain to the Canary Islands, 1300 km to the southwest. After taking on supplies there, he set out westward early in September, with good trade winds blowing from the northeast. By the middle of October he'd landed in the Bahamas—about 4800 km west of the Canaries. Here he made use of another aid to navigation—the lead and line. That's a heavy weight attached to a long rope. A sailor lowered it over the side until the lead hit bottom—or reached the end of the rope. Near land the water tends to be shallower so that the depth was some indication of the nearness of land.

From the Bahamas, Columbus explored southward, along the coast of Cuba to the northern coast of Haiti. He left Haiti in January of 1493 and headed northward, under what were mostly easterly winds, until the end of January. Then Columbus was far enough north to be in the region of the prevailing westerlies. So he had a good following wind astern for his eastward trip back to Spain.

At this stage, Columbus needed to have some idea of his latitude—how far north or south he was. After all his exploring, his dead reckoning could've been considerably out. So, he made a sighting on the North Star—Polaris—one of his infrequent uses of celestial navigation. Using an astrolabe, he measured the angular height of Polaris above the horizon. That gave latitude directly. When Polaris was 35° above the horizon, then Columbus was at 35° north latitude. But estimating the angle from the deck of a heaving ship was no easy task. Columbus did well to be within about two degrees, an error of over 200 km on the earth's surface, but very good according to the standards of the time.

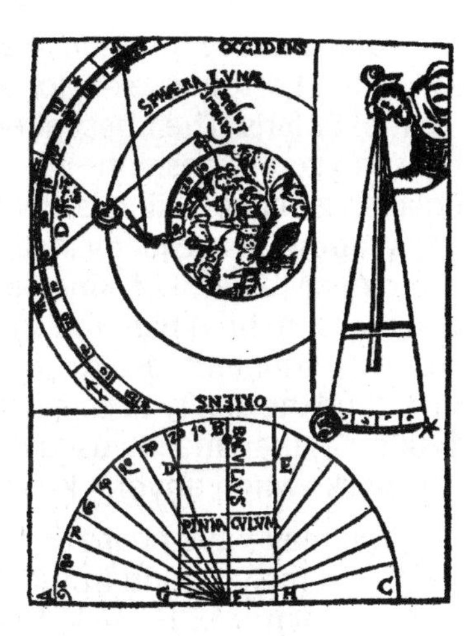

*A cross-staff such as Columbus would have used to find the angular height of a star.*

Columbus returned to the European coast in early March, 1493. Unfortunately, he landed at Lisbon. It took him more than a week to extricate himself from the Portuguese and return to the Spanish port of Palos, which he'd left seven months before. As we all know, his voyage was a tremendous achievement—even though he hadn't found the route to China he'd been looking for. Despite the crudeness of his instruments, Columbus was a superb navigator.

In the hands of skillful mariners, dead reckoning was a perfectly adequate method, especially for sailing east and west, as long as they didn't have to worry too much about their longitude, that is, their exact position east and west. Generally, the idea was simply, keep on sailing till you hit land. Navigators had great difficulty in using astronomy to determine longitude, even though in theory it's quite simple. All they had to do was to know the time in Lisbon when it was noon by the sun at their location. That's what we do now, except we use Greenwich instead of Lisbon. The position of the sun at noon on the earth's surface moves westward at 15° per hour. When the sun is due south of York University in Toronto (local noon), the time in Greenwich is 5.18 pm. Noon is five hours and 18 minutes later at York University than at Greenwich. The sun goes 75° in five hours, and 4.5° more in 18 minutes. That makes the longitude of York University 79.5° west of Greenwich.

Simple in theory, but in practice you need an accurate clock set to Lisbon time, and rugged enough to keep time on a bouncing ship. In Columbus's day, clocks didn't keep very good time, even on land. They could gain or lose five minutes an hour. And for navigation, every minute of error makes 15 miles error in position. As a matter of fact, clocks good enough for navigation weren't invented for another 250 years (see Chapter 13).

Polaris is fine for finding latitude as long as you're north of the equator. But as the Portuguese explored down the African coast, they eventually lost sight of Polaris. Their best alternative was to use the sun. But, unlike Polaris, the sun's zenith position—where it's highest in the sky—is different every day. Polaris's zenith position is always directly above the north pole. The *sun* is directly above the *equator* near the end of March and September—and 23° north near the end of June, and 23° south near the end of December. So, if you measure how high the sun is at noon, and know what day it is, you can calculate your latitude. But, to do that, you need tables compiled by astronomers, and the ability to calculate. These were both rare among seamen in the 1400s. So, they mainly used dead reckoning, with an occasional astronomical check—when anyone knew how.

You may have heard the legend that the Columbus venture was considered extremely daring in the 1480s because Europeans believed the earth was flat; hence, if he sailed out into the Atlantic he'd fall off the edge. Although a few unlettered peasants might have believed such a thing, no educated person did. The legend is false! From before the time of Aristotle, the Greeks had understood that the appearance of ships at a distance ("hull down at the horizon,") and the relation of the altitude of Polaris to the latitude of observing it, indicated that the earth is round. As we saw in Chapter 6, Sacrobosco's *Sphere* had made this knowledge available in Europe from the middle of the thirteenth century. Investors interested in the Columbus adventure would have been advised by persons familiar with Sacrobosco's book.

Columbus wanted to get to China by sailing westward. From numerous tales of seamen blown off course, he was sure there was more land to the west of the Canary Islands and the Azores. About 1480 he corresponded with Paolo Toscanelli (recall "perspective" in Chapter 7), who'd found that China in Ptolemy's *Geography* was 190° east of Lisbon (60° farther than is actually the case). Marco Polo had made it 30° even farther and then claimed that Japan was another 30° beyond that. As a result, Toscanelli concluded that the distance from Lisbon to Japan was about 110° instead of the actual 210°. Then, using a length for a degree about 60% of its real value, he concluded that the distance was less than 4000 nautical miles (about 7400 km).

Columbus was greatly encouraged by this, but used other "information" (and some fudging) to make the trip seem even shorter. He decided that if he started from the Canary Islands, he'd only have to sail 2400 nautical miles (about 4400 km) to get to Japan, where he could re-provision before going on to China. His actual distance when he made landfall at the Bahamas was only about 400 km beyond his estimate. Of course, Columbus had no way of knowing the Americas intervened between the Canaries and Japan. Lucky Columbus!

The first sea trip that finally made it to China was around Africa and across the Indian Ocean. By the Portuguese. They got to India in 1498 and China in 1517.

## *Vasco da Gama (1460–1524)*

Thirty years before Columbus's voyage, the Portuguese were using their fine new ships to explore down the Atlantic coast of Africa. By 1487, Bartholomeu Dias (1450–1500) had rounded the Cape of Good Hope and reached the Indian Ocean. Only then did the Portuguese realize they could get round Africa.

Meantime, the Portuguese had sent spies to the Middle East. One of these men sailed between Arabia and India, and sent back valuable information about the Indian Ocean. So, with the knowledge gained from their spies and from explorations, the Portuguese were ready to start trading with India. Vasco da Gama set out for India in July 1497.

His two ships were larger than Columbus's—about thirty meters long. Da Gama carried goods for trading and would be at sea for a much longer time than Columbus was. So Da Gama needed bigger ships. It took him 13 weeks out in the middle of the Atlantic just to sail as far as the southern tip of Africa.

That first stage of da Gama's voyage took him along the western coast of Africa to the Cape Verde islands. After taking on water there, he headed out into the Atlantic—southward with the north-east trade winds. Once he passed the equator, he encountered south-east trade winds and angled his course toward the southwest. That made him pass within a couple of hundred miles off the tip of Brazil. He got an occasional check on his latitude by sighting the sun at noon. When he judged that he'd gone far enough south, he headed due east. By then he was far enough south to pick up the prevailing westerlies, which would take him eastward toward Africa. He landed on the African coast 150 km north of the Cape of Good Hope—a relatively small error in his navigation.

Da Gama then sailed south along the coast, around the cape, and eastward into the Indian Ocean. Then he sailed north along the eastern coast of Africa. After he picked up an Arabian pilot on the east African coast in Kenya, he headed across the Indian Ocean toward India.

Vasco da Gama arrived at Calicut in India, in May of 1498—a voyage of 10 months. He returned triumphantly to Lisbon the next year. During the following twenty years, the Portuguese sailed aggressively into the Indian Ocean and the South China Sea. They established trading posts and permanent settlements along the way. Soon, much of the coastal shipping in south east Asia was Portuguese.

For, not only did da Gama's ships have trade goods in their holds, they also carried cannon on their decks. The new Portuguese design not only made ships efficient. It also made them fine platforms for guns. Warfare at sea was revolutionized. If you think of the thrilling pirate movies you've seen, you'll have some idea of sea warfare in the 1500s.

## *Europeans Conquer the World*

Thus began the European domination of Asia. Europeans had learned a lot of technology from the Chinese—the stern post rudder, gunpowder, the magnetic compass. Europeans put all these elements together and then sailed them back to China. Europeans with their armed ships overcame all opposition at sea and along the coasts. A Chinese official recalled one event some years after it happened. He wrote,

> The foreigners with the big noses came suddenly to the city of Canton and the noise of their cannon shook the earth.

It shook more than the earth. It shook Asian civilization. For the Europeans didn't just take their technology to Asia; they took the whole baggage of their European culture—their religion, their commercialism, their nationalism, their spirit of adventure. All so different from eastern attitudes.

Consider the simple acting of greeting someone. When two Europeans meet, with their heads held high, they thrust their right hands forward—to prove there's no weapon there. By contrast, when two Asians meet, they bow their heads gently and clasp their own hands together. A symbol of devotion that Europeans consider fawning or menial. Europeans have the right words for that gesture—like "your obedient servant, sir"—but I doubt they mean it.

Of course, while the Portuguese were conquering the east, the Spaniards were pillaging the Americas. Shortly to be followed by the British, Dutch, and French. American civilization was not merely shaken; it was utterly destroyed. It's a sad, oft-told tale.

## *Conclusion*

So Europeans conquered the world with their superior technology and their aggressive attitudes. Guns and ships provided the means. Religion provided the pretext. But gold and spices were the motive. One Portuguese reported that he and his buddies had gone abroad

> to serve God and the king, to give light to those who were in darkness, and to grow rich as all men desire to do.

Arrogant, aggressive, acquisitive, accomplished, the Europeans used the rest of the world as their supply house: gold and spices, slaves and cotton—the list goes on and on. They also used the world as a battle ground and as an extension of their own egos. You only have to think of the names of some of the regions: the Dutch East Indies, British North America, French Indo-China.

For four centuries, Europeans ruled the world. From 1550 to 1950, until the decolonization that followed World War II. Only recently have the Asians thrown off the yoke of European imperialism. Think of Gandhi in India and Mao in China. Sometimes the Asians used European technology to do it. One striking example occurred during the Russo-Japanese war in 1905. A Russian battle fleet left Europe the previous fall and sailed for more than seven months before engaging a Japanese fleet. The Russians discovered to their chagrin that the Japanese had built a modern battle fleet, even equipped with radio communications. In less than a day, the Japanese utterly destroyed the Russian fleet. Today, European imperial domination has declined, but European technology keeps on going, often in non-European hands.

And that means that European ideas still dominate Asia even though the political control is gone. Asians have had to learn European ways of thinking and doing in order to take control of their own lands. And not just technology and science. For the material side of European culture is thoroughly mixed in with its spiritual side. European civilization is based on a forward moving spirit, which has produced science and technology—the conquest of nature. When Asians want to share in those material advantages, they cannot separate them from the spirit that produced them. Which creates a dilemma for the Asians, who may want western technology, but who have a very different spirit: a will to contentment, harmony, and the golden mean.

This Asian dilemma was expressed with feeling by the Chinese author M. Chiang in 1947. He wrote about the impact of Europeans on his culture over the centuries:

> Since we were knocked out by cannon balls, naturally we became interested in them—thinking that by learning to make them we could strike back. We could forget for the time being in whose name they had come...

That is, in the name of Christ.

> ...since for us common mortals, to save our lives was more important than to save our souls. But history seems to move through very curious ways. From studying cannon balls we came to mechanical inventions, which in turn led us to political reforms; from political reforms we began to see political theories, which led us again to the philosophies of the West. On the other hand, through mechanical invention we saw science, from which we came to understand scientific method and the scientific mind. Step by step we were led farther and farther away from the cannon ball—yet we came nearer and nearer to it.

Even though the European invaders have gone home, they've left behind an indelible mark on Asia.

By 1500, Europeans, with their improved ships and cannon, with their national pride, their commercial greed, and their Christian zeal, were on their way to conquering the world. And they found the world very different from what they'd expected. Old maps had suggested that China was only about 4000 miles west of Spain. It's actually more than 10,000 miles. And North and South America block the route. During the three years from 1519 to 1522, Magellan's ships sailed right round the world and found out how broad the Pacific Ocean really is.

Soon, news of the enlarged world was spreading across Europe in printed books.

NICOLAI COPERNICI

net, in quo terram cum orbe lunari tanquam epicyclo contineri
diximus. Quinto loco Venus nono menſe reducitur., Sextum
deniq; locum Mercurius tenet, octuaginta dierum ſpacio circũ
currens. In medio uero omnium reſidet Sol. Quis enim in hoc

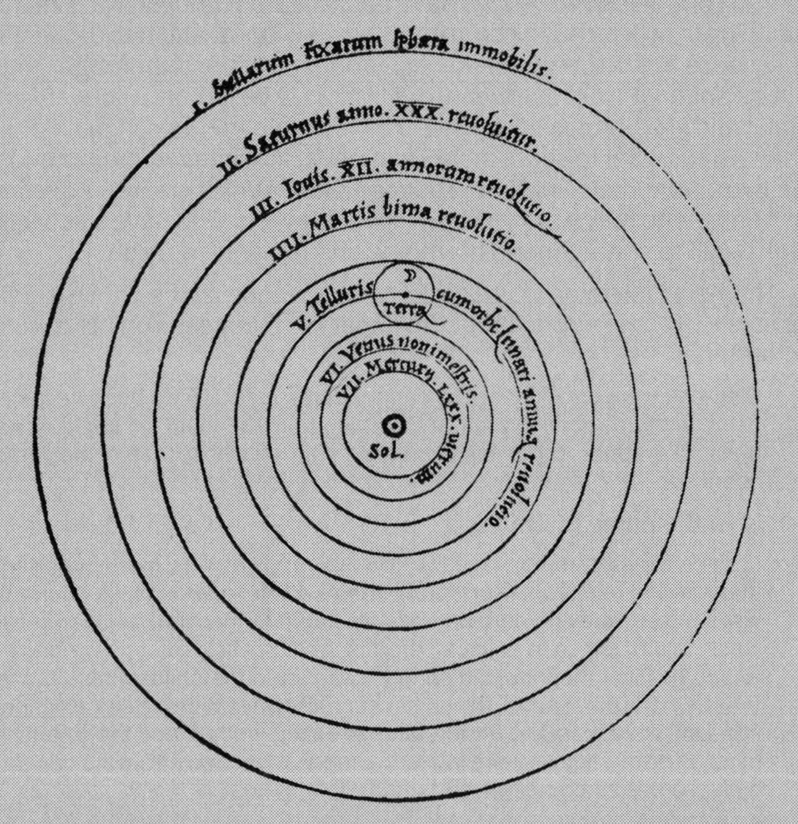


pulcherrimo templo lampadem hanc in alio uel meliori loco po
neret, quàm unde totum ſimul poſsit illuminare? Siquidem non
inepte quidam lucernam mundi, alij mentem, alij rectorem uo-
cant. Trimegiſtus uiſibilem Deum, Sophoclis Electra intuentẽ
omnia. Ita profecto tanquam in ſolio regali Sol reſidens circum
agentem gubernat Aſtrorum familiam. Tellus quoq; minime
fraudatur lunari miniſterio, ſed ut Ariſtoteles de animalibus
ait, maximã Luna cũ terra cognationẽ habet. Concipit interea à
Sole terra, & impregnatur annuo partu. Inuenimus igitur ſub
hac

*A page from Copernicus'* On the Revolutions of the Heavenly Spheres, *showing his sun-centered system. Compare it with the Aristotelian system, shown on page 59.*

CHAPTER NINE

# Enlarging the Universe

While European mariners were enlarging the earth with their overseas exploits, scholars were enlarging the universe with their astronomical investigations. From 1510 to 1610, astronomy was revolutionized at the hands of Copernicus, Tycho Brahe, Kepler, and Galileo. In 1510, Copernicus was in the midst of the studies that would lead him to propose the earth as a planet with others in orbit about the sun. From 1570 to 1600, Tycho Brahe made the most precise observations of the positions of stars and planets possible with the naked eye. In 1609, Kepler published the results of his analysis of Tycho's observations of Mars, concluding that Mars travelled in an elliptical orbit with the sun at one focus. In 1610, Galileo reported the innumerable array of stars visible through the telescope.

Among them, these four men transformed the science of astronomy, as well as the view of the heavens that Europeans could hold. For Kepler, the distance to the stars was 3000 times greater than the ancients had thought. In this chapter we'll investigate the reasoning that led to such a fantastic enlargement of the universe.

## *Nicolaus Copernicus (1473–1543)*

As Europeans recovered from the Black Death, their wealth increased, and the pope's authority declined. Some people turned to astrology for guidance. ("Is this a propitious day to sink my capital into that trading venture?") To improve their horoscopes, astrologers were keen to have better astronomical tables. They sought help at the University of Vienna, founded in 1365, as a center of astronomical revival.

In the 1450s, Georg Peurbach (1423–1461) was professor of astronomy and mathematics at the University of Vienna. He prepared two books to simplify Ptolemy's analysis and make it accessible to more people than could follow the complexities of the *Almagest.* He died before he was forty, leaving his work incomplete. However, it was completed and published by his pupil Johann Müller (1436–1476) of Königsberg (literally "King's Mountain"), who was known as Regiomontanus.

Regiomontanus set out to bring astronomy completely up-to-date, but he also died young. Nonetheless, his work, *The Epitome of the Almagest,* was printed in Venice in 1496. This work presented Ptolemaic astronomy in a clear, straightforward manner, a model textbook that was influential for much of the following century. Regiomontanus's *Epitome* was a major starting point for Copernicus.

*Nicholas Copernicus.*

Nikolaj Kopernik, known more commonly by the Latin form of his name, was born of German parents living in Torún, Poland, about 150 km south of the city that's now called Gdansk. His father died when he was ten, so Copernicus was raised by his maternal uncle, a priest who soon became a bishop. His uncle saw to it that Nicolaus was well educated—both in church law and in medicine. From 1490 to 1506, Copernicus attended universities at Cracow in Poland, and at Bologna and Padua in Italy. From then on, he lived as a church official in Poland near the Baltic Sea—a post obtained for him by his uncle the bishop—though Copernicus himself was never ordained.

His formal training in mathematics and astronomy was not extensive, but historians suspect he was studying them privately during his days at Cracow. His earliest knowledge of astronomy came from Sacrobosco's *Sphere*. Later, he studied Regiomontanus's *Epitome*. Some time after 1506, Copernicus began to study planetary astronomy in earnest. By 1515, he had recorded his ideas in a brief 24-page manuscript, which contained the startling hypothesis that the earth moves—both around its own axis and in orbit round the sun. In this little treatise Copernicus analyzed planetary motions with his new hypothesis to show that he could make predictions about as good as Ptolemy's.

But if his predictions weren't any better than Ptolemy's, why was Copernicus bold enough to move the earth? He didn't move the earth to improve the predictions but to improve the model. Copernicus had the common preconception that the planets revolved on actual spheres, and what bothered him about Ptolemy's scheme was the equant. Remember the equant is the off-center point from which planetary motion seems uniform. Copernicus said a real sphere just can't move uniformly around an off-center axis. It can only rotate evenly about an axis that goes exactly through the sphere's center. Otherwise it would wobble, like a wheel with its axle off center. So Ptolemy's equant was an affront to Copernicus's preconception about mechanical regularity in the heavens.

How could Copernicus get rid of Ptolemy's equant? Back in 150, Ptolemy had to explain two things with his models: the variations in the speeds of the planets and their retrograde motion.

Ptolemy had explained retrograde motion by using epicycles—having the planets move in small circles themselves as they moved around the earth on the circumference of a larger circle. And he explained speed variations by the equant.

With his preconceptions, Copernicus couldn't accept the equant. And he discovered in Regiomontanus's *Epitome* that an epicycle could explain planetary speed variations just as well as the equant could. But if Copernicus replaced Ptolemy's equant with another epicycle, it would mean having two epicycles for each planet. But having an epicycle on an epicycle would make calculations too difficult—that's why Ptolemy had chosen the equant in the first place. So, what Copernicus did was to replace Ptolemy's equant with an epicycle to explain speed variations. And then he went looking for a different way to explain retrograde motion.

He did it with another device he found in the *Epitome*. Copernicus made the planets revolve round the sun while the sun went round the earth. So, in this model he had the earth in the center, the sun in a circular orbit around the earth, and each of the planets (on their epicycles) circling the sun. This was Copernicus's first new model and it gave almost the same predictions as Ptolemy's. But notice—Copernicus hadn't yet moved the earth.

In fact, the model with the earth in the center of the sun's orbit, and the sun in the center of the planets' orbits is the one that the later astronomer Tycho Brahe would eventually adopt. We'll call it Tycho's model, even though it was already an option for Copernicus. But Copernicus didn't choose it. His reason was very old-fashioned, even if the consequence seemed newfangled. It happens that the radius of Mars's orbit round the sun is only 1.5 times larger than the radius of the sun's orbit round the earth. So in the Tychonic model, Mars's circle around the sun cuts across the sun's circle round the earth. As we'll see, that didn't bother Tycho. It bothered Copernicus a lot.

Copernicus couldn't allow orbits to intersect because he assumed that the planets were carried on actual spheres. You can have spheres revolving within spheres, but you can't have the body of one sphere crossing the body of another. So, because of his preconception that the spheres were real, Copernicus had to do something else: he had to move the earth. All he did was keep the planets orbiting the sun, but make the sun motionless, and put the earth into orbit around the sun too. That puts the orbit of the earth entirely *inside* the orbit of Mars, and then there's no intersection.

So Copernicus arrived at his newfangled idea for an old-fashioned reason. By clinging to Aristotle's concept of spheres, Copernicus destroyed Aristotle's idea of the central motionless earth. And Copernicus knew it. He knew that people would object, because there was no direct evidence of the earth's motion.

### *Copernicus's* ***Revolutions***

After Copernicus completed his earth-shaking treatise, he continued his regular duties as an official of the cathedral. In the following years, Copernicus described his model to friends, and they urged him to publish it. These friends even included a cardinal and a bishop. Eventually, Copernicus produced the book, called *On the Revolutions of the Celestial Spheres*. It was published in 1543, the year that Copernicus died at age 70.

By "revolutions" Copernicus simply meant "rolling again and again," signaling that his book was about continuous circular motions in the heavens. His use of the word "spheres" demonstrates his commitment to the traditional view of the celestial realm. So, there is nothing in his title to indicate the "revolutionary" nature of his work—indeed, the word was used in the sense of a great reversal of conditions only infrequently over the next couple of centuries.

The title page of Copernicus's *Revolutions* actually contains a brief notice of its nature:

> In this newly created and published book, you have, diligent reader, the motions of the stars, whether fixed or wandering, restored from both old and new observations; and moreover, you are provided with new and wonderful hypotheses. You also have the most convenient tables from which you can with the greatest ease calculate stellar positions for any time. Therefore buy, read, and profit.

Early in Book 1 (of six comprising his *Revolutions*) Copernicus argued that it was all right for the earth to move because it was a sphere—modifying Aristotle's principle by saying that spherical motion was *natural* for spherical objects. Just as it's natural for a ball to roll. That didn't convince many people, because while it may be natural for a ball to roll around on its own axis, it's not natural for it to roll around in circles.

A major advantage of Copernicus's scheme over Ptolemy's was that it explained retrograde motion without using epicycles. Mars doesn't really move on a retrograde path. It only seems that way to us on the moving earth. Because the earth is closer to the sun than Mars is, it moves around its orbit faster than Mars. The earth goes past Mars once every two years. As the earth overtakes Mars, we see it (so to speak) first ahead of us and then behind us, which makes its motion in the sky appear to be a loop. So, Copernicus explained the apparent retrograde motion of Mars and the other planets as simply a natural consequence of the earth's motion round the sun.

The Copernican scheme had other advantages over the Ptolemaic one. The main one is that Copernicus created a single *system*. Ptolemy had created seven separate systems—one for each planet. For Copernicus, the planets (including the earth) were all part of the solar system, revolving about the sun in orbits whose radii could be calculated. Copernicus established the definitive order of Mercury, Venus, Earth, Mars, Jupiter, and Saturn. This explained the differences between the "inferior" planets (Mercury and Venus are only seen relatively close to the sun) and the "superior" planets (Mars, Jupiter, and Saturn can appear due south at midnight). Other apparently arbitrary features of Ptolemy's models were similarly explained by Copernicus.

Copernicus explained the planetary orbits better than Ptolemy had. But his new model created as many problems as it solved. Besides moving the earth, which was an affront to the Aristotelian and Biblical notion of a "rock steady" earth, Copernicus's scheme had two other difficulties. The first was how Venus should look. In the Copernican model, Venus is closer to the sun

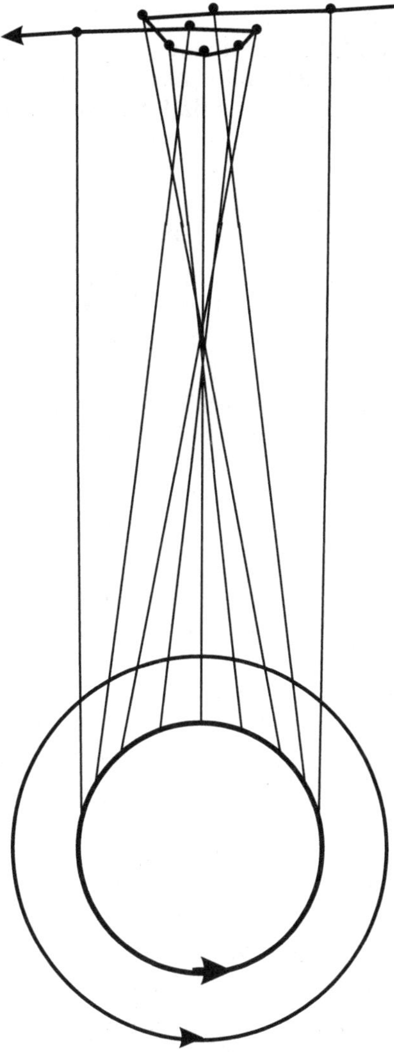

*The mysterious retrograde motion of the planets that was so difficult for ancient astronomers to explain can be shown to be an optical illusion in the Copernican system. In this diagram, the inner circle represents the orbit of the Earth around the Sun, while the outer circle represents the orbit of Mars. The straight lines represent lines of sight from Earth to Mars at different times over about six months. Because the Earth completes its orbit in about half the time that Mars does, the Earth "gets ahead" as it goes around the Sun. The top of the page represents the stars that would be seen in the background. As the diagram shows, Mars appears to move counter-clockwise, then stop, then go clockwise, then stop, and then resume its counter-clockwise motion, relative to the background stars.*

than the earth is. Venus's orbit is *within* the earth's orbit. If Venus shines by light reflected from the Sun, then, when it's between us and the sun, its bright side is away from us, and we shouldn't be able to see it. Venus should, in fact, show phases, like the moon. No one had ever seen phases in Venus, and no one would until telescopes were invented in the next century. Second, the motion of the earth from one side of its orbit to the other should cause a small parallactic shift in the positions of the stars. They should appear at slightly different angles during the course of the year. No one had ever seen that either.

Now, Copernicus had answers for these problems, but they sounded more like rationalizations than good answers. He thought that Venus always looked the same because maybe it had light of its own, or maybe it was transparent and refracted sunlight through its body. And he explained the lack of parallax in stellar observations by saying that the stars were so far away that their shifting positions were too tiny to be observed. That put the stars much farther away than people could possibly imagine. Yet Copernicus was right. The tiny shift in stars' positions does exist, but because the stars are so far away, it's less than a ten-thousandth of a degree. It was finally observed in 1838, 300 years after Copernicus.

In Copernicus's own time, his model had three fatal flaws: the appearance of Venus, the distance of the stars, and the moving earth. Although the Copernican model would provide an essential framework for future work, it didn't improve astronomy right away. In fact, except for moving the earth, Copernicus worked largely within Ptolemy's framework, even to using his observational data. He simplified some calculations, but his predicted positions of planets were hardly improved.

Another revolutionary feature of Copernicus's model was that he changed the meaning of the word **planet**. The ancients had seven planets—heavenly bodies that wandered through the sky, including the moon and the sun. For Copernicus, planets were bodies that orbited the sun. They included the earth, but not the sun or the moon. In his system the sun is motionless in the middle, and the moon orbits the earth. So for Copernicus there were only six planets. Making the earth a planet challenged Aristotle's distinction between heaven and earth. Historian Alexandre Koyré put it dramatically: "Copernicus wrested the earth from its foundations and hurled it into the sky." Another revolutionary feature was the tremendous enlargement of the universe. Copernicus put the stars a hundred times farther away than people had thought. He should've said a million, or more.

Copernicus's *Revolutions* was a technical mathematical treatise. He resisted simpleminded objections to his model by saying "mathematics is written for mathematicians." He dedicated his book to the pope of the time, Paul III. He did that because (he said in his letter of dedication to the pope)

> you are held to be most eminent both in your dignity and in your love of letters and even of mathematics; so, by the authority of your judgment you can easily provide a guard against the bite of slanderers.

But that didn't prevent some wits from making fun of his model. A poem written in France in 1578, as translated into English at the time, shows the kind of reaction that was voiced:

> Even so some brain-sicks live there nowadays
> That lose themselves still in contrary ways;
> Preposterous wits, that cannot row at ease
> On the smooth channels of our common seas.
> And such are those (in my conceit at least),
> Those clarks that think (think how absurd a jest)
> That neither heav'ns nor stars do turn at all,
> Nor dance about this great round earthly ball;
> But th'earth itself, this massy globe of ours
> Turns roundabout once every twice-twelve hours;
> And we resemble land-bred novices
> New brought aboard to venture on the seas,
> Who, at first launching from the shore, suppose
> The ship stands still, and that the ground it goes.

On balance, Copernicus's startling model produced very few ripples on the seas of scholarship for seventy years. Mostly people either dismissed it as foolish, or just didn't take it seriously. Protestants were sure that the Copernican scheme contradicted the Bible, while for many years the Catholic Church didn't see it as any threat to people's faith in God.

## *Tycho Brahe (1546–1601)*

Tycho was a member of the Danish nobility. He was also raised by his uncle, who, being childless, had kidnapped him from his parents. They had other children, so the uncle figured they could spare one. Tycho's uncle treated him well and supported him in university studies from age 13 to age 26. Maybe uncles haven't been given their rightful place in history!

*Tycho Brahe's observatory, Uraniborg, on the island of Hveen in Denmark.*

Tycho studied mathematics and astronomy at Copenhagen, Leipzig, Rostock, and Augsburg. His interest in astronomy began early when he learned that astronomers could predict events like eclipses of the sun. That seemed to him to be like reading God's mind. Tycho was attracted by astrology, too, and soon delved into astronomical predictions. When he bought printed copies of astronomical tables and checked their predictions by observation, he found they were often wrong. So, he resolved to make better observations and improve the tables.

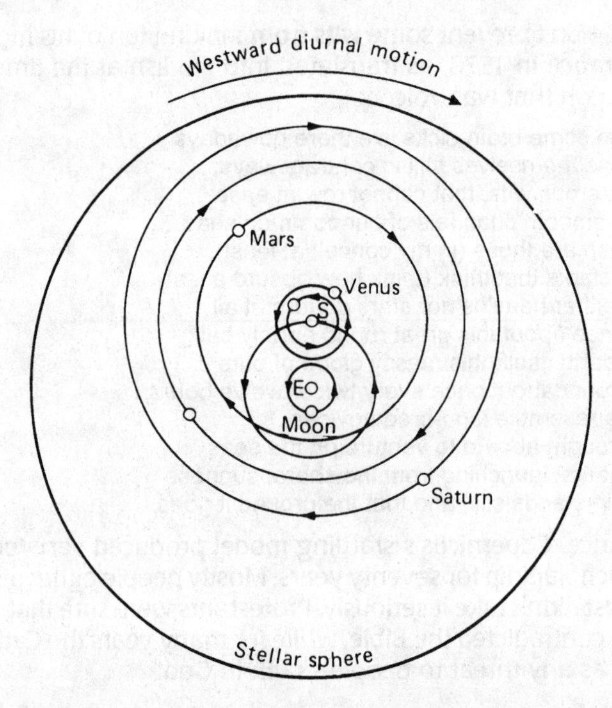


*The Tychonic system. A compromise between Ptolemy and Copernicus. The planets orbit the Sun, but the Sun orbits the Earth, which remains stationary.*

In fact, Tycho devoted the rest of his life to just that—making more accurate observations of planetary positions. He did this for over 25 years, until he died at age 55 in 1601. Most of the time he worked in an observatory on a small island in the Baltic, sponsored by the King of Denmark. Tycho designed the buildings, as well as the observing instruments, which he made very large—several meters across—so he could read angles to less than a sixtieth of a degree. Remember, these weren't telescopes, just sights for aiming at the stars. With these instruments, Tycho produced a catalog of star positions far more accurate than any before. It gave him the framework for measuring many different positions of planets.

Tycho also observed comets and estimated their distances. He discovered that comets moved along paths that would make them crash through the spheres of the planets—if those material spheres really existed. So Tycho gave up the Copernican prejudice of real spheres. Instead, he went back to the older prejudice that the earth couldn't move. As a result, Tycho decided that the sun orbited the stationary earth, carrying the planets in circles round the sun. As mentioned earlier, Copernicus had rejected that model when he moved the earth into orbit around the sun. This was a backward step for Tycho to take, especially because his model gave the same looping paths for the planets that Ptolemy's did. So, Tycho lost Copernicus's natural explanation of retrograde motion.

Tycho's most important contribution was to provide more accurate observations of the positions of the planets. Our next astronomer, Kepler, used those observations to effect another revolution in astronomy.

## *Johannes Kepler (1571–1630)*

Kepler was born in southern Germany and spent most of his life there and in nearby Austria. His parents weren't well off, but he was identified as bright and was trained for the Lutheran ministry at state expense. Kepler was a thoughtful and religious student, attracted by ideas of world harmony and unity. When he heard about the Copernican system from his mathematics professor at the University of Tubingen, he was struck by its order. Kepler immediately began to promote the Copernican model with youthful enthusiasm.

*Johannes Kepler.*

Before Kepler had completed his studies, he was sent to a seminary in Austria to teach mathematics. He never completed his theological training, and spent his life as a teacher and mathematical consultant to various governments. One of his main duties was to make astrological predictions. Governments were anxious about the future and expected their astrologers to predict the weather, plagues, and political events. Early on, Kepler gained fame for correctly predicting a cold winter and the Turkish invasion of Austria.

At age 25, Kepler made what he considered to be a great discovery. It was a scheme to explain the sizes of the planetary orbits. Kepler based his scheme on what are called the regular solids—space-filling shapes formed with regular figures like equilateral triangles and squares; each solid has all its edges of equal length. The simplest regular solid is the tetrahedron made of four equilateral triangles. Next is the cube, made of six squares. The octahedron is composed of eight equilateral triangles, and the icosahedron, of twenty. The dodecahedron is composed of twelve regular pentagons.

Now, there's a theorem in Euclid's *Geometry* proving that the number of convex regular solids cannot exceed the five listed above. Even God couldn't make more (at least not in the three-dimensional space He chose for the world). So, with his notions of universal harmony, Kepler decided that the five-ness of the regular solids must somehow be related to the six-ness of the

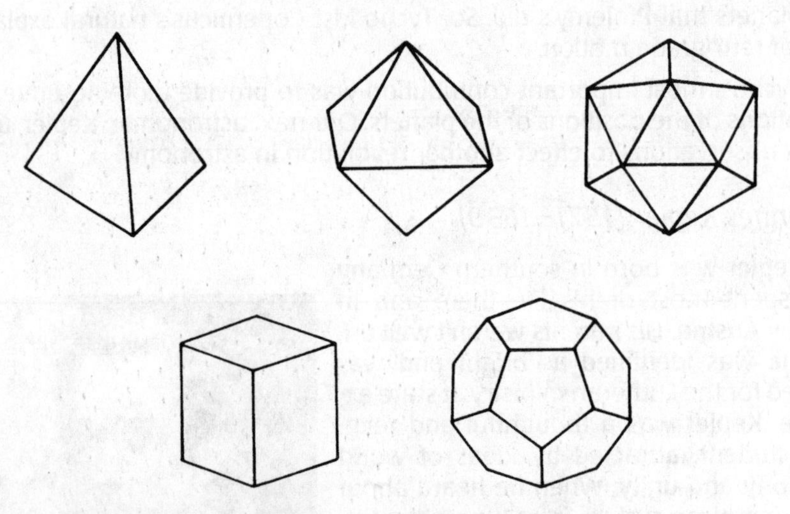

*The five regular solids: the tetrahedron, the octahedron, the icosahedron, the cube, and the dodecahedron.*

planets. (Copernicus's great achievement had been to show that six was a more important number than the traditional seven.)

Kepler imagined the regular solids to have been the basis for God's design of the planetary orbits. He thought that God had used the regular solids as separators (so to speak) between the spheres of the orbits of the planets. Start with the orbit of Saturn and build a cube that just fits inside it. Then put the orbit of Jupiter closer to the sun, just touching the inner sides of that cube. Build a tetrahedron within Jupiter's orbit, and put Mars' orbit against the sides of that tetrahedron. And so on. Mercury's orbit fits exactly within the fifth hollow solid, which happens to be the octahedron. Five separating shapes for the orbits of six planets. So, Kepler said, that's the reason there are only six planets. He published his theory in 1596 in *The Cosmographical Mystery*.

This book brought Kepler to the attention of Tycho Brahe, who in 1599 became mathematician for the emperor in Prague. About that time the archduke ruling Austria imposed Catholicism on his people and expelled Protestants like Kepler. He went to Prague to work with Tycho, succeeding him as imperial mathematician in 1601. This gave Kepler access to all of Tycho's records, and he began to use them to perfect his own model of the heavens—one similar to Copernicus's—that is, with the sun in the middle of the six planets. For that, he needed precise knowledge of the distances of the planets from the sun. Kepler used Tycho's observations to improve Copernicus's model for the planetary orbits.

The planet that had always given astronomers the most problems was Mars. No combination of circles, epicycles, or equants ever seemed to work as well for Mars as they did for Jupiter or Saturn. Mars always seemed to have

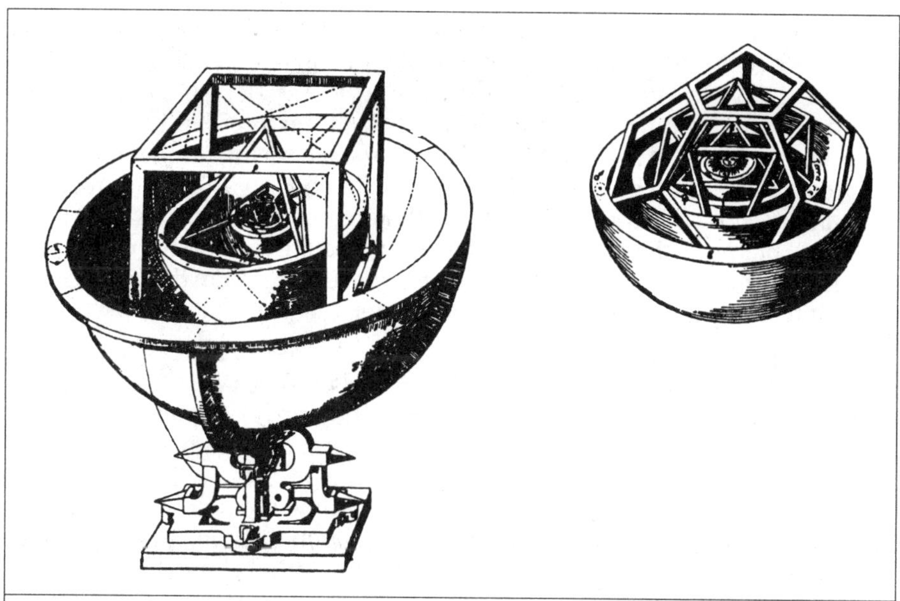

*Kepler's model of the universe from his Cosmographical Mystery (1596). Within the basic arrangement of the Copernican system, Kepler accounts for the sizes of the six planets' orbits by asserting that they are separated by the five regular solids. At left is the whole planetary system beginning with Saturn's orbit on the outside, separated by an inscribed cube from Jupiter's orbit, which is separated by a tetrahedron from the orbit of Mars. At right is a detailed view of the continuation toward the Sun; the orbit of Mars is separated from the Earth by the dodecahedron; the Earth is separated from Venus by the icosahedron, and Mercury is separated from Venus by the octahedron.*

an abnormal orbit. Kepler was motivated by a mystical conviction that the heavenly system must be perfectly orderly. But he was also convinced that Tycho's observations were more accurate than any others. So Kepler decided that *he* was the man to find God's perfect order in Tycho's observations of Mars. He used Tycho's data, but started with Copernicus's model: the sun motionless in the middle, and both Mars and the earth in orbit round it.

Kepler began by trying to calculate the best circle that would fit the observed positions of Mars. Very soon he made important changes to the methods Copernicus had used. Kepler decided to try to predict the orbit of Mars by calculating it from the exact center of the Sun. When he did that, he found that Mars's distance to the sun varied around its orbit, meaning that the sun was not in the very center of Mars's orbit. So Kepler reinstated Ptolemy's equant, from which Mars's speed would *look* uniform. With the equant Kepler found he could calculate Mars's speeds and distances more easily than when he used Copernicus's epicycle.

This was Kepler's first model of Mars's orbit. But, it still wasn't good enough. He realized that since observations were being made from a moving earth, he'd have to know the earth's orbit as well. He calculated it and found that the earth's orbit needed an equant too. With an improved orbit for the earth, Kepler returned to analyzing Mars. He made a second model for Mars's orbit that matched Tycho's observations to within an eighth of a degree, better than any previous model had done. Ptolemy and Copernicus had been content to blame such small differences on inaccurate observations. But Kepler was convinced that Tycho's observations were better than that. Which impelled him to try to improve his second model.

At this point Kepler introduced an entirely new method of calculating. He did it with a brand new hypothesis. He'd just read that William Gilbert had shown that the earth was a vast magnet (see Chapter 11). Kepler thought the sun and planets might all be magnets too, and that the magnetic force from the sun could explain the speed variations of the planets. He thought of the magnetic force spreading out from the sun, getting weaker at greater distances. The magnetic force was only strong enough to change their speeds, not to draw them into the sun. So Kepler made a rule for the speeds of the planets—that the planets moved more slowly when they were farther from the sun. Then he began to calculate the whole series of Mars's positions round its orbit, using its distances from the sun derived from his rule for speeds. Kepler tried to simplify this very tedious calculation by calculating the areas of narrow sectors between closely spaced distance lines from sun to Mars—a crude form of integration more than 50 years before the calculus was invented.

That led Kepler to his law of areas for orbits. For any two successive positions of Mars, the area formed by the line joining the sun to those two positions represents the time Mars takes to move from one position to the other. When Mars is farther from the sun, it moves slower and the area is long and skinny. When Mars is closer and faster, the area is short and fat. And Kepler found that this law fitted the observations very closely.

Kepler also found that Mars's orbit wasn't perfectly circular. At first he didn't recognize any regular shape. All he saw was that some parts of the calculated orbit fell within the circumference of his reference circle. After many months of laborious calculation—and some lucky breaks—he eventually realized that the shape of the orbit was an ellipse. An ellipse! All of a sudden, two thousand years of conviction that heavenly motions had to be perfectly circular came crashing down. The discrepancies between observations and predictions from the model vanished. Kepler proudly announced his discovery in a book with the double-barrelled title of *The New Astronomy, or Celestial Physics.* Though the book was completed by 1606, it wasn't published till 1609.

During the rest of his life, Kepler calculated new tables of predictions for the planets and published several more books. In 1619, he discovered that there was an important connection between how far each planet is from the

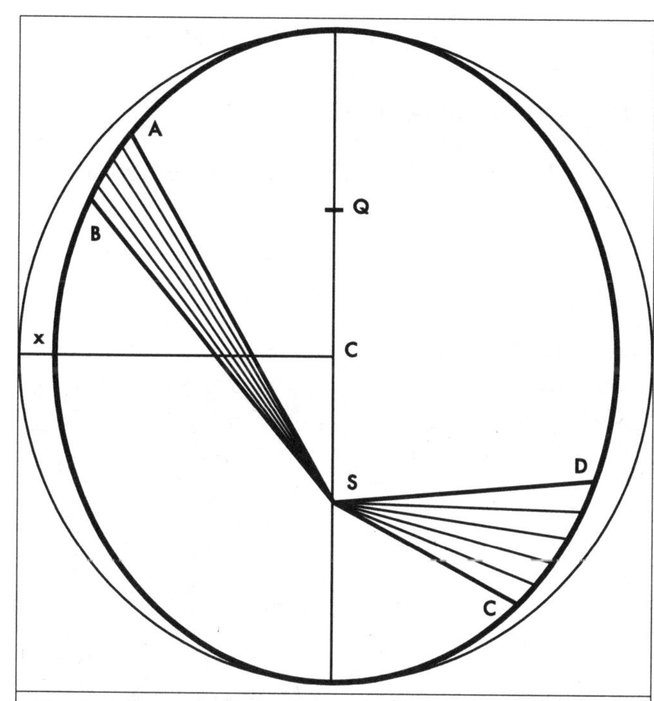


*Kepler's law of areas. As a planet moves in its orbit, a line drawn from the sun to the planet sweeps out equal areas in equal times. Here, if the time for the planet to move from A to B is the same as the time it takes to more from C to D, the wedges ABS and CDS will have the same area.*

sun and how long it takes to complete one orbit round the sun—what's called Kepler's harmonic law (see Chapter 12).

Kepler may appear to be a paradox. He was bound both by reason and by experience. You know how we say, "that's all right in theory, but experience shows otherwise." Not for Kepler! He was convinced both that Tycho's observations were accurate *and* that a reasoned explanation could be found for them. It was in trying to find reasoned explanations for accurate observations that modern science was born. We'll see this same attitude in Galileo's work too.

However, sometimes Kepler's reason ran away with him. Throughout his life he continued to improve the details of his regular solid model. It seems that his theory of elliptical orbits was only intended to make that original model more precise. Kepler believed that he had probed the mysteries of the harmonies of the heavens. He even calculated a musical scale to describe God's planetary harmony.

Kepler said that his goal was

> to show that the heavenly machine is not a kind of divine living being, but similar to a clockwork insofar as almost all the manifold motions are taken care of by one single absolutely simple magnetic force, as in a clockwork all motion is taken care of by a simple weight.

Weight-driven mechanical clocks had been invented about 300 years before Kepler's time. Some, like Dondi's *astrarium*, were marvelous mechanisms that showed the motions of the planets. They went from the heavens to clocks. Kepler went the other way—from clocks to the heavens. For him the heavens *were* a divine clockwork.

Kepler was always dependent on governments for his salary and often went unpaid. When he died in 1630 at age 59, he left four children under ten by his second wife and two older children by his first wife. They weren't any more successful than Kepler in obtaining his back pay and suffered badly in succeeding years.

## *Galileo Galilei (1564–1642)*

Galileo holds a special place in the history of science. If, as some say, Newton is physics, then Galileo is the father of physics, and Archimedes is the grandfather (see Chapter 10). Yet in the minds of the general public, Galileo is noted more for his astronomical discoveries and his troubles with the church in Rome.

*Galileo Galilei.*

Galileo called himself a mathematician and philosopher, not an astronomer. Nevertheless, his achievements, especially in observing novelties in the sky, were so great they must enter here. And once we start on him, we should also examine his troubles with the church. This is specially necessary, because the famous "Galileo affair" was not as deeply a conflict between religion and science as is still widely believed.

### *Galileo's Early Life*

Galileo was born in Pisa, Italy, in the same year that Shakespeare was born in England. Galileo's father was Vincenzio Galilei—which means "of the family of Galileo," the name of a famous ancestor. Pisa is a city in the state called Tuscany, whose capital was Florence. Vincenzio was a musician who played his lute for wealthy families in Tuscany—an uncertain employment at best.

Galileo was educated first in Pisa and then in a monastery school near Florence. In his early teens Galileo almost enrolled for the priesthood. But his father wouldn't allow it. Vincenzio was a good enough Catholic, but he didn't trust priests. Besides he wanted his son to restore the family fortunes, and priests didn't earn enough. In 1581, he sent Galileo to the University of Pisa to train for medicine. The family was living in Florence then, so Galileo stayed with an uncle in Pisa—another uncle!

After a couple of years at university, Galileo became more interested in mathematics and studied it privately instead of his regular courses. He left university before graduation and resolved to devote his life to mathematics. That annoyed his father, because mathematicians didn't make much money. Galileo was good enough at mathematics to become a professor at Pisa in 1589—aged 25. But he was only paid a tenth of what professors of philosophy got. Galileo couldn't support his aging parents that way!

Now, mathematicians in the 1500s were often involved in engineering. Galileo sometimes associated with engineers and developed a good reputation among them. One of them was a nobleman, who recommended Galileo for a better job, teaching mathematics at the University of Padua. That's in northern Italy, in the republic of Venice. Galileo moved there in 1592 for triple his previous salary. His father had died the year before, and Galileo could now afford to send home some money for his mother and his brothers and sisters.

Galileo made extra money by taking private pupils—teaching practical mathematics to young noblemen who wanted to learn about fortifications and other engineering topics. At the university, Galileo's teaching load was light. All he had to teach were the easy parts of Euclid's geometry and elementary astronomy based on Aristotle and Ptolemy.

Galileo knew the work of Copernicus and was impressed by its mathematical simplifications of Ptolemy; but initially he saw no reason to adopt the Copernican scheme. Then, about 1595, he had an inspiration. While travelling across the Venetian lagoon on a water barge, Galileo noticed how the water rocked back and forth when the speed of the barge changed. In his imagination, he let the barge represent the earth, and its water cargo, the oceans. The rocking of the water represented tidal motions. It occurred to Galileo that the tides could be explained mechanically (rather than mystically) by means of the Copernican motions of the earth.

He suggested to a friend that the combined effect of the daily and yearly motions of the earth would mean that parts of the earth travelled at different speeds at different times. That is, when Africa faced the sun its rotating motion would be subtracted from the earth's orbital motion, and 12 hours later would be added to it. So, as the speeds of parts of the earth changed, they might well rock the oceans.

Galileo's idea for using the Copernican motions to explain the tides gave him another reason for preferring Copernicus to Ptolemy. But it was not sufficient to make him a confirmed Copernican or to revise his teaching of elementary astronomy.

Galileo just tucked his tidal theory away and went on to his main interest—practical mathematics. He applied mathematics to simple machines like the lever and the pulley. And he looked for mathematical ways to describe the motions of balls rolling down inclines or falling straight down. That led Galileo to his revolution in physics (see Chapter 10). From 1592 to 1609, Galileo continued these various activities with only an occasional thought for astronomy.

*Invention of the Telescope*

In 1609, Galileo was 45, a well-established professor of mathematics in the Venetian republic, with a mistress who'd borne him three children. His salary had increased over the years, and he was reasonably well off. But he was still in debt back in Florence for the dowries of his two sisters. And then—out of the blue—came the telescope.

In the summer of 1609, Galileo heard a rumor that a Dutchman had invented a spyglass—a tube with two lenses in it that made distant objects appear three times larger—about as good as modern opera glasses. Immediately Galileo set out to duplicate that telescope in his workshop. Within two weeks he did even better; he made a telescope that magnified eight times. Near the end of August, Galileo presented his new telescope to the senators in Venice. Realizing how helpful it would be for seeing ships afar off, the senators confirmed Galileo's university post for life and promised to double his salary the following year. How pleased Vincenzio would have been!

Galileo returned to teaching in the fall and on the side improved his telescope even further. Soon he'd produced combinations of lenses that magnified 20 or 30 times. For several years, Galileo's telescopes were the best available.

Late in 1609, Galileo turned his telescopic gaze toward the night sky. In a very short time he observed three features that contradicted Aristotle's ideas about the heavens. When Galileo saw the moon in all her glory for the first time, he noticed small bright and dark spots changing in size as he watched. Something never seen with the naked eye. He soon concluded they were lighted peaks and shaded valleys—changing in appearance with the changing angle of the sun's rays—just like on earth. Very different from Aristotle's view that the moon was a perfectly smooth heavenly sphere. Instead, Galileo said, the moon had mountains and plains—its surface was rough. Score one against Aristotle.

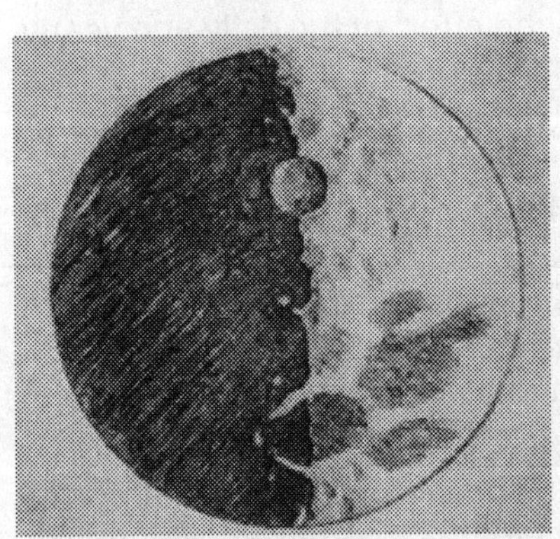

*Galileo's watercolor drawings of his observations of the moon, from his* Starry Messenger, *1610.*

Then Galileo swung his telescope through the stars. Planets like Mars and Jupiter looked like small discs, enlarged the way you'd expect. But the stars themselves were not enlarged. Indeed, the stars looked even smaller through the telescope because it cut out the sparkling effect we get when we look at the stars with our naked eyes. The telescope made the stars brighter but not larger. That could only be because the stars were an immense distance from the earth—very much farther than the planets—farther than Aristotle could ever have imagined. Score two against Aristotle.

And there were innumerably more stars visible now through the telescope than anyone had seen before. Not the fixed number of 7,000 Aristotle had claimed. Score three against Aristotle.

Then, early in January 1610, Galileo made the most startling discovery of all. One night he saw three or four tiny stars near Jupiter, along a straight line passing through the planet. Over the next few nights he noticed that as Jupiter took its wandering path through the fixed stars, those four little stars continued to stay near Jupiter—they wandered with Jupiter. Galileo soon decided they must be moons circling Jupiter, just like our moon circles the earth. He named Jupiter's moons the Medicean stars, in honor of the Medici family that ruled Tuscany.

Now Galileo rushed to get his discoveries into print before anyone else beat him to it. In March of 1610, he published a 50-page pamphlet titled *Sidereus Nuncius*—that's Latin for *Starry Messenger.* It was an immediate sensation, and people clamored for telescopes so they could see for themselves.

Now, what difference did all this make to astronomy? That is, the astronomy of making models to predict planetary positions. Not much really. Remember that Copernicus's revolution itself hadn't caused much of a stir back in the 1500s. But, Galileo's discoveries did become the foundation of a new *cosmology*, supporting the Copernican view in three ways.

First, Galileo felt his observations destroyed what had been considered the unique difference between the earth and the planets. Everybody had considered the planets to be more like the stars than like the earth. Through his telescope Galileo saw the planets as little round balls, not just points of light like the stars. So maybe, like the moon, the planets were also rough—earthlike. Galileo thought the planets were more like the earth than like the stars. If the planets are earthlike, then maybe the earth is planetlike—that is, also in orbit round the sun.

Second, in Copernicus's model the stars had to be very far away to explain the lack of stellar parallax—why the earth's motion didn't cause tiny changes in the angles at which the stars were seen. Galileo's observations supported that great distance to the stars.

Finally, the moons of Jupiter supported Copernicus by destroying a major objection to his model. The Aristotelians had said that there could be only one center of rotations in the universe—the earth. Copernicus had the sun in the center of the planets' orbits, but kept the earth as center for the moon. So, the Aristotelians objected to Copernicus because he had two centers. Now, Galileo said, that's okay—because we can even see a third center. At Jupiter with its moons.

These arguments weren't conclusive—but they gave aid and comfort to the Copernicans. Galileo mentioned these arguments in his *Starry Messenger*, but didn't insist on them. His opponents, the professors of philosophy, weren't impressed. Some of them suggested the telescope showed things that weren't really there. One of Galileo's colleagues at the University of Padua just refused to look through the telescope. He said it would give him a headache. Some headache!

Galileo was famous overnight. He applied for the post of mathematician to the Duke in Florence, Cosimo Medici, and got it for the same pay he'd been promised in Venice—and with no teaching duties! Now he could spend full time in research. Besides, the Duke paid off Galileo's debt for his sisters' dowries. Galileo returned to Florence in the fall of 1610. He left his mistress in Venice with their infant son and brought his two daughters home with him to grandma. Soon his two daughters were put in a convent where they spent the rest of their lives. Eventually Galileo had his son legitimized; he later got a minor government job in Florence.

In 1611, Galileo travelled to Rome, where he was treated like a hero. Astronomers at the Jesuit college there confirmed some of his discoveries, though they refused to go against Aristotle. Many Jesuits settled for Tycho's model with the sun orbiting the earth, and the other planets orbiting the sun. They had no trouble with the moons of Jupiter, but some of them refused to believe the moon was rough. They argued that the shadows of mountains on the moon were merely appearances, or else that there was a smooth invisible sphere of crystal encasing the moon. Galileo replied caustically, "If the crystal's invisible, how do you know *it* isn't rough too?"

Over the next couple of years Galileo observed spots on the sun whose motion showed that the sun rotated on its axis about once a month. He also observed that Venus showed phases like our moon. Remember the stories Copernicus had made up to explain why you couldn't see Venus's phases? With the telescope you could. That meant Venus must orbit the sun. Which was contrary to Ptolemy's arrangement, but fitted Copernicus and Tycho equally well. Galileo rejected Tycho's scheme because by then he had refuted all the arguments intended to prove that the earth was stationary. Though Galileo still couldn't prove that the earth moved by any experiments on the earth itself.

Until 1614, Galileo and his ideas were well received by the leaders of the Catholic church. Protestants had no serious objections either. Most people just treated the new discoveries as facts and left it up to philosophers and theologians to decide whether these new facts should affect the old beliefs. Besides, Catholics and Protestants at that time were busy struggling among themselves for control of various governments so they didn't have much time for astronomy or cosmology. You may have heard of the St. Bartholomew's Day massacre of Protestants in Paris in 1572; or the struggles in England between Catholic Mary and Protestant Elizabeth; or Guy Fawkes and the gunpowder plot in London in 1604; or the religious wars that wracked Europe from 1618 to 1648. All much more important to people than Jupiter's moons or Venus's phases.

Galileo was strongly opposed by academic philosophers who still clung to their Aristotle. And Galileo took great glee in combatting them. What person in his right mind would take the opinion of a Greek philosopher, dead for 2000 years, over the evidence of his own eyes? The more ridiculous Galileo made the philosophers look, the more they tried to get back at him. Finally, in 1614, they found a way.

*Galileo's First Troubles*

A philosophy professor at Pisa claimed that motion of the earth was impossible because it contradicted the Bible. Especially the passage where Joshua commanded the sun to stand still. If the earth revolved, the philosopher claimed, you'd expect that Joshua would have asked God to command the *earth* to stand still. Galileo had a reply to that argument. For, according to Aristotle, all the stars and planets revolve around the earth once a day with the sun. And they all move together because they're pushed by the outermost sphere—a sphere Aristotle called the *prime mover,* which he said was controlled by God. So, Galileo said, if philosophers believed Aristotle, *they* should expect Joshua to have asked God to make the prime mover stand still. The Aristotelian theory didn't fit the Bible either. Galileo wrote this in a letter to a friend, arguing that theology and science couldn't conflict and shouldn't be mixed. And that philosophers had no business in either science or theology. But just by answering the philosophers, Galileo had been drawn into theological controversy.

The letter Galileo wrote was copied by others. A priest in Florence got a copy and sent it to the Inquisition in Rome. He said that Galileo's arguments were stirring up dangerous controversy. In 1615, the Inquisition heard witnesses, examined Galileo's letter, and concluded it conformed to good Catholic doctrine. Meantime, Galileo got wind of the investigation and decided to go to Rome himself. He had two reasons for going. One was to look after his own reputation, to convince the authorities that he was a faithful Catholic. The other was to persuade church leaders that if they started to get involved in scientific decision making, it would only embarrass the church.

In Rome, in early 1616, Galileo found that his own reputation was secure. But he also found that cardinals who had once been friendly to him were now reticent. Galileo feared the question of the earth's motion might be decided by theologians, not scientists, which would damage both the church and Italian science. So he took every opportunity to argue that motion of the earth was purely a scientific question, which could only be answered by more research. Whenever he could find listeners, Galileo gave all the arguments and evidence against Aristotle and Ptolemy, and in favor of Copernicus.

Galileo stirred up enough controversy to annoy the pope, Paul V. He appointed a committee of theologians to make a final ruling on the earth's motion. Just what Galileo had opposed!

Near the end of February 1616, the committee reported to the Inquisition that motion of the earth was "foolish and absurd in philosophy, and an error in faith." That is, the theologians sided with Aristotelian philosophers, and based their theological decision on what can only be called outmoded science. Let's be clear about this. They didn't decide a scientific question by interpreting the Bible; they decided how to interpret the Bible by choosing to follow Aristotle. This was the very error Galileo wanted the church to avoid. He acknowledged the church's supreme authority in faith, but was sure the church would be in trouble if it authorized one brand of science over another. He didn't want the church to authorize Copernicus either, just to leave science free to follow its own methods. And to keep the church from committing itself to ideas that might later be proved false. It was not to be!

The church authorities took two actions following the decision of the theologians against the earth's motion. First, they issued an edict regulating books that discussed the idea that the earth moves. They prohibited books that interpreted the Bible according to the Copernican system. Although they permitted technical books that treated the motion of the earth hypothetically, they suspended Copernicus's *Revolutions* until passages that called the earth a star were removed.

In their second action, the church authorities dealt directly with Galileo. Cardinal Bellarmine warned Galileo to abandon his Copernican notions; if he didn't he'd be turned over to the Inquisition. Galileo acquiesced, and Roman affairs settled back into their normal course.

Galileo returned to Florence to continue his studies of motion and his work as the Grand Duke's mathematician. He was often ill with rheumatic complaints, which lasted till the end of his life. In the early 1620s, Galileo engaged in a battle of words—in print—with a Jesuit Aristotelian mathematician in Rome. Galileo wrote vigorously against the errors he found in the Jesuit's Aristotelian views of nature, but never mentioned the motion of the earth.

### *Galileo's* ***Dialogue***

In 1624, Galileo visited Rome to honor a new pope, Urban VIII, who was an old friend of his. Urban had enjoyed Galileo's anti-Aristotelian writings and encouraged Galileo to continue publishing. Galileo suggested writing a dialogue in which he'd compare the Copernican system with Aristotle and Ptolemy, and the pope agreed. Galileo finished the book six years later and took the manuscript to Rome to be checked by the censors. The censors edited and authorized the book. In February of 1632, the *Dialogo di Galileo sopra i due massimi sistemi del mondo* was published in Florence.

Galileo wrote his *Dialogue* in Italian to encourage intelligent non-academics, like doctors, lawyers, and businessmen, to think for themselves, not simply accept someone else's authority, especially not Aristotle's. The book is usually called "Dialogue on the two chief world systems: Ptolemaic and Copernican," though Galileo's original title had been "Dialogue on the Tides." It was written in the form of a conversation between three men. Galileo had them discussing all the topics that had engaged him over the years. They reasoned about the motions of the earth, the arrangement of the planets, and many other things. Over and over Galileo poked fun at the Aristotelians and showed that Copernicus's system made more sense than Ptolemy's in the light of reason and all the available evidence. In the preface to the *Dialogue*, Galileo said it was a review of arguments, not a statement of his own beliefs.

Galileo's *Dialogue* was an instant success. The edition of a thousand copies sold out in a few months. And then the ax fell! Enemies in Rome brought charges that Galileo had done more than review arguments in the *Dialogue*, that he had really defended the earth's motions and actually believed the earth moved, contrary to the order given him back in 1616.

In August of 1632, the Inquisition ordered Galileo's printer in Florence to suspend sales of the *Dialogue*. In October Galileo was summoned to Rome. His doctors' certificates of illness were ignored, and Galileo finally made the trip in January 1633. He stayed in the home of the Tuscan ambassador until April, when he was moved to a small apartment in the Inquisition building. Inquisition officials questioned Galileo about his interview with Cardinal Bellarmine back in 1616. Galileo was charged with strong suspicion of heresy although motion of the earth had never officially been ruled heretical!

What had happened was this. The authorities had looked up the file on Galileo in the Inquisition office. They found a stenographic report of that meeting with Cardinal Bellarmine. Since the report hadn't been signed, it clearly had no legal standing. But the report claimed that in addition to the cardinal's warning, the commissary of the Inquisition had commanded Galileo never again even to *discuss* the motion of the earth. Bellarmine had later told Galileo to ignore the command, but the report didn't mention that. So, just by writing the *Dialogue* Galileo appeared to have disobeyed the special command the report said he'd been given. He was now suspected of heresy, for having disobeyed that command. Galileo pleaded that the censors had licensed his book, surely that proved there was no heresy in it. But he admitted he hadn't told the censors about any special command not to discuss the earth's motion.

Galileo was convicted of second-degree heresy. His *Dialogue* was put on the list of prohibited books. He was forced to recant his belief in the motion of the earth and was sentenced to prison. That was soon commuted to confining Galileo to his own home near Florence for the rest of his life. Visits by others to his home were restricted, and he couldn't leave it without permission from the Inquisition. He lived just a short walk from the convent where his daughters were and couldn't even visit them without permission.

Galileo was crushed publicly as well as personally. His sentence of conviction was widely reported within church circles. In fact, the pope ordered inquisitors throughout Europe to gather local mathematicians together and read Galileo's sentence to them. Galileo was held up as an example and warning to anybody who might ever again try to use science to challenge the church's authority.

Galileo's *Dialogue* was translated into Latin in 1635, and 300 copies were sent to Paris. An English translation was published in London in 1665. Although Luther (around 1550) had called Copernicus a fool, Protestants now supported the new astronomy as a mark of their opposition to the pope and to Catholicism.

As for Galileo himself, he gradually overcame his grief. His indomitable spirit won out against great miseries. His favorite daughter died in 1634, and he soon became blind. Yet he continued to work on his mathematical analysis of motion: his book on *Two New Sciences* was published in Holland in 1638 (see Chapter 10). Galileo didn't have to smuggle that book out of Italy. It's true it couldn't be published there. But the Inquisition didn't prevent Galileo from writing or receiving letters. And they did even allow some visits. A Dutch publisher visited him in 1636 and took the first part of the manuscript of the *Two New Sciences* away with him. The rest was sent by mail, through a friend in Venice.

John Milton also came from England to visit Galileo. A few years later, Milton mentioned the visit in his *Areopagitica*, a defense of freedom of speech. He wrote eloquently about the plight of Italian scholars who felt tyrannized by the Inquisition. As he said,

> they did nothing but bemoan the servile condition into which learning amongst them was brought; that this it was which had damped the glory of Italian wits; that nothing had been there written now these many years but flattery and flummery. There it was that I found and visited the famous Galileo, grown old, a prisoner to the Inquisition for thinking in astronomy otherwise than the Franciscan and Dominican licensers thought.

Galileo died early in 1642, just short of his seventy-eighth birthday. He was buried in a back crypt of the family church in Florence. The pope forbade any elaborate ceremonies. Only a hundred years later was it possible to erect a monument to Galileo in the church of Santa Croce, where it can be seen today.

## *Science and Religion*

From time to time Galileo's difficulties with the Church have been used to symbolize a conflict between science and religion. Although you may wish to develop your own opinion about this, you should not allow yourself to be easily misled by authors with an ax to grind. If you investigate the issue for yourself, you should find a number of facts that must be accounted for. In the first place, Galileo always tried to be an obedient Catholic, even if he came from a family that resented any abuse of power by priests. Second, Galileo had numerous friends at various levels within the hierarchy of the church, including cardinals. For example, at the end of his trial in 1633, he spent the first five months at the home of his friend, the Archbishop of Siena. In addition, some of his closest colleagues (like Benedetto Castelli) were themselves priests.

That Galileo was severely disciplined by papal authority cannot be denied. Nor can it be denied that Catholics were restricted from reading the works of Galileo and Copernicus for about 200 years. However, papal authority was not absolute in these matters. Galileo was not excommunicated: when he visited his son in Florence in 1638 to be close to doctors, he was allowed to attend mass, as long as he kept to himself so as not to become the focus of rebellious talk. Moreover, there were numerous priests, especially in France, who continued to pursue science and astronomy quite openly, even if some chose not to put their true opinions into print.

The atmosphere of the times was by no means free and open, but it is a mistake to think that science and religion were diametrically opposed. Besides, restricting intellectual freedom was not a monopoly of Catholics. Calvinists, Lutherans, and Anglicans also had strong feelings about what ideas should be allowed to be published in the areas where they were in control. Some of those restrictions were gradually lifted in later years, partly because scientific ideas began to seem to be useful for the conduct of human affairs.

So, while religious authorities were concerned to protect their traditions and their power, they did not oppose every new idea just because it was scientific. Also, most scientists in the seventeenth century were professed Christians, not at all interested in toppling religious authority. To be sure, many scientists were avidly opposed to Aristotelian ideas, which did bring them into conflict with the traditional philosophers of their time. And occasionally, some church groups (like the Jesuits) did seem to make Aristotelian dogma into articles of faith. But that was more to combat religious heterodoxy than scientific novelties.

The "Galileo affair" was much more of a local dispute than an example of any general warfare between science and religion. Once Galileo's conviction had been obtained, the Jesuits did use it in their ongoing battle to lure German Protestant princes back to the Roman fold. However, except in Italy, there is little evidence that Galileo's conviction had much effect on the continuing growth of science. It may have hastened the shift in scientific activity towards the north and west, but that shift was already underway for other reasons—including the economic decline of Italy, and the growth of political and economic power in France and England.

## *Conclusion*

Between 1510 and 1610 three men transformed technical astronomy. They did it by starting where others had ended, but then changing the preconceptions implicit in their starting point. Copernicus started with Ptolemy's models and transformed them by eliminating the equant and making the earth move. He did that to save his preconception of solid celestial spheres.

Tycho Brahe destroyed solid spheres by his observations of comets. By destroying the spheres, he could keep his preconception of the motionless earth. Tycho also observed the positions of the planets much more accurately. Both Tycho and Copernicus kept the preconception that planetary orbits were perfectly circular.

Kepler had his own preconceptions—that the universe is harmonious and that men can find the harmonies by a mathematical analysis of careful observations. Kepler used Tycho's observations of Mars, Copernicus's arrangement of the planets, and Ptolemy's equant. But, in the process, his ellipses destroyed the longstanding preconception of perfect circular orbits.

I've summarized a hundred years of astronomy like this to show that science doesn't progress in a straight line. Ptolemy had the equant; Copernicus threw it out; Kepler brought it back. Ptolemy had the earth stationary; Copernicus made it move; Tycho made it motionless again; then Kepler moved it again. Copernicus had solid spheres; Tycho destroyed them; Kepler almost brought them back. In addition, Kepler introduced ellipses over the perfect circles of all his predecessors. Yet, despite all Kepler's improvements, his idea of a *magnetic* force between sun and planets would have to be changed. And his favorite idea that the regular solids determine planetary orbits quickly became an antique curiosity.

In 1610, observational astronomy was revolutionized by the telescope. In Galileo's *Starry Messenger*, people learned about the telescope and the heavenly wonders it revealed.

Galileo was not a mathematical astronomer the way Copernicus and Kepler were. He was more interested in combatting Aristotle's cosmology than Ptolemy's astronomy. Galileo preferred Copernicus because the system with the earth moving made more sense scientifically and fitted in better with his mathematical analysis of motions on earth. Although Kepler's theory of elliptical orbits gave better predictions than Copernicus's circles, Galileo paid no attention to them. They were just a minor technical detail. Of course he liked Kepler for being a Copernican.

Galileo's *Dialogue* was really more about cosmology and natural philosophy—the physics of things—than about technical astronomy. In fact, Galileo chose to attack Aristotle in the realm of cosmology rather than in the dry mathematics of falling bodies, because he knew it would attract more attention. His main concern was to drag the study of nature out from under the dead hand of the tradition of Aristotle. He did that for popular consumption in the *Dialogue* by dealing with cosmology. But in his own research, he would do best when he studied the motion of falling objects. Next chapter!

*Frontispiece of Galileo's* Dialogue on the Two Chief World Systems. *From the 1642 edition published in Leiden. Pictured are Aristotle and Ptolemy, representing the old world system, and Copernicus, representing the new.*

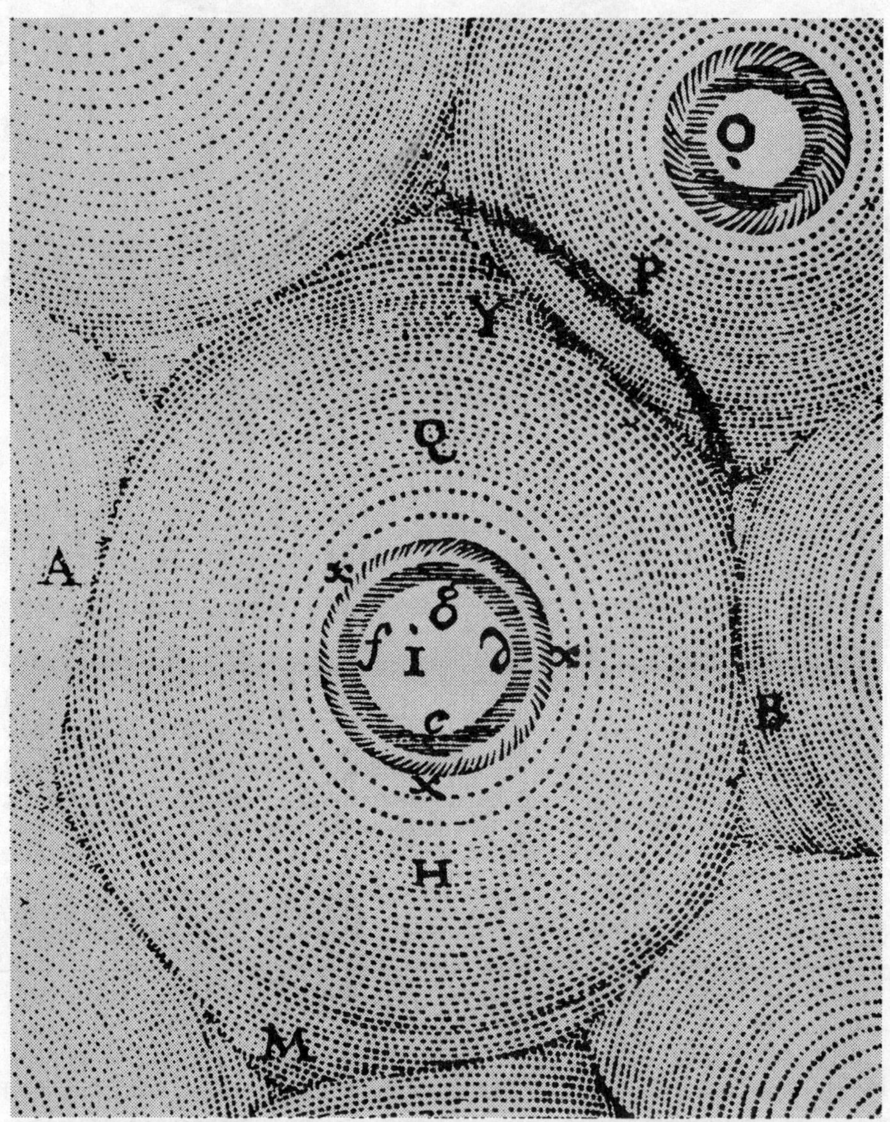


*Descartes' idea of how inertial motion is transformed into circular motion by swirling in vortices, ultimately carrying the planets around the universe. From Principles of Philosophy, 1644.*

## CHAPTER TEN

# Revolution in Motion

Until Galileo's time, motion was studied as a branch of natural philosophy. A lot of it derived from Aristotle's *Physics,* and some from his work *On the Heavens*. The philosophers of Galileo's time followed Aristotle in examining the causes of motion, assuming the effects to be well known. Galileo, on the other hand, resisted the search for causes. Instead, he used mathematics to analyze the effects, that is, how objects actually move. In his *Two New Sciences* (1638) he wrote:

> *Now doesn't seem the right time to investigate the causes of the acceleration of natural motion, on which many philosophers have produced many opinions...It'll be enough for us to investigate and demonstrate a few properties of accelerated motion.*

Galileo stole the study of motion from the philosophers and handed it to the mathematicians. His contemporary in France, René Descartes, claimed to be revising the whole of philosophy that way, but he couldn't achieve such an overly ambitious objective. More modestly, Isaac Newton in England soon showed the way to a full mathematical treatment of all kinds of motions.

## *Motion Studies before Galileo*

Aristotle had divided matter into four elements—earth, water, air, and fire—a common sense way of accounting for the gross differences in materials. Aristotle also used his theory of elements to explain ordinary motions. He said that earth and water fall because they're heavy, and air and fire rise because they're light. He considered these vertical—straight up and down—motions to be natural for the elements. They moved like that because...well..."What else would you expect them to do?" It's not so much an explanation of motion as a place to start. You don't have to explain why a book sits on a table: "What else would you expect it to do?"

Aristotle said it's natural for a stone to fall down, but it's unnatural for a stone to move up, or sideways; you have to throw it. Then he thought he had to explain what keeps the stone moving after it leaves the hand—why it continues to move unnaturally. Aristotle considered that something must be in contact with the stone to keep pushing it. His best guess was that the stone was carried along by the air that's also pushed by the hand. As early as 600 CE, some scholars began to think that was a pretty weak explanation.

Well into medieval times, philosophers began saying that if you threw a stone horizontally, the hand actually transferred its force to the stone. Then that force gradually leaked out, until the stone did indeed fall down.

Interesting, but not very helpful. Continuing horizontal motion was a problem right into the 1600s. Galileo mostly ignored the imaginary medieval force. But Isaac Newton was still puzzled by continuing horizontal motion. He finally solved the puzzle by creating a new science—dynamics—the science of force and motion.

Aristotle's authority was so great that it lasted 2000 years, even though the Greek mathematician Archimedes had taken a different approach back in the third century BCE, a century after Aristotle.

Archimedes belonged to the mathematical tradition of Aristotle's teacher, Plato. Archimedes didn't consider that air and fire were absolutely light, but only less dense than earth and water. He used the idea of density to explain why some solids can float on water. And he did it with great mathematical precision. Archimedes showed by geometrical reasoning that if water was more dense than a particular solid, like wood, the water would exert an upward pressure on the wood—a buoyant force—great enough to support the weight of the wood. The wood remains at rest, with only part of it submerged in the water. Archimedes' analysis is called **hydrostatics**. *Hydro* is the Greek word for water. *Static* means the wood's at rest—in equilibrium in the water.

Archimedes wrote a treatise on hydrostatics, and two others on ordinary statics—the conditions needed to keep balances and levers in equilibrium. He calculated the center of weight—or balance point—of irregular shapes. We call it the **center of gravity**, since "gravity" is just the Latin word for heaviness.

Archimedes' work on statics and hydrostatics didn't affect the study of motion until after 1500. Some medieval Arabic and European mathematicians did know Archimedes' work, but the philosophers who taught Aristotle's theory of motion weren't mathematicians. The best that can be said for the medieval mathematicians is that they kept alive Archimedes' tradition of statics; they didn't extend it to motion. Then, engineers began to find Archimedes' mathematics useful.

*Niccolo Tartaglia (1500–1550)*

Tartaglia made his living as a mathematics teacher. His pupils were mostly men who had a practical need for mathematics—especially military and civil engineers. He also did some important work in the growing field of algebra and acted as consultant to noblemen who wanted engineering advice. His practical consulting was different from Kepler's use of mathematics to make astrological predictions for noblemen (see Chapter 9).

In 1537, Tartaglia published a short book titled *La nova scientia* ("The new science"), which we should distinguish from Galileo's *Two New Sciences*, published a hundred years later. Tartaglia analyzed the path of a ball fired from a cannon. His elaborate geometrical reasoning was not correct. But he did de-

L'altro disse che molto piu tiraria a dui ponti piu basso di tal squara (laquale era diuisa in. 12. parte) come di sotto appare in disegno.

*An illustration from Tartaglia's* The New Science, *showing how to aim a cannon.*

duce correctly that a cannon would shoot farthest if it was elevated at an angle of 45 degrees.

Now just because Tartaglia called his book *La nova scientia* doesn't mean he'd actually succeeded in making the study of motion scientific. He didn't test his theories with careful measurements, although he did make a few rough trials. And he'd made a start in the right direction in that he used mathematics to analyze certain motions.

Tartaglia did more than just use and teach the mathematics of the ancient Greeks. He also put it into print. In 1543, Tartaglia published a Latin translation of Archimedes and an Italian translation* of Euclid. His Italian *Euclid* made geometry available to students who knew no Latin.

## *Galileo's Studies in Motion*

One of Tartaglia's pupils became Galileo's mathematics teacher. In the 1580s, Galileo studied Euclid and Archimedes privately while he was a university student. Galileo's first original mathematics was a study of centers of gravity, extending Archimedes' ideas. That brought him to the attention of a nobleman and mathematician who helped Galileo land a job teaching mathematics.

In 1589, Galileo, aged 25, became professor of mathematics at the University of Pisa, where he'd been a student only five years earlier. There he embarked on the career that would make him the father of physics. And he started by choosing Archimedes over Aristotle.

Galileo had a colleague at Pisa—a professor of philosophy—who was imbued with Aristotle's principles of motion. In 1591, this man, named Francesco Buonamico, published a gigantic thousand-page volume (in Latin) titled simply, *On Motion*. This book was entirely devoid of mathematics. The arguing was ponderous, scholastic, medieval. Here are a few sentences:

> Why do things which move according to nature move more quickly at the end than at the beginning of their motion? Many things have been said about this question, as much at the time of Aristotle himself as since then down to our own times. Many causes have been proposed: on the one hand are the intrinsic causes such as the nature of the object, or of the place where it is; and on the other hand, the accidental causes, such as

> the intervention of obstacles, the rarification of heat, a certain adventitious heaviness, and whether those causes act together or separately. Such explanations may sound reasonable enough, but without the eyes of Argus one could easily be mistaken—and we must examine all the particular causes with greater attention.

Buonamico then went on to describe the opinions of everyone who'd ever written on the subject: Greeks, Arabs, medieval Latins, and contemporaries.

To Galileo the mathematician, Buonamico's approach was a total waste of time. So, in the 1590s, he tried to do better by applying Archimedes' hydrostatic principle to motion. He replaced Aristotle's idea that earth is heavy and air is light, with the idea that air is simply less dense than earth. From that he concluded that denser objects fall more quickly than less dense ones because lighter objects are buoyed up by the air. He transferred Archimedes' principle of hydrostatics from water to air. Later, Galileo concluded from this argument that in a vacuum a feather would fall as fast as a stone!

Buonamico, however, like Aristotle, thought weight determined the speed of falling. So he said a large cannon ball would fall much faster than a small bullet. Galileo said they'd fall at the same speed even though the cannonball was heavier, because they both have the same density. Density, not weight was what counted in speed. That's not exactly true, but it's close. When Galileo tested this by dropping iron balls of different weights from the leaning tower of Pisa, he found that the heavier one landed just a tiny bit sooner. According to Buonamico, a ball that was ten times heavier should fall ten times faster, not just a little faster.

Galileo knew that Aristotle and Buonamico were wrong, but he still didn't have a good general rule to replace theirs. In 1592, Galileo moved from Pisa to Padua and took up other studies. He invented a scaling ruler for gunnery calculations and wrote a mathematical analysis of levers, pulleys, and other simple machines. Then, about 1602, Galileo started studying motion again, and within six years had figured out the basic ideas we still use today.

Galileo was determined to dispense with all arguments and opinions about the causes of motions. Instead of asking *why* objects fall, he would just concentrate on *how* they fall. He decided to study motion in a new way. He'd make actual, careful measurements and analyze them mathematically. But just what should he measure?

When an object moves, it covers a certain distance in a certain time. So, Galileo began to measure distances and times. Distances were easy; he used a very accurate ruler, good to the nearest millimeter. But it was difficult for him to measure the short time it takes a stone to fall a few meters. He had no stop-watch, and the mechanical clocks of his time didn't even give accurate minutes, let alone seconds.

So Galileo created a kind of "slow motion" by rolling a ball down a slightly tilted board with a groove in it. Then he marked that slow motion off into distances travelled in equal times. Since the motion was slow, he could use fairly simple timing devices. All he needed was a series of equal short time-intervals, about half a second each.

Galileo knew that a pendulum swinging back and forth keeps regular time even as it swings through shorter and shorter arcs. He'd discovered that by watching two equal pendulums about 2 m long swinging back and forth in step, even when one was started with a larger arc than the other. So he used a pendulum as his timer—"tic toc tic toc." Sometimes he may have sung a tune with a regular beat (recall that his father was a musician), such as:

*Fra Martino, campanaro,*
*Dormi tu, dormi tu,*
*Suona le campane, suona le campane,*
*Din, don, dan; din don dan.*

As the ball rolled down the plane Galileo marked its position at every beat.

In that experiment Galileo found that the distance the ball travelled in each half a second increased in a regular way: the successive distances were in the continuing ratio of 1 : 3 : 5 : 7 etc. (the odd numbers starting at 1). Then he used mathematics to calculate distances and times in vertical fall. But, he also wanted to relate those distances and times to speeds, which he couldn't measure directly. Galileo filled many pages with calculations and eventually figured out what we now call Galileo's law of uniform acceleration—acceleration simply means increasing speed. To create that law, Galileo had to invent the concept of what's called **instantaneous speed**: how fast the ball travels at any one point. That's not something you can measure, because there's no distance at a point.

Eventually, Galileo figured out how to handle instantaneous speeds, and how to relate them to the distances and times he *could* measure. With all that finished, Galileo could now say, "When a body falls from rest, its speed increases in proportion to the elapsed time." That's his **law of uniform acceleration**. He had it by 1608, six years after he'd started. A proper definition of speed had been the hardest thing to pin down. Then, being a mathematician, Galileo analyzed speed relations in various situations, such as on planes inclined at different angles.

Also in 1608, Galileo made another great discovery—that the path followed by a thrown object has the shape of a parabola. Aristotle had said only vertical falling was a natural motion. Galileo had the new idea that continuing horizontal motion is also natural. So if asked why a ball keeps rolling along the floor?—he could answer, "What else would you expect it to do? Don't even try to explain it. It's natural." In theory, the ball would keep moving forever. Since it doesn't, you may have to explain why the ball stops, but *not* why it keeps moving. The ball stops because it bumps into something, or it's slowed to a stop by friction with the floor and the air. This new principle of continued horizontal motion transformed Aristotle's idea of what motions are natural. Not just vertical motion, but horizontal motion too. Galileo combined these two motions to give a parabolic path, also a natural motion.

Consider a ball rolling along a table and off the edge. Once past the edge of the table, the ball continues to move forward at constant speed, while it also starts to fall downward with increasing speed. Galileo made measurements and calculations of these two independent and simultaneous natural motions—one of constant speed going forward, the other accelerating downward. And he found that when a ball rolled off the edge of a table, these two simultaneous motions combined to give it a parabolic path through the air—till it hit the floor. Cannon balls shot at low speeds follow parabolic paths. And you can deduce they'll go farthest horizontally when aimed at an angle of 45 degrees. Tartaglia had previously deduced that correct result, though from faulty principles.

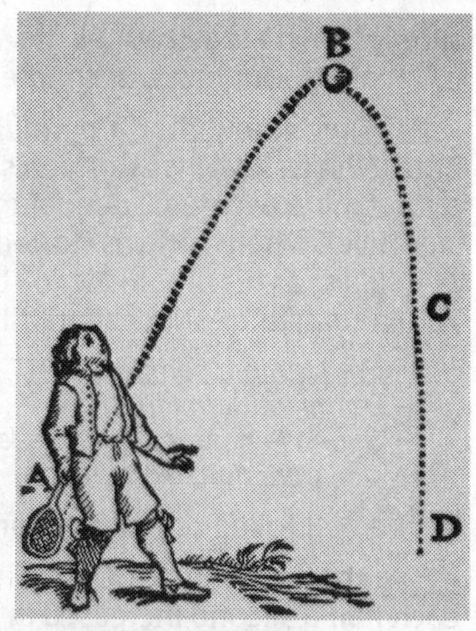

*An illustration of the parabolic arc of a tennis ball.*

Galileo could now make a whole series of deductions about many kinds of motions. In 1608, he began to put them together in a mathematical treatise he could publish. Then, in the summer of 1609, the telescope intervened (see Chapter 9). Galileo's treatise on motion wouldn't be completed for another thirty years.

## *Galileo Makes Physics Mathematical*

In 1591, Buonamico had said the last word on the philosophy of motion. Motion belonged to natural philosophers with their long-winded arguments and opinions about intrinsic causes and essences. Within 20 years, Galileo had stolen the science of motion from the philosophers and delivered it to the mathematicians. Like Robin Hood, he stole from the rich and gave to the poor. In the process, he invented the mathematical way to do physics.

Galileo brought physics into mathematics alongside astronomy by modeling his way of studying motion on the methods in astronomy. In 1602, he wrote a memorandum on how to investigate nature, stating the procedure in four parts:

1. start with the phenomena, the sense observations we see every day;
2. then make hypotheses, suppositions about the underlying structure of things;
3. use geometry to demonstrate how various events can follow from the hypotheses;
4. finally, use the geometrical model to calculate particular values for the phenomena.

Although Galileo didn't say so, his own methods in following years used an additional step:

> 5. compare calculations from the model with measurements of the phenomena. And where they compare poorly, revise your hypotheses (step 2), and then repeat steps 3 and 4.

Galileo devised this procedure by elaborating on some statements that had appeared in the preface of Ptolemy's *Almagest.* Although Galileo stated these steps originally in the field of astronomy, he was soon applying them in his study of motion. By our standard definition of science, Ptolemy had made astronomy scientific in 150; Galileo now made the study of motion scientific.

Galileo also learned from Archimedes. In fact, his very first original mathematics was done to extend some theorems of Archimedes on finding centers of gravity. But Galileo slightly changed Archimedes' basic approach by making his assumptions physical instead of purely mathematical. For example, in analyzing the equilibrium of a balance or a lever, Archimedes began with the assumption that "equal weights at equal distances are in equilibrium." There and in the following theorems, Archimedes didn't even mention the horizontal beam or lever arm. He must have assumed them to be weightless. But when the theorems are applied to real cases the beam cannot be weightless. Therefore, you must also assume it to be uniform. As Galileo described these theorems, he did apply them to real cases, but stated his uniformity assumptions only some of the time. His style of expression may be seen in this phrasing from his *Two New Sciences*:

> ...prisms and cylinders of constant thickness increase in moment beyond their own resistances as the squares of their lengths...

Galileo assumed the materials he discussed, like wood and iron, to be uniform (i.e., isotropic) without saying so.

We might say that Archimedes took a third of a step from mathematics to rational mechanics (a branch of mathematical physics). Galileo took a second third of a step. It was only the school of European mathematicians after Newton (men like Euler and the Bernoulli family) who took the final third of a step.

Galileo didn't get around to producing his book on motion until he returned to Florence after his condemnation in 1633. Its full title is *Discourses and mathematical demonstrations on two new sciences.* It was published in Holland in 1638.

Galileo's chief new science was **kinematics**—the study of motion by the mathematical analysis of measurements of accelerated motions, parabolic trajectories, and the like. In his other new science, Galileo invented a mathematical analysis for the strength of materials—especially for beams, the supporting structures of buildings. It wasn't as thorough as his analysis of motion, but it provided a sound foundation on which others would build. Galileo also described briefly some work he'd done on sound and vibrations.

## *René Descartes (1596–1650)*

*René Descartes.*

Descartes was born into the minor aristocracy in France. That gave him an independent income, so he never had to work for a living. Descartes got a classical education in a Jesuit college and later was a soldier for a few years. Then in the 1620s, he decided to devote himself to reforming mathematics and philosophy. He spent most of the rest of his life in Holland, writing out his philosophical ideas.

Most of Descartes' work belongs to the history of philosophy. His most lasting contribution to science was to invent a new branch of mathematics—analytic geometry. Unlike Euclid's geometry, it uses $x$ and $y$ axes to plot the shapes of curves, making it possible to describe them with algebraic equations. Analytic geometry would make the analysis of motion much simpler for Newton than it had been for Galileo.

Galileo hadn't even used algebra, only the proportion theory and geometry of Euclid. Galileo got about as deeply into describing motion as Euclidean mathematics would allow. Indeed, some historians think the Greeks themselves didn't study motion mathematically because their mathematics made it very clumsy to do. They used ratios of quantities. Ratios are fine for constant values. For things like weights on levers—but not for motions where distances and times are always changing. It was only because Galileo was so ingenious that he was able to handle instantaneous speeds and accelerations with ratios. With algebra and analytical geometry, Galileo would've found relations between distance, time, and speed much easier to express.

Here's a simple example. Suppose you're driving along at a constant speed of 100 km/h. It's easy to calculate how far you'll go in three hours—just multiply speed by time—100 times 3 means you'll go 300 km. If you write the equation, distance equals speed times time, $d = vt$, that's algebra. If you put it on a graph, that's analytic geometry. But Galileo didn't have analytic geometry and distrusted algebra. Following the prescriptions of Euclid, he was restricted to ratios of like quantities. So he'd have said one hour is to three hours, as 100 km is to the distance travelled in three hours. And then used the rules of ratios to calculate the distance to be 300 km. One is to three as 100 is to 300. That's how Euclid's proportion theory works.

Now, you might think that Descartes would have used his analytic geometry to take Galileo's analysis of motion farther. But he didn't. Descartes claimed that all reasoning should be as clear as mathematics, but he didn't actually use mathematics for everything. When he studied nature, Descartes tended to get wordy and philosophical. He said you shouldn't deal with things until you had clear, distinct ideas of them. Like Aristotle, he thought physics had to begin with clearly understood causes. So, when he read Galileo, Descartes said Galileo's work was without foundation because he hadn't begun by defining weight, and explaining how weight causes motion. Galileo measured weights and related them to motion; he didn't want to get tied up in words. His motto was: find out how things behave first. Then you can decide about definitions. But not Descartes!

In 1644, Descartes published a book called the *Principles of Philosophy*. It was a complete system that he meant to replace Aristotle's system. Like Aristotle, Descartes built up his picture of the world very carefully from what he considered to be first principles. So Descartes began by defining such things as thinking, god, space, matter, and motion. Then, in his **first law of nature**, he made a very general statement implying that everything remains constant unless something makes it change. He treated uniform motion in a straight line as a natural state that would continue unabated until it was caused to change. Descartes' actual statement of the law was:

> Any particular thing in so far as it is simple and undivided remains always to the best of its ability in the same state—nor is it ever changed from that state except by external causes.

This doesn't seem to say much more than that "things stay the way they are except when they don't." But Descartes included in it an explicit statement that uniform motion was a state of existence. This contrasted with Aristotle, who seemed to require a continuing cause to maintain an object in motion. Isaac Newton would later convert Descartes' first law of nature into his first law of motion.

In picturing the universe, Descartes tried to explain everything mechanically. For him the universe was full of particles swirling about and colliding. There were no empty spaces—only particles of different sizes—no spirits or essences. But even without spirits, Descartes managed to find a role for God. He said God had started the whole thing off by (so to speak) giving it a kick. And, somehow or other, God continued to preserve the universe by guaranteeing its laws—the laws Descartes had discovered, of course. Descartes then used his idea of gigantic whirlpools of particles to explain gravity and the tides and the motion of the planets around the sun. (See illustration on page 164.)

As you can imagine, Descartes' picture of the universe got him interested in collisions and circular motion. The rules he derived for the speeds of colliding bodies were wrong, and there's no evidence he ever tried to correct them by measurements. For circular motion, Descartes recognized that a stone swung in a sling exerts a force that tries to move the stone away from the center. That concept (the *endeavor* away from the center) would later come to be called centrifugal force. Descartes didn't analyze circular motion beyond that—he just described it.

Analytic geometry, collisions, straight and circular motions. These are the Cartesian ideas that influenced Newton.

## *Isaac Newton (1643–1727)*

*Sir Isaac Newton.*

Three hundred years ago, the name Newton provoked as much wonder as the name Einstein does in our own time. People might have respected Galileo, but they practically worshipped Newton.

Newton graduated from Cambridge University in 1665 at age 22. His education was strictly traditional—Aristotle and the Latin classics—with only elementary mathematics. Yet in 1664, on his own he began to study the most up-to-date mathematics books available, particularly algebra and Descartes' analytic geometry. Newton worked through those books diligently, and within two years he was inventing a whole new branch of mathematics—calculus.

All you need to understand about **calculus** is that it's a method for calculating the values of quantities that are continually changing. With calculus you can make equations to find how directions change along curved paths, or calculate the area enclosed by curved lines. Today we can see calculus as a natural extension of analytic geometry, but back then it took a genius like Newton to invent it.

Newton used his new rules of calculus to analyze forces in circular motions. Where Descartes had only described them vaguely, Newton calculated them exactly. He made an equation for the acceleration of an object traveling in a circular path. Now, Newton knew from Galileo's *Dialogue* that when a pendulum swings, it's a kind of falling motion, *down* along the arc of a circle. By timing the swings of a pendulum and using his new equation, Newton found a way to calculate the acceleration of vertical fall from the length and period of a swinging pendulum. He got a much better value for it than the one he'd taken from Galileo's *Dialogue*. Newton also used calculus to describe instantaneous speeds more clearly than Galileo could.

Then Newton investigated Descartes' laws of motion for colliding bodies. He corrected Descartes' mistakes, and, in the process, established his own general **laws of motion**. As published formally by Newton, they read (in an English translation made in Newton's time):

> 1. Every body perseveres in its state of rest, or of uniform motion in a right line, unless it is compelled to change that state by forces impressed thereon.
> 2. The alteration of motion is ever proportional to the motive force impressed; and is made in the direction of the right line in which that force is impressed.
> 3. To every action there is always opposed an equal reaction: or the mutual actions of two bodies upon each other are always equal, and directed to contrary parts.

With his laws of motion and his calculus, Newton could analyze all kinds of motion with great precision. Starting with clues from Galileo and Descartes, Newton founded the science of dynamics. By **dynamics** is meant using the ideas of force and momentum to write equations for the motions and interactions of moving objects. **Momentum** is our word for the quantity of motion in a moving object. It depends both on the mass and speed of the object. And **force** is equal to the change in an object's momentum.

To make Newton's ideas more concrete, imagine two people playing catch. Suppose Anne throws a tennis ball to Bob. Her moving hand applies a force to the ball. That gives the ball a certain speed—a certain momentum. The momentum will stay constant till the ball gets to Bob. Then, Bob has to exert a contrary force when he catches the ball—to reduce its momentum to zero. Anne'll have to exert more force if she wants to throw the ball faster. The faster the ball goes, the more momentum it has. And Bob'll have to exert more force to stop the faster tennis ball with its greater momentum. If Anne switches to a baseball, heavier than a tennis ball, she'll need more force to throw it at the same speed as the tennis ball. If the two balls go at the same speed, the heavier baseball has more momentum—and Bob'll have to exert more force to stop it.

## Conclusion

In the 1590s, the study of motion was still wandering through Buonamico's wordy thicket of the Aristotelian philosophy of motion: "What's the nature of this? Why does that happen?" Early in the 1600s Galileo came along and said, "You'll never answer all those whys of motion, let's find out how." By making measurements, geometrical demonstrations, and arithmetic calculations, Galileo found some simple rules for how things move, whether accelerating straight down or following parabolic paths. Galileo used mathematical theorems instead of scholastic arguments and experiments instead of opinions. In his own words, Galileo claimed that knowledge should be sought in *dimostrazioni necessarie e sensate esperienze*, which may be loosely translated as "conclusive mathematical proofs and careful experiments."

In the 1630s, Descartes invented analytic geometry and built a philosophy of motion based on particles whirling and colliding. That gave Newton two things. First, he advanced Descartes' mathematics to the calculus, and found out more about how things move than Galileo could with his limited mathematics. Second, by correcting Descartes' analysis of circular motions and collisions, Newton invented the science of dynamics, with rules we still use today.

Newton did his major work in calculus and dynamics in only three years—before he'd reached his twenty-fourth birthday. During that same time he also made important experiments in optics (Chapter 11). Most important of all, as he himself wrote "the same year [1666] I began to think of gravity extending to the orb of the moon."

By inventing the principle of universal gravitation, Newton made the rules of dynamics apply throughout the universe. Both in the heavens and on earth. How he thereby broke the celestial/terrestrial dichotomy that had ruled western thought since the time of Aristotle (if not before) will be discussed in Chapter 12.

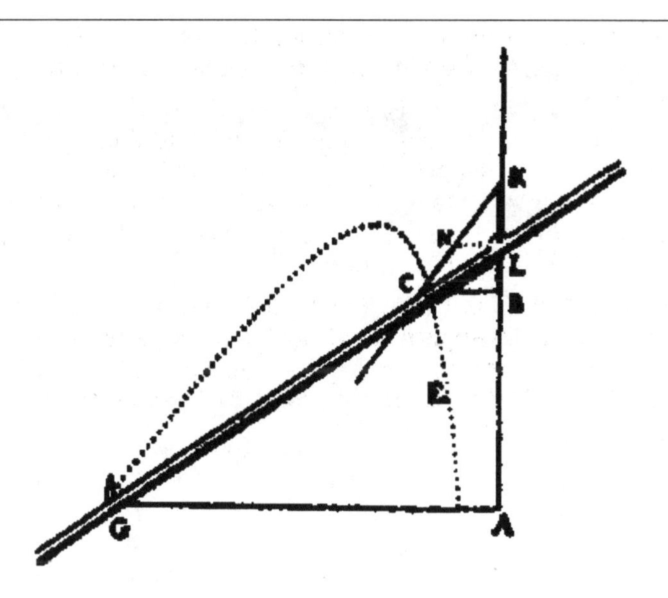


Aprés cela prenant vn point a diſcretion dans la courbe, comme C, ſur lequel ie ſuppoſe que l'inſtrument qui ſert a la deſcrire eſt appliqué, ie tire de ce point C la ligne C B parallele a G A, & pourceque C B & B A ſont deux quantités indeterminées & inconnuës, ie les nomme l'vne $y$ & l'autre $x$. mais affin de trouuer le rapport de l'vne à l'autre; ie conſidere auſſy les quantités connuës qui determinent la deſcription de cete ligne courbe, comme G A que ie nomme $a$, K L que ie nomme $b$, & N L parallele a G A que ie nomme $c$. puis ie dis, comme N L eſt à L K, ou $c$ à $b$, ainſi C B, ou $y$, eſt à B K, qui eſt par conſequent $\frac{b}{c}y$: & B L eſt $\frac{b}{c}y - b$, & A L eſt $x + \frac{b}{c}y - b$. de plus comme C B eſt à L B, ou $y$ à $\frac{b}{c}y - b$, ainſi $a$, ou G A, eſt á L A, ou $x + \frac{b}{c}y - b$. de façon que mul-

Sſ tipliant

*Page from Descartes'* La Geometrie *(1637) demonstrating an algebraic description of a parabola.*

*Isaac Newton experimenting with light.*

## CHAPTER ELEVEN

# Revolution in Science

The two preceding chapters dealt with some aspects of what has been called by historians, the scientific revolution. The study of motion was revolutionized by Galileo when he transformed it from philosophy to mathematics (Chapter 10). After him, Descartes and Newton added the concept of force to Galileo's kinematics, and created the new science of dynamics.

That revolution represented a change in the methods scientists used to probe nature. As we saw, Galileo got a major clue to his transformation of motion studies from astronomy. Astronomy had already been made fully mathematical by Ptolemy back in the second century. So, the revolution in astronomy wrought by Copernicus and the others (Chapter 9) did not represent any significant change in method. The Copernican revolution was much more a transformation in **cosmology** (how people thought about the universe) than in science (how people measured and calculated).

The mathematical sciences were strengthened and extended in the sixteenth and seventeenth centuries in various ways. But changes in Europeans' general view of nature and the universe were not limited to mathematical sciences. The expansive spirit of the Renaissance—in exploration, trade, art, religion, and literature—carried over into all the sciences.

In 1948, a general historian, Herbert Butterfield said of the scientific revolution:

> It outshines everything since the rise of Christianity and reduces the Renaissance and the Reformation to the rank of mere episodes, mere internal displacements, within the system of medieval Christendom. Since it changed the character of men's habitual mental operations even in the conduct of the non-material sciences, while transforming the whole diagram of the physical universe and the very texture of human life itself, it looms so large as the real origin both of the modern world and of the modern mentality that our customary periodization of European history has become an anachronism and an encumbrance.

In this chapter we'll look at some of those other sciences—anatomy, aspects of earth science (but geology is postponed to Chapter 16), and optics. In addition, we'll look at a couple of broader features of the scientific revolution.

## *Harvey and the Circulation of the Blood*

During the middle ages, medicine had been taught mostly out of books written back in Greek times. Some of the Greek anatomy had been based on the dissection of apes rather than humans. So the texts were often wrong. But medieval teachers mostly accepted the authority of the Greeks, and didn't bother to check whether the texts were right. Often the teachers had religious scruples about performing dissections of human corpses.

During the 1400s, physicians gradually recognized the errors in the Greek texts and began to correct their ideas about the human body. They were assisted by the new attitude among renaissance artists—like Leonardo da Vinci—who wanted to portray nature realistically, as *they saw* nature, not as some Greek had imagined it.

Leonardo, for example, made drawings of anatomical dissections so he'd know exactly how muscles were put together. By the early 1500s, physicians too were performing more dissections. They began to see individual organs and parts of the body as part of systems. Eventually, one of them, Andreas Vesalius (1514–1564), made a complete atlas of the human body—muscles, organ systems, skeleton. Vesalius performed the dissections and had an artist create the elaborate, precise illustrations. (See examples on pages 2 and 102.) Vesalius's book, *On the Construction of the Human Body*, was published in Latin in 1543, the same year Copernicus and Tartaglia published their books. Printing was spreading science abroad!

For a while Vesalius taught anatomy at the University of Padua. Its famous medical school was attracting students from all over Europe. Later, William Harvey (1578–1657) from England attended the Padua medical school and absorbed the new spirit of careful observation. After returning home in 1602, Harvey was puzzled about how the blood moved through its vessels, the arteries and veins. The Greek texts had said the blood only flows out from the heart and liver, carrying nourishment to the rest of the body. But Harvey combined careful observation with reasoned speculation to conclude that the heart pumps blood through the arteries to the veins and back to the heart in a continuous circulation.

Harvey published the results of his research in *An anatomical disquisition of the motion of the heart and blood in animals* (1628, in Latin), containing 16 brief chapters. His dedication to the newly enthroned king of England, Charles I, began:

> The heart of animals is the foundation of their life, the sovereign of everything within them, the sun of their microcosm, that upon which all growth depends, from which all power proceeds. The King, in like manner, is the foundation of the kingdom, the sun of the world around him, the heart of the republic, the fountain whence all power, all grace doth flow.

The analogies of the heart and the king with the sun betray the influence on Harvey's thinking of the Copernican hypothesis of the solar system. That should make us suspect that Harvey would combine his experimental evidence with an inspired speculation about how the whole system of heart and blood vessels was put together.

In human and animal bodies the arteries and veins run very close together and are often difficult to distinguish clearly. They're clearer when a corpse is dissected; for, the arteries are frequently nearly empty, though there may still be blood in the veins. When Greek physicians observed that, they speculated that the arteries carried air (the "breath of life") from the lungs to the body, while the veins carried nutrients in the blood from the liver. They thought of the heart more as a mixing chamber than a pump.

In early chapters, Harvey described his observations on the beating of the heart in live fish, frogs, snakes, and pigs. He noted the muscular nature of the heart, and the relation of its contractions to the pulsing of blood into the arteries. In chapter five he generalized that

> the sole action of the heart is the transmission of the blood and its distribution, by means of the arteries, to the very extremities of the body; so that the pulse which we feel in the arteries is nothing more than the impulse of the blood derived from the heart.

In the next two chapters, Harvey described the pulmonary circulation (that is, between the heart and the lungs). Here, he owed much to the earlier researches of the Paduan anatomist Realdo Colombo (1559).

In chapter eight, Harvey began to argue the consequences of his anatomical observations. With the heart continually pumping blood into the arteries, he realized that a large quantity of blood was involved:

> But not finding it possible that this could be supplied by the juices of the ingested aliment without the veins on the one hand becoming drained, and the arteries on the other getting ruptured through the excessive charge of blood, unless the blood should somehow find its way from the arteries into the veins, and so return to the right side of the heart; I began to think whether there might not be a MOTION, AS IT WERE, IN A CIRCLE. Now this I afterwards found to be true; and I finally saw that the blood, forced by the action of the left ventricle into the arteries, was distributed to the body at large, and its several parts, in the same manner as it is sent through the lungs, impelled by the right ventricle into the pulmonary artery, and that it then passed through the veins and along the vena cava, and so round to the left ventricle in the manner already indicated. This motion we may be allowed to call circular, in the same way as Aristotle *says that the air and the rain emulate the circular motion of the superior bodies [the stars]; for the moist earth, warmed by the sun, evaporates; the vapours drawn upwards are condensed, and descending in the form of rain, moisten the earth again.*

To clinch his argument, Harvey used explicitly quantitative reasoning. He measured the volume of the left ventricle of the heart of a human corpse to be about two ounces (weighing 57 g). He supposed as a minimum that an eighth of the blood is expelled at each contraction of the ventricle; i.e., 7 g. If the living heart beats 64 times a minute, you get 450 g. of blood expelled each minute, 27 kg in an hour—about four times the total amount of blood in the body. It *has* to circulate!

In fact, Harvey's estimate of the weight of blood expelled per stroke of the heart pump was far too low. For the sake of his argument, he erred on the right side.

In later chapters Harvey described other important anatomical features, such as the valves in the heart and in the veins. (See illustration on page 195.) All going to support his contention of the complete circulation of the blood.

Harvey used both careful observation and speculation to describe the motion of the heart and the blood. He observed as much as he could, but he wasn't able to observe the tiny microscopic blood vessels—the capillaries—which connect the arteries to the veins. So, all he could do was speculate that the blood must somehow get from the arteries to the veins. Capillaries were observed for the first time only 30 years later, after the invention of the microscope. That confirmed Harvey's speculation by showing red cells passing through the capillaries. In the same way, Galileo's observation of the phases of Venus with his telescope confirmed Copernicus's speculation that Venus orbits the sun. Instruments to extend the senses, like telescopes and microscopes, made a great contribution to the scientific revolution.

## *Francis Bacon (1561–1626), Prophet of the New Science*

A contemporary of Harvey's in England was Francis Bacon, born into a noble family three years before Shakespeare and Galileo. Bacon was trained in law, and from time to time was a member of the House of Commons and later the House of Lords. He was deeply embroiled in the many political intrigues of the time. Bacon was knighted in 1603 and became Lord Chancellor to James I in 1617, but was disgraced four years later.

During his rocky political career, Bacon found lots of spare time for thinking and writing. In his writings he caught the expansive spirit of the Elizabethan age. The power of England was growing, and Bacon proposed to enhance it with science. Though he wasn't a scientist, Bacon promoted science in his writings. He was bothered that his contemporaries speculated about nature using ideas taken from ancient Greeks like Aristotle. Instead Bacon wanted men to gain their knowledge directly from nature herself. He believed human life could be enriched by combining science with technology.

When Bacon was about 30, he wrote an essay that displays the spirit of all his later writings. He began by lamenting the uselessness of learned philosophy.

> Are we any the richer by one poor invention from all the learning we've had these many hundred years? The industry of artisans does make some small improvement in invented things; and chance experimenting sometimes makes us stumble on something new; but all the disputation of the learned never brought to light one effect of nature previously unknown.

Bacon blamed this on the barren methods of philosophers and alchemists.

> All the philosophy of nature we have now is either the philosophy of the Greeks or of the Alchemists. That of the Greeks is founded in words, in ostentation, in confusion, in sects, schools, and disputations. That of the Alchemists is founded in imposture, in clandestine tradition, and obscurity. The one never fails to multiply words, the other ever fails to multiply gold.

Instead of those vanities, Bacon proposed paying attention to technology. He identified the three great inventions of his age: printing, gunpowder, and the magnetic compass. He said:

> What a change have these three made in the world of our times; printing in the state of learning, artillery in the state of war, the compass in the state of navigation and treasures. And those, I say, were but stumbled upon by chance.

Stumbled on by chance! Bacon thought that even more benefits could be brought to humans by systemic investigation.

> Now we govern nature in opinions, but we are in thrall to her in necessity. If we would be led by her in invention, we should command her in action.

*The title page from Bacon's Great Instauration shows a ship sailing out between the pillars of Hercules. This image of setting sail for a new world is frequently used by Bacon to represent his break with the ancients.*

Bacon also wrote, in another place, "Nature cannot be conquered but by obeying her." That is, to achieve the goal of controlling nature, we have to understand nature's rules. And that, he said, was "not an opinion to be held, but a work to be done."

Well, you may ask, what work? Bacon answered that with a little parable:

> Men of experiment are like the ant—they only collect and use; the reasoners resemble spiders, who make cobwebs out of their own substance. But the bee takes a middle course—it gathers its material from the flowers of the garden and the field—but transforms and digests it by a power of its own. The true business of philosophy is like that.

An admirable doctrine! But Bacon wasn't clear about the processes his scientist-bees should use to transform and digest knowledge. All he proposed was that teams of investigators should simply collect data—what he called natural histories—and compile catalogs of careful observations of related phenomena. He thought that by comparing different phenomena, investigators would be able to pick out common features that would show nature's rules. This technique is sometimes called the Baconian method—or the method of **induction**. That is, start with particulars, and build up to generalities. It's the opposite of **deduction** —where you start with a general statement and deduce particulars from it. The logic of Aristotle and his followers was largely deductive.

Now, modern scientists actually use both deduction and induction. Bacon rather missed the mark by disparaging deduction, and also didn't see that one could understand complicated experiences by analyzing simpler, more abstract situations. Like Galileo rolling balls down inclined planes—where you get simple rules that have to be modified in everyday experience—for example, to take account of friction. Since Bacon distrusted abstract rules, he insisted on sticking to everyday situations with all their complexities. In fact, Galileo's methods often work better than Bacon's. Nonetheless, Bacon's writings did encourage scientists to make detailed observations, as Galileo, Kepler, and Harvey were already doing. Bacon's writings also inspired succeeding generations—inspired them to observe nature, to consider that human life could be improved by science, and to cooperate in investigating nature.

Actually, Bacon didn't inspire Harvey much. Harvey once said derisively of his illustrious compatriot that Bacon "writes philosophy like a Lord Chancellor."

### *William Gilbert (1544–1603) and Magnetism*

Another Englishman, 17 years older than Bacon, was William Gilbert, sometime physician to Queen Elizabeth. In 1600, Gilbert published a book in Latin, *On the magnet.* In this work he described experiments he'd performed, which led him to the startling conclusion that the earth itself is a gigantic magnet.

Before 1300, Europeans had learned from the Chinese that a magnetized needle will point north when it rotates freely. No one knew why. Some people speculated it was because of some kind of magical influence from the stars. But even without understanding magnetism, European navigators began to use magnetic compasses on their voyages. As they travelled the world, they noticed their compasses often deviated from due north, sometimes pointing as much as 50 degrees east or west of north. Eventually they coped by making surveys of compass variations and putting them on their maps.

Gilbert wondered why a magnetic compass points more or less northward. He rejected the theory of "magical influence from the stars" and chose instead to concentrate on magnetism itself. He made many experiments on the attractions of magnets. This led him to speculate that the behavior of compasses is also a magnetic attraction—that the earth itself is a gigantic magnet.

Gilbert tested his speculation by another experiment. He got a small ball of magnetic iron ore to use as a model of the earth. He called it a "little earth"—a *terrella*. Then Gilbert made little magnetic compass needles out of short iron wires. He laid the little compasses on the terrella at various locations and observed that they pointed more or less northward—the same way a ship's compass did round the world. But Gilbert also observed that his little compasses stuck out at angles to the surface of the terrella. Except at the equator, where they lay flat. Midway between equator and pole the needles stuck out at 45 degrees. At the poles, they stood straight out.

Now this was something else Gilbert could test—would a compass behave like that on the real earth? He mounted a regular magnetic needle so it rotated up and down—in the vertical plane—instead of back and forth, horizontally, as was usual. This new form of compass is called a **dipping needle**. He asked navigators to take dipping needles on their voyages. They confirmed that indeed, they were horizontal near the equator. When they sailed north or south from the equator, the needles dipped one end downward, at greater angles the farther north or south. So, on the real earth the needles behaved as they had on the model magnetic earth (the *terrella*), confirming Gilbert's speculation that the earth is a magnet.

Gilbert's work provides another example of combining careful observations with inspired speculations.

One of the readers of Gilbert's book was Kepler. You may recall he suggested magnetism to explain his elliptical orbits of planets round the sun—using one speculation to support another. Galileo also read Gilbert's book, but Galileo concentrated on trying to make magnets stronger—using Gilbert's techniques. Galileo didn't like Kepler's speculation that some mysterious attraction reached out across space from the earth to the moon, or from the sun to the planets.

## *Discovery of Atmospheric Pressure*

You shouldn't get the idea Galileo avoided all speculations. Here's one he made—and was wrong. Engineers had found that their pumps couldn't lift water more than about thirty feet, or 10 m. Galileo thought: maybe a column of water behaves like a weak rope and breaks under its own weight when it's more than 10 m high. It's not true! But a pupil of Galileo's figured out what is true, using reasoned speculation and careful experimenting.

Evangelista Torricelli (1608–1647) was in his early 30s when he lived with Galileo, in the last few months before Galileo died. Torricelli then succeeded Galileo as the Duke of Tuscany's mathematician. In 1644, Torricelli speculated about that limitation on water pumps. Knowing that mercury is about 14 times denser than water, Torricelli decided to test the limitation with mercury. He figured the height limit using mercury should be 1/14 of the limit for water. Instead of working with a column of water 10 meters high, he could use a column of mercury less than a meter high—much easier to handle.

Torricelli then built what we now call a **barometer**. He got a narrow glass tube about a meter long, filled it with mercury, and put his finger over the open end to hold in the mercury. Then he turned the tube upside down with his finger at the bottom. He dipped the end of the tube into a jar of mercury and removed his finger. Immediately, the level of the mercury in the tube dropped to a height of only three-quarters of a meter—about 30 inches. The upper portion of the tube was empty. Really empty! Not even any air.

To explain this effect, Torricelli turned Galileo's reasoning upside down—almost literally. Instead of asking how much liquid can hang down without breaking (Galileo's idea) Torricelli asked, what's holding the mercury up? What can hold up 75 cm of mercury—or 10 m of water—and no more? Here's Torricelli's speculation—in his own words:

> We live submerged at the bottom of an ocean of the element air...On the surface of the mercury in the jar there weighs a mass of air 50 miles high...[and] the mercury rises [in the tube] to such an extent as to come into equilibrium with the weight of this outside air which presses upon it.

Fantastic! An ocean of air and only 50 miles high. You mean the universe isn't full? Neither Aristotle nor Descartes could have handled that one. Descartes tried—with what he called "subtle matter"—so fine it could pass through the pores in the glass. That didn't help at all.

But how do you actually test the "ocean of air" speculation? Well, if the mercury is held up in Torricelli's barometer by the pressure of the ocean of air above us—then if you go up a mountain you'll have less air above you. The pressure of the atmosphere should be less; and the mercury won't rise as high in the tube. That test was devised in 1647 by a brilliant young Parisian, Blaise Pascal (1623–1662). He asked his brother-in-law Périer to make the test, since he lived near a mountain in southeastern France. A mountain about a thousand meters high. Périer made the test, and found that the mercury in the barometer on the mountain top stood about 8 cm lower than at the foot of the mountain. As you go higher, pressure decreases, and the height of mercury is less. So, by reasoning and experimenting, Torricelli and Pascal established the idea of **atmospheric pressure**.

Also, right from the start, Torricelli realized there was no air in the space above the mercury in his barometer—it was a vacuum. That's why he could explain the barometer only by the external action of the atmosphere. Since the space above the mercury was empty, there was nothing inside the tube to push down on the mercury. Torricelli had demolished the Aristotelian idea that a vacuum is impossible.

Soon, other men devised machines for taking the air out of containers—vacuum pumps. And then they found that sound wouldn't travel through a vacuum, but light would. Sound needs air; light doesn't. Also they could test Galileo's earlier speculation that in a vacuum, a feather and a stone would fall at the same rate. They do!

An intriguing combination of Harvey's idea of the heart as pump with the idea of air pressure was made in the early 1700s by Stephen Hales (1677–1761). This English clergyman used a barometer arrangement to make a direct measurement of the blood pressure of a horse. He inserted a tube into the horse's main artery to see how high up in the tube the heart could push the blood. Poor horse! She had to be sacrificed. Fortunately for horses, it was an experiment that only needed to be done once. But the experiment confirmed the *indirect* method for measuring blood pressure, which is still an important tool in medical practice today.

*Stephen Hales measuring the blood pressure of a horse with a barometer-like device.*

## *Scientific Communications*

How did these scientists with their experiments, speculations, and arguments communicate with one another? Since printing was well established, they read each others' books and reacted to them. Sometimes they wrote pamphlets to denounce opponents' ideas. Galileo, for example, engaged in a couple of these pamphlet wars. All in all, printing contributed greatly to the spread of scientific ideas.

Other forms of communication were helpful too—regular meetings and (eventually) periodicals. Scientists also wrote letters to congratulate someone for a new idea—or to argue against it.

### *Marin Mersenne (1588–1648)*

One of the most diligent letter writers of the seventeenth century was Marin Mersenne. In the 1620s, this French priest began writing letters to keep his scientific friends up-to-date on what the others were doing. Mersenne became, what one historian called the "secretary of scholarly Europe," or you might say the "letter box." He also organized regular meetings of scientists and mathematicians in Paris, where they discussed the latest reports of the work of their colleagues across Europe. These informal gatherings led to the formation of the *Académie Royale des Sciences* in 1666.

Of peasant stock, Mersenne nevertheless obtained a good education. His early schooling was at the same Jesuit college that Descartes attended. Mersenne was ordained a priest in 1613 and taught philosophy for a few years. From 1619, he lived the rest of his life at his order's monastery in Paris. There, Mersenne conducted his far-flung correspondence and published a number of books on the science of his time.

Mersenne's career gives the lie to those who claim to see in the seventeenth century a great conflict between science and religion. He took upon himself the role of demonstrating that the new science need not conflict with religious beliefs. That was largely because scientists like Harvey and Galileo were relatively modest in their claims about scientific knowledge. They did not attempt the creation of great philosophical or political systems that would challenge religion's supremacy in its own field. (When Giordano Bruno did that, he was burned at the stake, though his contemporary, Tommaso Campanella, merely spent many years in prison.) Of course, they could get into trouble from time to time with religionists who attempted to control scientific activities. Mersenne was opposed to that.

In the 1630s, Mersenne published several books that made Galileo's ideas available to Frenchmen. One was the first appearance in print of Galileo's work on *Mechanics* (levers and pulleys, etc.), which had been circulating in manuscript for thirty years. In another book, Mersenne included brief summaries of Galileo's *Dialogue,* despite its condemnation in Rome. (Although Frenchmen were mostly devoted Catholics, they tended to be less fearful of the power of Rome than Italians were.) In 1638, Mersenne published a French condensation of Galileo's *Two New Sciences*, a scant couple of months after that work issued from the press in Leiden.

Mersenne was a great popularizer of science. In books, letters, and meetings, he promoted the new scientific ways of investigating nature. As Bacon had demanded, and as Galileo and Descartes were doing, Mersenne encouraged people to see and think for themselves, instead of blindly following the traditions of Aristotle. The spirit of science was spreading abroad in Europe—Catholic Europe.

And in Protestant England too.

### *The Royal Society of London*

Englishmen read the European scientific books and travelled abroad to meet the authors. Englishmen also read Sir Francis Bacon. Some of them were inspired by his vision of collective action to promote science. Bacon had said he was "ringing the bell to call the wits together." And in the 1640s the wits—or scholars—did come together. Some of them began meeting regularly in London and Oxford to discuss the new science. In 1662, these men founded the Royal Society, with a charter from Charles II. Its full name is the *Royal Society of London for Improving Natural Knowledge.* It started with about a hundred members; some were scientists, but many were just interested amateurs. The Royal Society held weekly meetings and hired Robert Hooke (1635–1703) to perform experiments at each meeting. The Royal Society also began the publication of the very first scientific periodical called the *Philosophical Transactions*.

*Frontispiece of Thomas Sprat's* History of the Royal Society, *making much of the patronage of King Charles II.*

The early Royal Society actually tried to follow the precepts of Francis Bacon. They established teams to study various aspects of science and technology. One team, for example, investigated what they called "the natural history of trades." Using Baconian principles, their objective was to collect information about all the crafts, like glassmaking and woodworking, in order to improve them and contribute to the welfare of the English economy. But it was premature. Many practices in the crafts had been perfected by long experience. It would take more than a catalog to improve them. And it would take a much deeper knowledge of the workings of nature to apply science to technology than a handful of inspired amateurs could muster. Cooperative research would need ample funding and the full-time dedication of professionals. This wouldn't come for another couple of hundred years.

In the meantime, the Royal Society was more successful in supporting the work of gifted individuals. For example, they sponsored the publication of Newton's *Principia* (see Chapter 12) in 1687. Even before that, Newton's very first publication had been a letter in the *Philosophical Transactions*. In that letter Newton described an experiment on what he called "the celebrated phenomenon of colours."

## *Newton's Optical Researches*

In Greek times, Euclid and Ptolemy had written brief treatises on optics. They dealt largely with the geometry of light transmission, reflection, and so on. Their work was added to by Arab scholars, and picked up from them by medieval mathematicians. Under the influence of the theology of light, a couple of the medievals made further additions, including a start on explaining rainbows.

Kepler wrote a couple of books on light, dealing particularly with the ray optics of lenses. While Galileo was making telescopes, Kepler explained their operation. However, the correct law of refraction was not published until 1637 (by Descartes), although it had also been found earlier by a Dutch scientist named Willebrord Snel (1591–1626).

Through all this work, investigators had considered white light to be simple and primary. Then, Isaac Newton performed an experiment with an equiangular prism, which led him to propose a different hypothesis. After becoming professor of mathematics at Cambridge University in 1669, Newton sent a letter describing his experiment and conclusions to the editor of the *Philosophical Transactions*, who published it in 1672. This letter might be considered the first scientific paper in a scientific journal. An excerpt of Newton's famous letter is reprinted on pages 192-193.

Newton observed the spectrum of white light (red, orange, yellow, green, blue, indigo, and violet) spread out on his wall with a length five times its breadth. He could only conclude that white light was composed of those colors, which had been differentially refracted by his prism. Indeed, he performed a number of tests of that hypothesis and found them confirming it.

Newton's report raised a storm of controversy. In the following three years, eight critical letters were published in the *Philosophical Transactions*. Chief among Newton's critics was the curator of the Royal Society, Robert Hooke. He'd written about light and colors in a book published in 1665, and seemed unwilling to abandon his views for Newton's. The novelty and simplicity of Newton's ideas shattered traditional views and the results of Hooke's own hard work.

Newton's critics attacked him on every possible basis, both experimental and theoretical. Newton composed a response to every one of them. For a while he answered patiently, "Why could they not just do what he did, and see what he'd seen?" Newton considered his idea of the different degrees of refrangibility of the component colors to follow directly from his experiments. But his critics called the idea an unwarranted hypothesis. Finally, Newton could take no more. He withdrew into his shell at Cambridge and refused further communication with the Royal Society. He was particularly annoyed by Hooke's attitude.

Newton continued his optical studies at Cambridge and created a pretty complete analysis of physical optics. Yet as long as Hooke lived, Newton refused to share his further work with any but a small coterie of young admirers.

Only in 1704, the year after Hooke's death, did Newton finally publish his *Opticks.* On the very first page he struck again at the heart of the criticisms of 30 years before:

> My Design in this Book [i.e. in the first of three "books" comprising the Opticks] is not to explain the Properties of Light by Hypotheses, but to propose and prove them by Reason and Experiments.

At the end of the *Opticks,* Newton did allow his hypothetical imagination free rein in a series of what he called "queries." Several brief examples will show the tenor of Newton's speculations on issues for which he lacked the full support of "reason and experiments":

> 5. Do not Bodies and Light act mutually upon one another; that is to say, Bodies upon Light in emitting, reflecting, refracting and inflecting it, and Light upon Bodies for heating them, and putting their parts into a vibrating motion wherein heat consists?
>
> 13. Do not several sorts of Rays make Vibrations of several bignesses, which according to their bignesses excite Sensations of several Colours, much after the manner that the Vibrations of the Air, according to their several bignesses excite Sensations of several Sounds? And particularly do not the most refrangible Rays excite the shortest Vibrations for making a Sensation of deep violet, the least refrangible the largest for making a Sensation of deep red, and the several intermediate sorts of Rays, Vibrations of several intermediate bignesses to make Sensations of the several intermediate Colours?
>
> 28. Are not all Hypotheses erroneous, in which Light is supposed to consist in Pression or Motion, propagated through a fluid Medium? For in all these Hypotheses the Phænomena of Light have been hitherto explain'd by supposing that they arise from new Modifications of the Rays; which is an erroneous Supposition.

With Hooke gone from the Royal Society, Newton was elected its president, a post he retained till his own death in 1727. The period of Newton's presidency of the Royal Society did not mark any great advances in science. It was rather a time for consolidating and diffusing Newton's ideas. His disciples

## *Newton's Letter to the Editor of the Philosophical Transactions of the Royal Society of London*

*(3075) Numb.80*
*PHILOSOPHICAL TRANSACTIONS*
*February 19. 1671/72.*

*The CONTENTS.*

*A Letter of Mr.* Isaac Newton, *Mathematick Professor in the University of Cambridge; containing his New Theory about* Light *and* Colors: *Where* Light *is declared to be not Similar or Homgeneal, but consisting of difform rays, some of which are more refrangible that others: And* Colors *are affirmed to be not Qualifications of Light, derived from Refractions of natural Bodies, (as 'tis generally believed;) but Original and Connate properties, which in divers rays are divers: Where several Observations and Experiments are alledged to prove the said Theory. ...*

*A Letter of* Mr. Isaac Newton, *Professor of the Mathematicks in the University of Cambridge; containing his New Theory about Light and Colors: sent by the Author to the Publisher from Cambridge, Febr. 6. 1671/72; in order to be communicated to the* R. Society.

Sir,

To perform my late promise to you I shall without further ceremony acquaint you that in the beginning of the Year 1666 (at which time I applyed my self to the grinding of Optick glasses of other figures than Spherical) *I procured me a Triangular glass-Prisme, to try therewith the celebrated* Phænomena *of* Colours. *And in order thereto having darkened my chamber, and made a small hole in my window-shuts, to let in a convenient quantity of the Suns light, I placed my Prisme at his entrance, that it might be thereby refracted to the opposite wall. It was at first a pleasing divertisement, to view the vivid and intense colours produced thereby; but after a*

circulated his ideas far and wide in lectures and books on optics, calculus, dynamics, and the principle of universal gravitation (Chapter 12). It took a couple of generations for scientists to absorb all that and to learn how to apply Newton's principles in their analyses of phenomena other than the ones he'd illuminated so vividly.

*while applying my self to consider them more circumspectly, I became surprised to see them in an* oblong *form; which, according to the received laws of Refraction, I expected should have been* circular.

And I saw...that the light, tending to (one) end of the Image, did suffer a Refraction considerably greater than the light tending to the other. And so the true cause of the length of that Image was detected to be no other, than that *Light* consists of *Rays differently refrangible*, which, without any respect to a difference in their incidence, were, accordingly to their degrees of refrangibility transmitted towards divers parts of the wall....

Then I placed another Prisme...so that the light...might pass through that also, and be again refracted before it arrived at the wall. This done, I took the first Prisme in my hand and turned it to and fro slowly about its Axis, so much as to make the several parts of the Image...successively pass through...that I might observe to what places on the wall the second Prisme would refract them.

When any one sort of Rays hath been well parted from those of other kinds, it hath afterwards obstinately retained its colour, notwithstanding my utmost endeavors to change it....

I have refracted it with Prismes, and reflected with it Bodies which in Daylight were of other colours; I have interceted it with the coloured film of Air interceding two compressed plates of glass; transmitted it through coloured Mediums, and through Mediums irradiated with other sorts of Rays, and diversely terminated it; and yet could never produce any new colour out of it.

But the most surprising, and wonderful composition was that of *Whiteness*. There is no one sort of Rays which alone can exhibit this. Tis ever compounded, and to its composition are requisite all the aforesaid primary Colours, mixed in a due proportion. I have often with Admiration beheld, that all the Colours of the Prisme being made to converge, and thereby to be again mixed, reproduced light, intirely and perfectly white.

Hence therefore it comes to pass, that *Whiteness* is the usual colour of *Light*; for Light is a confused aggregate of Rays indued with all sorts of Colours, as they are promiscuously darted from the various parts of luminous bodies.

## *Conclusion*

By 1700 scientific practice had been completely transformed. As in Bacon's image of spiders and ants, scientists were no longer spinning beautiful webs out of nothing (like Aristotle), nor making haphazard piles of disconnected facts (like the alchemists). Instead, like bees, they were building a honeycomb of related ideas about nature. They observed and experimented where they could, but they also went beyond experience by speculating. That is, they proposed hypotheses about what they couldn't observe. They combined induction from experience with deduction from hypotheses. Then they tested their deductions by further observation or experiment.

They didn't hesitate to go beyond their direct evidence. Harvey was unable to observe the capillaries that carried blood from the arteries to the veins on its way back to the heart. Nonetheless, he concluded from his many observations and experiments that the blood must be circulating. He hypothesized the connection from arteries to veins and soon microscopists found it.

Gilbert deduced a prediction for the behavior of dipping needles from his hypothesis that the earth is a magnet. When dipping needles were found to behave as Gilbert had predicted, that supported his hypothesis.

Torricelli speculated from his barometer experiments that we live at the bottom of an ocean of air. From that hypothesis, Pascal deduced that barometric pressure should be less on a mountain top, and his experiment confirmed the deduction.

Newton experimented with a beam of sunlight and a prism, and saw the vivid colors of the spectrum. He performed a number of tests that convinced him that white light is composed of all the colors. As we've discussed, Newton was very distressed when Hooke and others said his conclusion was just a hypothesis. Yet Newton's ideas about light did go beyond direct evidence.

They had to in order to make good science.

*Otto von Guericke's demonstration of the power of the vacuum. Sixteen horses were unable to separate two bronze hemispheres that had been placed together and the air pumped out of the space inside. Magdeburg, Germany.*

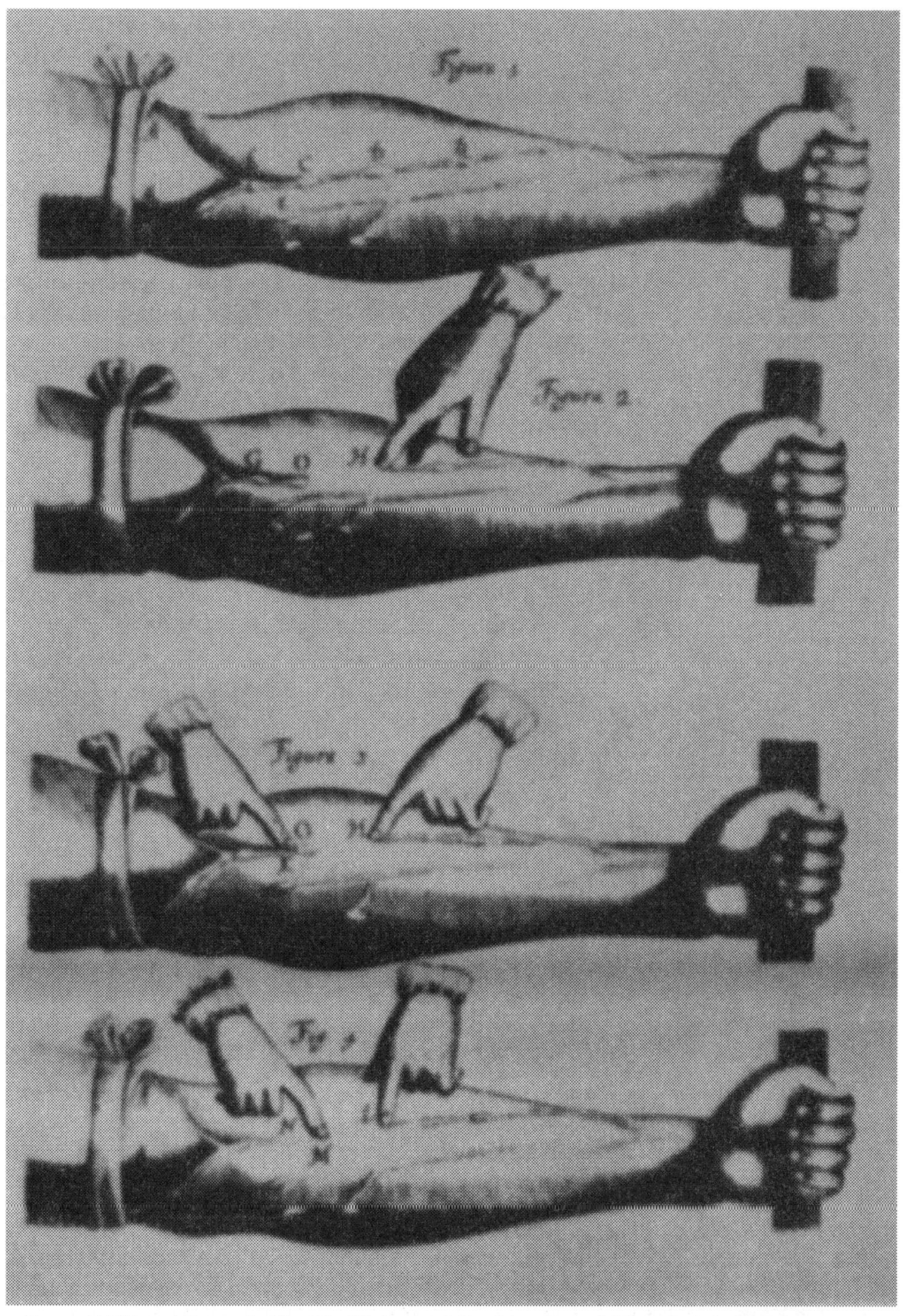

*From William Harvey's* On the Motion of the Heart and Blood *showing the valves in the veins that allow blood to flow only toward the heart.*

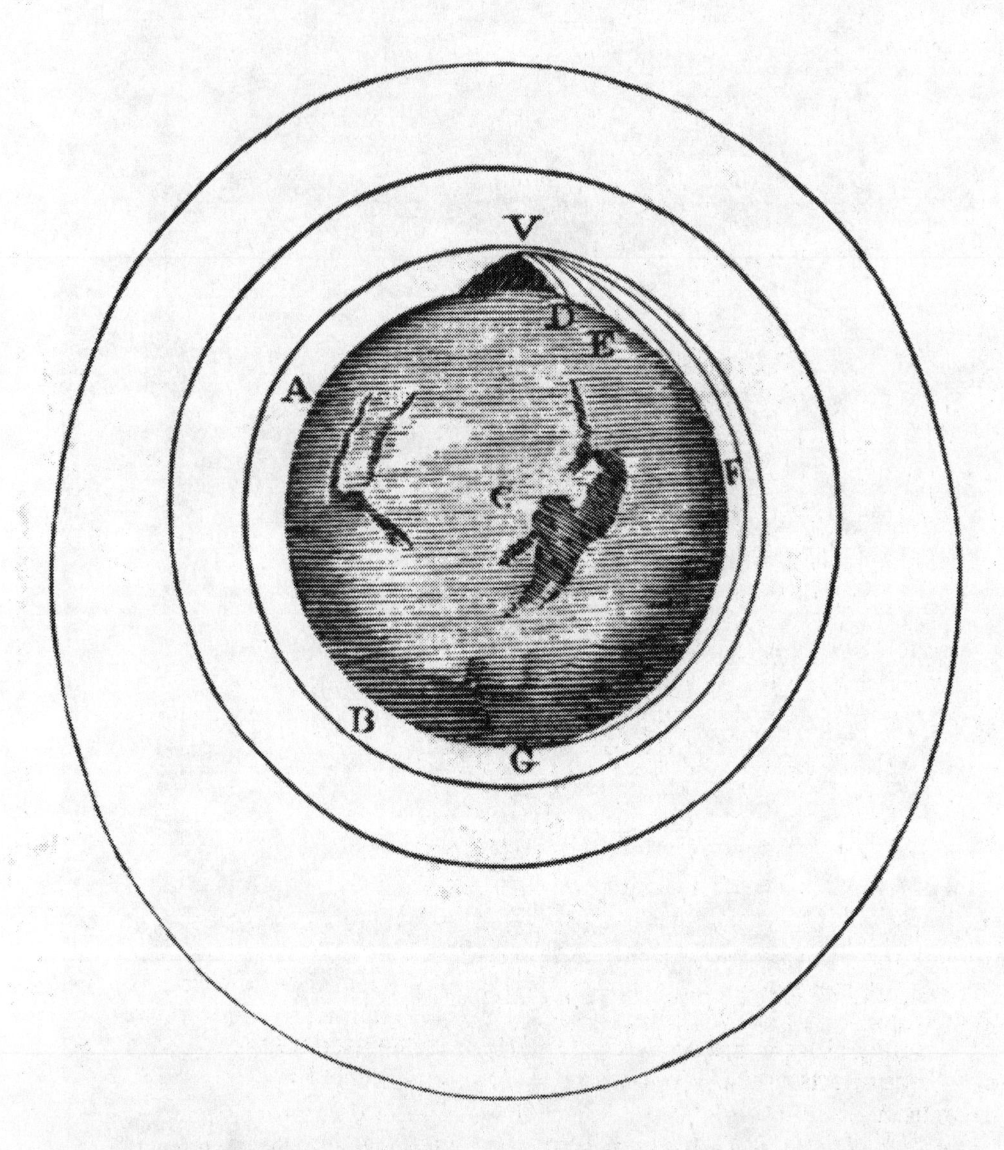


*Newton's illustration of how inertial motion plus gravity can produce an elliptical orbit. A cannon fired from atop a mountain will travel in a parabolic path (as shown by Galileo) before it comes to earth. But if the inertial force of firing is greater, the cannon ball will travel a considerable distance before it arcs down to earth. Finally, if the inertial motion from firing is great enough the cannon ball will never actually hit the earth, but instead will go into an elliptical orbit around the earth. From Newton's* System of the World *(1728).*

## CHAPTER TWELVE

# Gravity Embraces the Stars

Until the 1680s, scientists of the "new philosophy" dealt mainly with the world around them. Of course, they had the Copernican revision of astronomy, but they still considered the stars and planets to be heavenly objects, subject to different laws from those that applied on earth, despite Galileo's having shown the moon to be earth-like. Physics was still earthbound.

As we've seen, the new scientists had lots to investigate on earth: the earth itself, its numerous inhabitants, the motions of objects, the properties of light, and many other things. But the Galilean glimmers could encourage scientists to dream about a unification of heaven and earth.

When Kepler called his *New Astronomy* "celestial physics," he showed he was thinking in that direction. But his magnetic physics of the heavens could not be sustained. Descartes made a valiant attempt to produce a single universal system, but as we'll see, it was much more a philosophical speculation than a proper scientific theory.

The idea of unifying the sciences of heaven and earth was "in the air" in the seventeenth century. It was not achieved until Isaac Newton combined his inspired speculation of universal gravitation with a deep analysis of motions, confirmed in observations by both astronomers and physicists.

Before we get to Newton, we should review some of the steps that led to his great achievement.

## *Views of the Universe before Newton*

### *In Greek Times*

Back to Aristotle. Physics (φυσις) is the Greek word for "nature." In Chapter 4 we saw how the Greeks had invented the idea of nature. They separated themselves and their gods (subjective beings with feelings) from the world of nature (objects without feelings like rocks and trees). But they only got half way. Aristotle's nature didn't extend past the moon. His physics applied only to elements and causes and motions on or near the earth—the terrestrial realm. The moon and everything beyond it occupied the celestial realm—the heavens. Motions in the heavens were only circular, while the natural motions of earthly elements were only vertical—straight up and down. So Aristotle's universe had two radically distinct parts—earth and heaven. It was an almost perfect dichotomy—a schizoid universe.

Looking at Aristotle's views today, we can ask embarrassing questions like "do you suppose the light that comes to earth from the heavens somehow changes at the moon's orbit?" Although there's no indication such questions were ever asked at the time, we might expect Aristotelians to answer that we've absolutely no idea what light is like in heaven. Nonetheless, Ptolemy believed that his observations of the refraction of light from water to air also applied to light from the stars: that their observed positions could be affected by the refraction of light in the earth's atmosphere.

*In the Middle Ages*

The Aristotelian view was preserved and Christianized by medieval European scholars. The sharp distinction between heaven and earth seemed to fit well with their interpretations of Holy Scriptures, even though God was the ruler of both domains. Medieval scholars even considered light to be the medium of transmission of influences from heaven to earth (as in astrology for example), also apparently without letting that affect their view of the radical schism between the two realms.

*In the Renaissance*

In the early 1500s, Copernicus "threw the earth into heaven" by having the earth orbit the sun. Think what that did to Aristotle's schizoid universe. If the earth *has* circular motions—on its own axis and around the sun—that puts the earth into Aristotle's category of heavenly things. But the natural motion of a part of the earth (like a falling rock) is only straight. Aristotle had argued that what belongs to a part (the rock) must belong to the whole (the earth). So he said that only straight motions are natural for the earth.

Since Copernicus wanted to stay as close to Aristotle as he could, he was in a bind. The earth can't be both earthly and heavenly. The best Copernicus could do was to deny the basic logic that what's natural for the parts is also natural for the whole. He argued lamely that since the whole earth is a sphere, then spherical—or circular—motion must suit the earth, even though straight motion suits a rock—a piece of the earth. In a sense, Copernicus tried to make a heavenly rule fit the earth. And remember, Copernicus also clung to the old-fashioned idea that the planets, including the earth, were attached to real spheres—even though nobody could detect them.

Copernicus may have stunned Aristotle's schizoid universe, but he didn't kill it. A heavier blow came from Galileo's revolutionary mathematization of motions on earth—directly attacking Aristotle's physics itself. Galileo established mathematical laws for objects moving near the earth. A stone falls naturally with accelerated motion—but horizontal motion is natural too. Since Galileo considered these motions natural, he didn't think forces caused them—gravity was just the name for *heaviness*. Nor did he ever apply his laws to motions in the heavens. So, while Galileo transformed the earthly side of Aristotle's schizoid universe, he wasn't greatly concerned with the heavenly side. Of course, Galileo's telescopic observations of the moon, sun, and planets did help reduce the distinction between earthly and heavenly objects—with mountains on the moon and spots on the sun showing they're not pure, perfectly smooth celestial things. But Galileo didn't attempt to unify earth and heaven—to get a single universal system.

What Galileo was too modest even to consider, Descartes asserted boldly—the unity of the universe. He had the sun as merely another star, with each star surrounded by a gigantic whirlpool of subtle matter. (See picture on page 164). The whirlpool kept the planets in their places revolving round the sun. The materials of earth, stars, and whirlpools were all similar—and obeyed the same laws. So, in his philosophical system, Descartes rigorously rejected Aristotle's schizoid universe. But Descartes' universal laws of motion were too vague to be useful. And people couldn't detect his whirlpools any better than Copernicus's spheres. Descartes was long on generalities, but short on details.

Descartes described his system in a book published in 1644, titled *Principles of Philosophy*. A follower of Descartes, Jacques Rohault, published a version of the Cartesian system in 1671 and called it *La Physique*—physics. That title directly signalled Descartes' challenge to Aristotle, that nature is not only earthly, but heavenly too—a physics of the whole universe. Rohault's *Physics* was widely reprinted and translated, and became influential in philosophical circles in France, and England too. But it didn't last long—for, what Descartes and Rohault only dreamed of, Isaac Newton would soon make real!

## *Newton's Early Years*

Newton's father owned a small farm in the county of Lincolnshire in England. He died three months before his son was born. Isaac Newton was born on Christmas Day in 1642, on the Julian calendar still used in England, which hadn't yet switched to the Gregorian calendar. Catholic Europe had made the switch back in 1582 and lost ten days. So, when Newton was born, the date in Europe was actually January fourth, 1643. (This sounds pretty trivial, but historians who neglect it give Newton's life a whole extra year.) Newton's mother remarried when he was three, moved in with her new husband, and left Isaac in the care of his grandmother and uncle. Yet another uncle!

*Newton's birthplace. Woolsthorpe Manor.*

Isaac had a normal enough childhood and an ordinary university career at Cambridge in the 1660s. At age 22, he caught fire mathematically; on his own, he learned all the mathematics there was, invented calculus, and deduced a rule for acceleration in circular motion (Chapter 10). Newton stayed on at Cambridge after graduation, as a fellow of his college, Trinity.

Then the plague struck England, and the university was closed for two years. Newton spent that time back home in Lincolnshire. There in 1666, as he wrote later, he began "to think of gravity extending to the orb of the moon." Now, Newton considered gravity to be the force that caused a falling rock to accelerate, whereas Galileo had considered this accelerated falling to be a natural motion. Galileo calculated distances and times, avoided causes, and paid scant attention to forces. Newton, on the other hand, with his first law of motion, could have no acceleration without a force. Unless a force acts on an object, its speed stays constant.

## *Newton's "Moon Test" of 1666*

Here's where the famous story of Newton and the apple comes in. Many years later Newton recalled that the idea of gravity, as it related to the moon, had come to him in his garden, when a falling apple attracted his attention. But he never said what went through his mind. Papers he wrote at the time provide some clues to suggest a story like the following.

*An apple tree in Kew gardens in England. This tree was grown from a graft from an apple tree at Newton's home, Woolsthorpe Manor. The original tree was maintained until it blew down in 1820. Whether the story of Newton conceiving universal gravitation while watching an apple fall from this tree is true or not, "Newton's apple tree" has become a popular symbol for theoretical science. Similar trees have been planed at several universities, including York University in Toronto and the University of Wisconsin in Madison, and of course, Trinity College at Cambridge University (see illustration on page 203).*

Imagine Newton sitting in his garden one afternoon, a few days before full moon. He can see the moon in the southeastern sky. Between him and the moon he sees an apple hanging on the tree. He's daydreaming when the apple falls and snaps him out of his reverie. Newton then went through a thought process something like this: "The apple's going round the earth in a circle—around the center of the earth—4000 miles straight down below me and the apple. Because of its circular motion, that apple has a certain **centrifugal** force; it's trying to flee from the center. Using my law of circular motion I can calculate that force and compare it to the force of gravity. And...

"And the moon too! The moon's going round the earth in a circle, 250,000 miles out there. Because of its circular motion, the moon has a centrifugal force I can also calculate. Maybe the moon stays in its orbit because its centrifugal force balances the force of gravity, if it reaches out that far. I should compare the strengths of those three forces." That is, gravity at the earth's surface, and the centrifugal forces of the apple and the moon.

That's exactly what Newton did. He found that the centrifugal force of the apple was about one three-hundredth of the force of gravity. So, because the earth rotates, that makes the apple—and Newton (and all of us)—a third of a percent lighter than if the earth were at rest.

Newton also found that the centrifugal force of the moon, as it circles the earth 380,000 km away, was about 1/4300 of the force of gravity at the earth's surface.

On the same sheet of paper, Newton also calculated the centrifugal forces of the various planets away from the sun, using Kepler's third, or harmonic, law. He wrote:

> Since for the planets the cubes of the distances from the sun *are inversely as the square of the number of revolutions in a given time* [that's Kepler's third law], *then the endeavors of receding from the sun* [i.e., their centrifugal forces from the sun] *will be inversely as the square of the distances from the sun.*

Newton was saying that the farther a planet is from the sun, the less will be its centrifugal force—in double doses! That is, inversely proportional to the square of the distance. At double the distance, the force is only one-quarter; at triple the distance, one-ninth, etc.

That's what Newton figured out in the 1660s. But it still wasn't universal gravitation because he got it backwards so to speak. He didn't turn it around for 18 years. Turning it around means: If no force acts on a planet, then it'll travel in a straight line at constant speed, by Newton's first law. Since planets don't travel in straight lines, there must be some force acting on them to pull them into their orbits. In the 1680s, Newton called that a **centripetal** force—centripetal force, seeking the center—as opposed to a centrifugal force, fleeing from the center. Once Newton had made that switch in his mind, though, he found that his earlier calculations still worked. That is, planets' distances and speeds fit the idea that a centripetal force from the sun keeps them in orbit. And that centripetal force is inversely proportional to the square of the distance, exactly as he'd found earlier for centrifugal force.

Then Newton made a tremendous guess: that the centripetal force from the sun is supplied by gravity, the very same force that holds us on the earth. Then he speculated that this same force, gravity, keeps the moon in orbit around the earth. As he'd said, "The earth's gravity extends to the orb of the moon." Much later, when Newton looked back over his old papers, he fooled himself into thinking he'd had the right idea back in the 1660s. All he had then were the right calculations, but with the force backwards.

Recall his earlier moon calculation, and consider that gravity extends out from the earth, decreasing as the distance increases, according to the inverse square law. On the surface of the earth, we're one radius from the earth's center. Astronomers had long known that the moon's distance is sixty radii from the earth. Since $60^2 = 3600$, the earth's gravity at the moon's distance is 1/3600 of gravity at the surface of the earth. Almost 20% more than the 1/4300 Newton had calculated in 1666. But when Newton originally made that calculation, he was thinking about centrifugal force (the endeavor of the moon trying to flee from the earth), not the centripetal force of the earth's gravity (the earth pulling on the moon).

Historians now generally acknowledge that Newton didn't have universal gravitation before the 1680s. Yet the apple story of 1666 is valuable in showing us the development of Newton's thinking. Besides it's such a common story, we should get it straight. It is definitely not the story of "Newton inventing gravity." People had said for centuries that gravity causes falling. Galileo put it neatly in his *Dialogue:*

> *SALVIATI.* ...I'll tell you what makes the earth move, if you can tell me what it is that moves parts of the earth downward.
>
> *SIMPLICIO.* The cause of that effect is well known; everyone knows it's gravity.
>
> *SALVIATI.* You are wrong Simplicio. You should say everyone knows it's called gravity. But I didn't ask for its name, but the essence of the thing; of which essence you know not one whit more than you know the essence of what moves the stars around...But we really understand no more about it than about the principle or force that moves a stone downward, or upward after it's left the thrower, or what moves the moon around; except (as I said) the name.

When Newton invented the idea of the force of universal gravitation, he used the same calculations he'd used with the falling apple and the moon. All he had to do then was change his point of view. Instead of imagining himself on the moon trying to flee from the earth, he imagined himself on the earth keeping the moon from fleeing. It's not centrifugal force that's important, but centripetal force. Those two forces are equal, but opposite, so that Newton had just started at the wrong end, so to speak.

## *Mathematics Professor at Cambridge*

In 1667, Newton set aside those calculations on centrifugal forces and went on to other things, like calculus and optics. In 1669, he was appointed to the chair of the Lucasian Professor of Mathematics at Cambridge University. He was only 26.

Through the 1670s, Newton became almost a recluse at Cambridge, giving a few lectures on mathematical topics, but spending most of his time studying alchemy and biblical history. Then, the controversy over the spectrum of white light (Chapter 11) bothered Newton so much that he became even more of a recluse.

*Trinity College, Cambridge University, as it looked when Newton was there. In 1954 a graft from the apple tree at Woolsthorpe was planted in the garden (in the foreground to the left of the gate).*

In the meantime, the Dutch mathematician, Christiaan Huygens (1629–1695), published a book containing a law of force for circular motion. It was essentially the same as Newton's. Huygens had found the law independently of Newton. But by publishing it, Huygens made it available to everyone. Soon, others besides Newton were thinking seriously of something like a force of gravity spreading out from the sun. A force that kept the planets in their orbits. And the calculation that Newton had made in the 1660s, relating gravity to Kepler's harmonic law, wasn't hard for them to do–if they assumed that the planetary orbits were circular. But by then, many people were accepting Kepler's idea that the orbits were elliptical. And no one was able to relate elliptical orbits to gravity. The mathematics was just too difficult for them.

## *Newton's **Principia**, 1687*

In 1684, three men were discussing that problem in London: Robert Hooke, Christopher Wren (1632–1723), and Edmund Halley (1656–1742). Hooke was the experimenter for the Royal Society, Wren, the well-known architect of St. Paul's cathedral, and Halley, the famous astronomer. They wondered how they could show that gravity from the sun could produce elliptical orbits. When they realized they couldn't, Halley went to Cambridge to ask Newton. Newton said he'd already figured it out, but then he couldn't find his

*Robert Hooke.*

*Christopher Wren.*

*Edmund Halley.*

papers. He promised to do it over and send the results to Halley. Which he did before the end of 1684.

Apparently in 1679, by using his laws of motion and centripetal force, Newton had shown that a body could be made to travel in an elliptical orbit by a force spreading out from one focus of the ellipse— a centripetal force varying inversely as the square of the distance of the body from the focus. With the sun at the focus, the farther the body was from the sun, the less the force, and the less the speed of the body. Exactly what Kepler had found for the planets. Well, not *exactly*.

For, in 1684, when Newton repeated his demonstration, he now had the theory of universal gravitation, where *every* body attracts every other body. So, Mars is not only attracted by the sun, but by all the other planets too. The force of the sun is by far the greatest, because it's by far the largest body. But, Jupiter also exerts a small force on Mars. (Jupiter's maximum pull on Mars [when they're closest] is 0.016% of the Sun's pull, and the earth's maximum pull on Mars is 0.003%.) So Mars's orbit can't be a perfect ellipse. And as telescopes got more and more precise, people began to observe exactly the kind of deviations Newton's theory of gravitation predicted. An astounding tour de force!

In addition to his mathematical analyses, Newton had transformed his ideas about force in the period from 1675 to 1684. Initially, following Descartes, he had considered forces only mechanically, that is, as the results of collisions among particles, including particles of subtle matter. However, he was never able to construct a satsifactory theory of motion on such a basis. Recently, some historians have urged us to realize the contribution to Newton's ideas of force from his alchemical investigations. That work convinced him that the particles of gross matter did not merely respond passively to collisions; rather they were the site and source of active powers, such as attraction and repulsion. By extending these ideas into the cosmos, Newton developed the idea of a universal force of gravitational attraction.

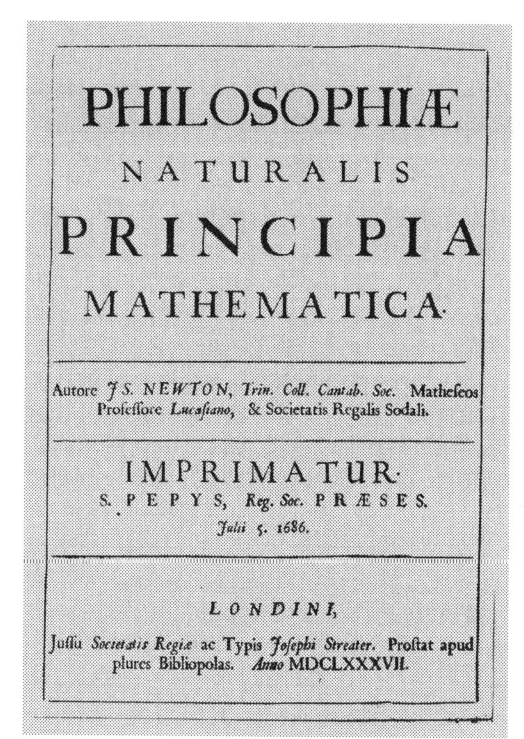

PHILOSOPHIÆ
NATURALIS
PRINCIPIA
MATHEMATICA.

Autore *J S.* NEWTON, *Trin. Coll. Cantab. Soc.* Matheſeos Profeſſore *Lucaſiano*, & Societatis Regalis Sodali.

IMPRIMATUR.
S. PEPYS, *Reg. Soc.* PRÆSES.
*Julii* 5. 1686.

*LONDINI,*
Juſſu *Societatis Regiæ* ac Typis *Joſephi Streater*. Proſtat apud plures Bibliopolas. *Anno* MDCLXXXVII.

*Title page of Newton's* Mathematical Principles of Natural Philosophy, *1687.*

At Halley's urging, Newton set down all his deductions in a book. Since Newton despised Descartes' *Principles of Philosophy*, which he called a "philosophical romance," Newton wanted a title for his book that would express his disdain for Descartes' system. So Newton called his book, *The Mathematical Principles of Natural Philosophy*. The book was published in 1687, in Latin. Halley paid for the printing. Though few people could understand Newton's work, everyone acknowledged his tremendous genius.

Newton's book is often called by the first word in its Latin title—the *Principia*. In the *Principia* he set forth a complete mathematical analysis of the dynamics of moving bodies. He also included a whole section on motions of bodies within fluids, and the motions of fluids themselves, like wave motions. This section was intended to demolish the idea of vortices of subtle matter propounded by Descartes. The fact that Newton had once been entranced by vortices and had devoted much fruitless effort to putting them on a sound foundation surely added to to the disdain with which he treated them and their French author.

Using his principle of universal gravitation, Newton explained a wide range of phenomena. Gravity keeps the planets in orbit round the sun, holds moons in orbit round their planets, holds us on the earth, causes objects to fall to earth, and causes the tides. At every stage, Newton referred to the evidence that confirmed his theories: the observations of astronomers and the measurements of experimenters.

Newton applied physics to the whole universe. Gravity, a previously earthbound force, now embraced the stars too; it acted on all matter. And the same laws of motion applied to all systems in the universe.

Aristotle's schizoid universe was finally dead. One single set of laws would cover motions in the heavens as well as on earth.

## *Newton's Later Life*

In 1687, almost overnight, Newton was acknowledged as the major genius of his age, even by Cartesians who didn't agree with him. Now, 1687 also marked a time of political controversy in England. King James II was a Catholic, and Englishmen were fiercely Protestant. Newton was elected to parliament. In that session, the MPs kicked James out and brought the Protestant William of Orange over from Holland to be King of England. A few years later, Newton was put in charge of the mint, responsible for the English coinage. In 1696, he moved from Cambridge to London and spent the rest of his life there.

Later, in the early 1700s, Newton was knighted by Queen Anne and became Sir Isaac Newton. And he became president of the Royal Society, the paramount organization of scientists, founded in 1662. Until his death in 1727, Sir Isaac ruled science with an iron hand. One biographer called him "the autocrat of science." Newton was a hero to young mathematicians, and he saw to it that his disciples got all the good university posts. To his enemies, Newton was an implacable foe.

All this makes Newton a fascinating subject for psychological study—and it's been done by Professor Frank Manuel in his book, *A Portrait of Isaac Newton* (1968).

Manuel used lists of words the youthful Newton wrote in his copy book. Isaac's feelings towards his mother and step-father show up in the list for *W*: "*Wife, Wedlock, Wooer, Widdow, Widdower, and Whoore*;" and for *F*: "*Father, Fornicator, Flatterer, Frenchman, and Florentine*." These last two convey, according to Manuel, "overtones of sin and sensuality to a young puritan."

Then, Manuel continues:

> At the end, under *Y*, *Yeoman* stands alone—the status of his true father, who died before he was born. Ambivalence toward his mother would color Newton's whole style of life.

Besides such personal colorings, Newton held the religious tenets of his time and country. He was morally a puritan and a militant anti-Catholic. Less typically, he was also an anti-Trinitarian, although covertly.

This combination of strong religious and personal feelings gave Newton a lifelong ambivalence. On the one hand, at age three, he was abandoned by his mother when she went to the home of her new husband and left him in his grandmother's care. That gave him feelings of rejection and deprivation. On the other hand, Newton considered himself to be among the elect, with direct access to God and the wisdom of the ages. After all, was he not born on Christmas Day while his father was in heaven? Manuel even attributes his anti-Trinitarianism to Newton's feeling that "he was the only son of God and could not endure the rivalry of Christ."

Nor could he endure rivalry in science. Robert Hooke was only one of the targets of Newton's venom. He was most vicious toward G.W. Leibniz

(1646–1716), a German scholar and inventor of a form of calculus different from Newton's. After being accused of plagiarism by one of Newton's admirers, Leibniz appealed to the Royal Society for redress. By the time the issue was dealt with, Newton was the society's president. In adjudicating the case, Newton chose the judges, provided the evidence, and wrote the judgment, all intended to vindicate himself, God's servant. Leibniz died in 1716, a broken man.

*Gottfried W. Leibniz.*

Despite Newton's great achievements in mathematics and physics, he spent more time on theology and alchemy than on science. He was an obsessive copier; millions of words in his manuscript remains contain long passages from the books he read.

None of these personal characteristics can detract from Newton's intellectual accomplishments. Whether they help to explain them is still a contentious issue among historians.

Newton died at 84. He was given a state funeral and buried in Westminster Abbey among the kings and admirals—the heroes of England. Quite an achievement for the farm boy from Lincolnshire. The poet, Alexander Pope, wrote an epitaph for Newton in these words:

> Nature and nature's laws lay hid in night:
> God said, Let Newton be! *and all was light.*

## *Mechanics after Newton*

Newton's *Principia* appeared about 50 years after Galileo's *Two New Sciences*. That fifty-year period marked not only the invention of mechanics, but also an unparalleled advance in analytical sophistication. While all competent mathematicians could read and understand Galileo, only a handful at first could handle the depths of Newton's work. Despite the giant leap Newton achieved, he did not by any means complete the subject. Indeed, as with most works of genius, the *Principia* initiated a whole industry within physics for several generations afterward.

*Newton's grave in Westminster Abbey.*

In particular, what Newton left for mathematicans was the task of converting the theoretical structure of the *Principia* into the new language of calculus. Although the infinitesimal ideas of calculus are contained on many pages of Newton's work, they are framed in a mathematical language that looks like Euclidean geometry. That kind of putting "new wine into old bottles" only made Newton's work more forbidding. So, beginning shortly after 1700, mathematicians on the continent began to translate Newton's work into the "algebraic" language of Leibniz's calculus.

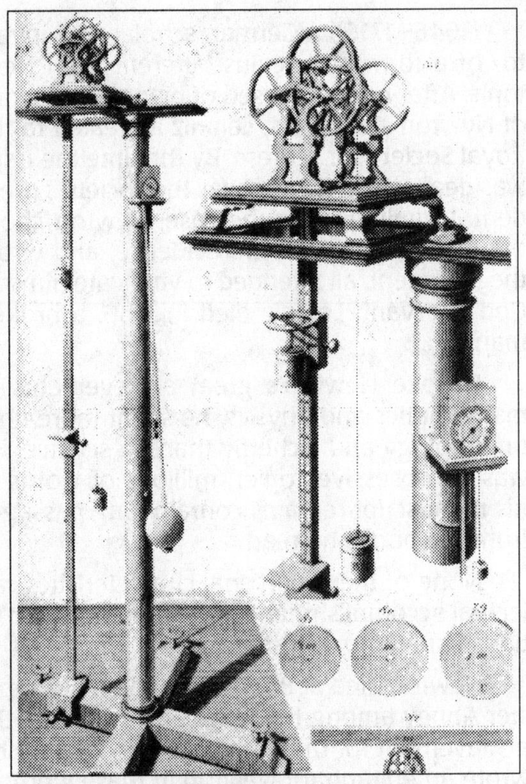

*An apparatus for testing Newton's second law of motion. From George Atwood's* Treatise on the Rectilinear Motion and Rotation of Bodies *(1748).*

There are several reasons for the shift of attention from England to the continent. First, Leibniz's symbolism for calculus was signficantly superior to Newton's. Second, people who were directly under Newton's gaze—in the Royal Society and in the British universities—were expected to conform to their master's style. So, while David Gregory (1659–1708) and Colin Maclaurin (1698–1746) produced pale imitations in England, Leonhard Euler (1707–1783) and three generations of Bernoullis (including Jakob 1654–1705, Johann 1667–1748, Daniel 1700–1782) converted the ideas of the *Principia* into an analytical instrument of exquisite finesse. They reformulated problems and solved them with calculus, creating the rational mechanics that lies at the foundation of engineering practice today, more than two centuries later.

At the same time, there were people both in Britain and on the continent who began to spread Newton's ideas to larger audiences who lacked any significant mathematical knowledge. This was done in England by Newton's disciples and by travelling lecturers in natural philosophy. In place of mathematics, they used mechanical devices to illustrate their master's principles (see Chapter 13). The most helpful of Newton's expositors in France were Voltaire (1694–1778) and his very close friend Emilie, Marquise du Châtelet (1706–1749). Together they wrote a treatise on the Newtonian system, *Éle-*

*ments de la philosophie de Newton*. Madame du Châtelet also produced a French translation of the *Principia, Principes mathématique de la philosophie naturelle* (1759).

*Henry Cavendish.*

Finally, a form of experimental confirmation of Newton's gravitational law was found by Henry Cavendish (1731–1810) and published in 1798. He used a torsion balance to measure extremely tiny forces. He hung small lead balls of about 775 g at opposite ends of a light rod about 2 m in length. He suspended the rod on a wire about 1 m long. By the twisting of the wire, Cavendish could measure the force on the small balls when two large lead balls of about 50 kg were brought near them, from opposite sides. From the measurements he obtained, Cavendish could calculate the density of the earth to be about 5.5 times that of water. Although this confirmed the attraction of masses for one another, it contributed nothing to relating the solar-planetary force to gravity.

## *Science and the Enlightenment*

*Immanuel Kant.*

The term "enlightenment" was applied to his own age by the German philosopher Immanuel Kant (1724–1804). His contemporaries in France called their century the *siècle des lumières*. In both cases, the light was the light of reason. They were signalling the replacement of the traditional world view of the Christian middle ages by a new view of the world and society—a view derived from the transformations in science and religion of the preceding two centuries.

In western Europe of 1700, three major systems of thought were competing for the adherence of intellectual laymen: orthodox Aristotelianism, gasping its last breaths, largely in Catholic countries; the rationalism of Descartes, still prominent in France, but losing ground in England; and the new experimental philosophy inspired by Newton's synthesis. Through the course of the eighteenth century, the first two lost out, and Newtonianism reigned supreme. Newton's acknowledged scientific successes provided a foundation for continuing the assault on religious and intellectual traditions that had begun with Luther, Bacon, Galileo, and Descartes.

We should not suppose that Newtonian science "caused" the enlightenment. A complex of developments in western Europe had created the environment in which science grew and prospered. The success of Newton's program simply added to that complex and assisted in its further development. In religion, the domination of Rome was broken and the nations of western Europe established themselves as masters in their own territory. Europeans' conquests in Asia and America added immensely to their economic well-being. Although there was a period of centralized authority in England under the Stuart kings, the English civil war gave more power to the middle classes. This eventually led to the dominance of Parliament after the exile of James II in 1688. In France, Louis XIV destroyed the power of the aristocracy by putting men of lower birth into positions of authority. All these factors contributed to a gradual enlarging of the realms in which rather ordinary men were able to achieve great things.

*John Locke.*

The success of the Newtonian program of describing natural phenomena mathematically encouraged intellectuals to extend the "geometrical spirit" into philosophy and politics. John Locke (1632–1704) in England created a view of human nature as the product of reason and experience, which helped to encourage people to take charge of their own lives. Locke quite consciously modelled his philosophy on Newton's science. Since Newton had shown that nature was rational, then there was no reason why society should not also be organized rationally.

*Benard le Bovier de Fontenelle.*

French intellectuals like Benard le Bovier de Fontenelle (1657–1757), Voltaire, and the mathematician Jean le Rond d'Alembert (1717–1783) spread these ideas vigorously among their fellow citizens. Among the most influential of their works was the *Encyclopédie* (1751–1772), subtitled *A reasoned dictionary of the sciences, arts, and crafts, published by a society of men of letters.* Its seventeen large volumes of text and eleven of illustrative plates were edited by d'Alembert and Denis Diderot (1713–1784). (For more information, see page 210 and an illustration on page 245.) Like Voltaire and Fontenelle, Diderot was a man of letters rather than a scientist. Such men put their considerable literary powers at the service of the new philosophy—the philosophy of reason and experience. You may recall Galileo's words (Chapter 10), "conclusive mathematical proofs and careful experiments."

*Jean le Rond d'Alembert.*

*Denis Diderot.*

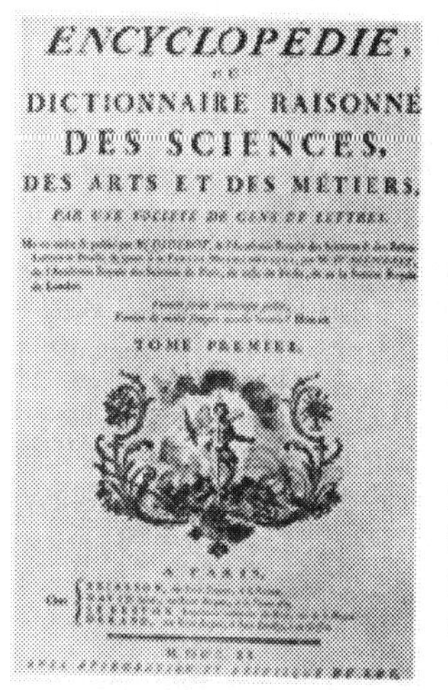

ENCYCLOPÉDIE,
OU
DICTIONNAIRE RAISONNÉ
DES SCIENCES,
DES ARTS ET DES MÉTIERS,
PAR UNE SOCIÉTÉ DE GENS DE LETTRES.

TOME PREMIER.

A PARIS,

*Title page of the* Encyclopédie.

*Voltaire (François Marie Arouet).*

Although a reformed Christianity was the goal of many philosophers before 1700, those of the enlightenment attacked the principles of "revealed" religion with considerable vigor. Not only were they opposed to the authority of priests and bishops, they also sought to rid religion of all taints of supernaturalism. While most enlightenment philosophers allowed a place for some kind of God, God was remote and inactive in their world. God may have established the foundations, but now everything was supposed to be running naturally. Moral rules were to be determined rationally, not handed down from on high. Indeed, a natural theology developed in England, where people felt they could deduce all the principles of religion and morality from their experiences in the world of nature.

In this way, religion, philosophy, science, and politics were determined to walk along together hand in hand. Human powers, according to the light of natural reason, would solve all problems. Such a comfortable view of life was wittily satirized by Voltaire in his *Candide* (1759).

Growing out of the foundations of the Reformation and the scientific revolution, the period of the enlightenment provided the stage for the revolutions that have formed our modern world—revolutions both political and industrial.

## Conclusion

Copernicus, Kepler, Galileo, and Descartes were vindicated, extended, and corrected. It was Newton's genius to pick out the kernels of truth from the chaff, grind them together with new insights of his own, to bake the incomparable loaf of the *Principia.*

In at least one instance, the process of correction was done right before our very eyes. Remember that Rohault's book, *Physics,* had spread Descartes' ideas far and wide. It had even become a university text in England. Once Newton's ideas began to be accepted, two of his disciples, named Clarke, produced a new edition of Rohault's *Physics.* And every place they found an error in the Cartesian system, the Clarkes added a footnote correcting the error according to Newton. Just imagine a textbook where almost every idea in the text is contradicted in the footnotes. Students likely caught on pretty soon and just read the footnotes. Eventually, of course, the text was dropped, and the footnotes became the text. But that didn't happen for a few decades.

Now, with all those contributors to the final synthesis of Newtonian physics, who should get the credit for destroying Aristotle's schizoid universe? Let's look at it from Newton's point of view. To make his discoveries he needed the idea that the earth and the other planets orbited the sun—credit Copernicus for that. And credit Kepler and Galileo for popularizing the idea of the earth's motion. Then Newton needed the idea that the planetary orbits were elliptical—credit Kepler for that. But Newton didn't credit Kepler! Because, claimed Newton, he had *proved* that the orbits weren't precisely elliptical. Kepler had merely found that ellipses fitted Tycho Brahe's observations. Newton had deduced the precise orbits from his fundamental principles of motion and gravity.

For the ideas of motions on earth, such as speeds and accelerations, Newton readily acknowledged Galileo's achievements. In fact, Newton gave Galileo credit for the idea of force, which he didn't really deserve. Descartes had some influence on the concept of force, but Newton refused to credit Descartes for anything. In fact, in the *Principia*, Newton produced some deductions specifically to prove that Cartesian whirlpools simply wouldn't work.

As for the idea of a force spreading out from the sun, it's clear that a number of Newton's contemporaries also had the same general idea. Chief among them was Robert Hooke, who'd even mentioned it in a letter to Newton in 1680. So, when the *Principia* was being printed, Hooke wanted Newton to give him some credit in the book. Newton was furious:

> Now is not this very fine? Mathematicians that find out, settle & do all the business must content themselves with being nothing but dry calculators & drudges & another that does nothing but pretend and grasp at all things must carry away all the invention…

So Hooke lost, and Newton's arrogance kept the number of credits pretty narrow.

Actually, we now tend to give credit like Newton did—that is, only to the person who constructs the grand theory. Merely guessing at the theory can't be enough. The winner must also show how all the parts fit together. Later we'll see how this notion attached Darwin's name to the theory of evolution.

Nevertheless, it's a distortion of the way science works to neglect all the others. Even the great theory-builder needs the work of observers and experimenters. Sometimes, it's particularly crucial for the theory-builder to have earlier theories to criticize. Newton didn't just need Descartes for analytic geometry and some ideas about force. We might also say that Descartes helped Newton by being wrong. There's a neat phrase for that: "truth comes more readily from error than from confusion."

*The octagonal clock room of the Royal Greenwich Observatory, founded 1675. Presided over by John Flamsteen, first astronomer royal.*

So, Newton put it all together. What Descartes had only dreamed, Newton made exact and consistent. He created a system for science that has served down to our own time. A system that helps even when it's wrong and needs to be corrected—as it would be by Einstein for instance. But that's a couple of hundred years farther on.

In 1931, Einstein himself had an opportunity to evaluate Newton. In a brief foreword to a new edition of Newton's *Opticks* he wrote:

> Fortunate Newton, happy childhood of science! He who has time and tranquility can by reading this book live again the wonderful events which the great Newton experienced in his young days. Nature to him was an open book, whose letters he could read without effort. The conceptions which he used to reduce the material of experience to order seemed to flow spontaneously from experience itself, from the beautiful experiments which he ranged in order like playthings and describes with an affectionate wealth of detail. In one person he combined the experimenter, the theorist, the mechanic and, not least, the artist in exposition. He stands before us strong, certain, and alone: his joy in creation and his minute precision are evident in every word and in every figure.

*The 210-ft reflector telescope at Parkes, New South Wales.*

# PART 4

# *The Modern World*

CONTENTS

Mevrouw Smit's world is a world we have largely lost—a comfortable world of family relations even in its business dealings. Oh, you can still find it in small pockets in the heartland of America: family firms in small towns, and family farms fast vanishing; but the action, and the centers of population are elsewhere. The faceless crowds in the bustling streets of Metropolis bespeak the faceless corporations that employ them. As Peter Lazlett put it in his book *The World We Have Lost* (1971):

> Who could love the name of a limited company or of a government department as an apprentice could love his superbly satisfactory father-figure master, even if he were a bully and a beater, a usurer and a hypocrite?

Of course, this transformation didn't happen in a day—not even in a generation. It began to happen in England in the eighteenth century. There had been occasional glimmerings of new social relations in employment before then. A few exceptional industrial enterprises could not be run by a single extended family. They needed a large crew of able-bodied men just to make them go—constructing a Gothic cathedral, delving in mines for precious minerals, handling a three-masted, hundred-gun man-of-war. A curious sign of the eighteenth century, then, was the remark of John Byng on seeing a cotton mill at night:

> These cotton mills, seven stories high, and filled with inhabitants, remind me of a first-rate man-of-war; and when they are lighted up on a dark night look most luminously beautiful.

Beauty is surely in the eye of the beholder who can so admire engines of war and of profit, which housed the miserable wage-slaves his mighty empire was built on.

Our modern world of changed social relations gathered momentum from the time of those early cotton mills. After thousands of years of living our whole lives mostly within the bosom of a family, we went out to earn our livings in the mills—and later in the myriad cubicles of the fifty-story towers that line the traffic jams of downtown Megalopolis.

Our world has changed—we've invented a whole technology of change—a "new improved" version of the world every day. In the last 250 years the world has seen more changes than in the preceding 25,000 years. To say change happens a hundred times faster now, can only be a figure of speech; maybe the number should be as little as ten, or more than a million. Messages travel ten million times faster by telephone than by horseback. Planes go thirty times faster than clipper ships. Supertankers carry up to a million times the cargo of a pack horse. North Americans use energy at rates forty times greater than neolithic hunter-gatherers. And the earth's current five billion population is more than a thousand times what it was back then.

Constant change is a historian's dream...or nightmare.

It all started innocently enough: some British entrepreneurs in the eighteenth century wanted to improve their competitive edge in textiles, while some enterprising inventors were harnessing coal to pump water out of mines. Coal, steam, iron, and cotton soon got together to spin and weave stronger fabrics faster and more abundantly than ever Leonardo could have imagined. The sites of industry changed from cottages to mills and factories; the control of industry passed from owner-operators to the corporate boards in their castles in the sky.

Stage coaches gave way to railroads and steamships, which in their turn gave way to cars, trucks, and buses, through DC–3s to Concordes, with sub-orbital rocket ships on the drawing boards.

As our technological juggernaut was roaring out of the starting gate, scientists were extending and elaborating the methods of Galileo and Newton to the far reaches of the universe—upward to the stars, down through the organic world thinly layering our globe, into geological strata, into the depths of atoms, into the very genetic code that reads the framework and history of life on earth.

While geology, chemistry, and biology each have their own unique principles, not derivable from physics, their major investigative tools, their techniques of observation, experimentation, and logical deduction are mainly those of our seventeenth-century physicists. And all the sciences benefit from advances in technology: telescopes, microscopes, vacuum pumps, electric centrifuges, -ray tubes, and on and on. Some of these instruments were based on prior scientific investigations, some were accidental discoveries, and some were the result of systematic improvement.

As we follow these various developments through 250 years and more, we'll see gradual changes in the processes of innovation in technology and science. In 1700, science was often in the hands of school-trained mathematicians. However, even as late as the 1820s, Michael Faraday, a self-educated son of a blacksmith, could make important discoveries in physics. Most of the inventors in the earlier eighteenth century were ingenious artisans. James Watt apprenticed to an instrument maker, but had solid contacts with scientists like Black and Priestley.

In the nineteenth century, many chemical developments came out of the PhD factories of German universities. But the biological innovations of Darwin and Wallace came from men whose special training was mostly of their own making—Darwin's degree fitted him to be a country clergyman; Wallace didn't attend university. In technology, inspiration and perspiration were still often more effective teachers than the schools.

At the turn of the century, radio inventors were sometimes self-taught, sometimes they had PhDs in physics or electrical engineering. Sometimes they applied scientific principles, sometimes even a PhD could have an inspired hunch. Now we have great research institutes for both science and technology—laboratories of universities, corporations, and governments. R & D is big business; yet as late as the 1970s, a couple of Steves could build the Apple Corporation out of an inspiration in their garage in Cupertino, California.

All in all, we've come a long way in less than 300 years, from steam-operated mine pumps in Cornwall to the Olympic Games in Sydney live on TV across the world.

### *In a Mind's Eye*

He pulled his V–8 off the road; got out and walked to the lake edge. On the shortest day of the year, Abe Lincoln Pratt stood watching the sun rise over Lake Erie. In normal times he'd already be stripped to the waist tending the rolling mill in Dearborn. But, times were not normal. His call for the picket line at Ford's Rouge plant wasn't till nine.

You don't change the habits of a lifetime over a little interruption in the daily routine. He'd risen at the regular time, though he admitted to dawdling over breakfast. Betsy was up before him, as usual, feeding the chickens, getting the fresh eggs for their meal. Dear Betsy—without her devoted labors they'd be living in Detroit's ghetto. Instead, they had that neat little cottage on two acres outside Monroe. It had always seemed the right place for immigrants from Monroeville back in Alabama, where Abe Senior still eked out his mean sharecropping existence.

With young Henry II now in charge, Abe thought back twenty years to when that first Henry Ford had hired him into the steel mill. He'd been proud to be part of the team that turned out the first Model A back in '27. He sure had more dignity as an auto worker than he'd ever get in the South. Not that he had any love for old Henry—he'd been there in May of '37 when those goons pommeled Walter Reuther and threw him bloodied down the steps of the overpass. At least strikes were a little more civilized now, thanks to Reuther and the CIO. He wondered if Henry II would be any different from his grandfather.

Was '37 also the year Orson Wells shocked the country with his *War of the Worlds*, "live from New Jersey" the announcer had said? It was Hallowe'en—that might've given listeners a clue to the hoax—but it was all so realistic. No, it must've been '38. Abe wondered if there really were little green men on Mars. He knew the astronomers were sure there wasn't life on the moon. But Mars…that might be a different matter. Didn't they say there were canals on Mars?

*War of the Worlds* was a hoax, but World War II had been real enough. Young Grover's bones under the sands of Iwo Jima were proof enough; Betsy was only now reconciling herself to the loss of her firstborn so far from home. Then Abe thought about the holocaust of Hiroshima, that other graveyard across the Pacific. Would Franklin D have dropped the Bomb as easily as Harry Truman had? Abe thought he probably would—after all, they said it saved the lives of a million other Grovers. That hadn't consoled Betsy—only time could heal her wounds.

As he walked back to the car he thought about the bleak Christmas in store for them. Well, we do what we can. He'd heard that Ford had lots of money in the bank, but if they didn't settle soon, GM and Chrysler would drive them out of business. "Solidarity forever," he hummed under his breath as the engine caught and roared into action.

## *For Further Reading*

Aitken, Hugh G. *Syntony and Spark: The Origins of Radio*. New York: Wiley, 1976.

———. *The Continuous Wave: Technology and American Radio, 1900-1932*. Princeton: Princeton University Press, 1985.

Bowler, Peter. *Evolution, The History of an Idea*. Berkeley: University of California Press, 1989.

Brown, Anthony Cave, and Charles B. MacDonald, eds. *The Secret History of the Atomic Bomb*. New York: Dell, 1977.

Bruce, Robert V. *Alexander Graham Bell and the Conquest of Solitude*. Boston: Little, Brown, 1973.

Cowan, Ruth Schwartz. *More Work for Mother: The Ironies of Household Technology from the Open Hearth to the Microwave*. New York: Basic Books, 1983.

Darwin, Charles. *On the Origin of Species* [1859]. New York: Atheneum, 1967.

Deane, Phyllis. *The First Industrial Revolution*. Cambridge: Cambridge University Press, 1980.

Eiseley, Loren. *Darwin's Century: Evolution and the Men Who Discovered It*. New York: Doubleday, 1961.

Flink, James J. *The Automobile Age*. Cambridge, MA: MIT Press, 1988.

Garson, Barbara. *All the Livelong Day: The Meaning and Demeaning of Routine Work*. New York: Penguin, 1977.

Greene, John C. *The Death of Adam: Evolution and Its Impact on Western Thought*. Ames, IA: Iowa State University Press, 1959.

Hallam. A. *Great Geological Controversies*, 2nd ed. Oxford: Oxford University Press, 1989.

Hughes, Thomas P. *Networks of Power: Elctrification in Western Society, 1800–1930*. Baltimore: Johns Hopkins University Press, 1983.

Ihde, Aaron J. *The Development of Modern Chemistry*. New York: Dover, 1983.

Josephson, Matthew. *Edison: A Biography*. New York: McGraw-Hill, 1959.

Klingender, Francis D. *Art and the Industrial Revolution*. London: Paladin, 1972.

Landes, David S. *The Unbound Prometheus: Technological Change and Industrial Development in Western Europe from 1750 to the Present*. Cambridge: Cambridge University Press, 1969.

Malthus, Thomas. *An Essay on the Principle of Population as It Affects the Future Improvement of Society*. 1798

Marcus, Alan I., and Howard P. Segal. *Technology in America: A Brief History.* New York: Harcourt Brace Jovanovich, 1989.

Mayr, Ernst. *One Long Argument: Charles Darwin and the Genius of Modern Evolutionary Thought.* Cambridge, MA: Harvard University Press, 1991.

Meadows, Donella H. *et al. The Limits to Growth: A Report for the Club of Rome's Project on the Predicament of Mankind,* 2nd ed. New York: Universe, 1974.

Mihajlo, Mesorovic and Eduard Pestel. *Mankind at the Turning Point: The Second Report to the Club of Rome.* New York: Dutton, 1974.

Musson, A.E., and P. Robinson. *Science and Technology in the Industrial Revolution.* Toronto: University of Toronto Press, 1969.

Purcell, Carroll W., ed. *Technology in America: A History of Individuals and Ideas.* Cambridge, MA: MIT Press, 1981.

Rae, John B. *The American Automobile: A Brief History.* Chicago: University of Chicago Press, 1965.

Sayre, Anne. *Rosalind Franklin and DNA.* New York: Norton, 1975.

Schumacher, E. F. *Small is Beautiful: A Study of Economics as if People Mattered.* London: Abacus, 1974.

Segré, Emilio. *From -rays to Quarks: Modern Physicists and Their Discoveries* . Berkeley: University of California Press, 1980.

Speyer, Edward. *Six Roads from Newton: Great Discoveries in Physics.* New York: Wiley, 1994.

Wachhorst, Wyn. *Thomas Alva Edison: An American Myth.* Cambridge, MA: MIT Press, 1981.

Watson, James D. *The Double Helix: A Personal Account of the Discovery of the Structure of DNA.* Critical edition edited by Gunther S. Stent. New York: Norton, 1980.

*Steam locomotives of George Stephenson's design in coal mines in Hetton around 1820.*

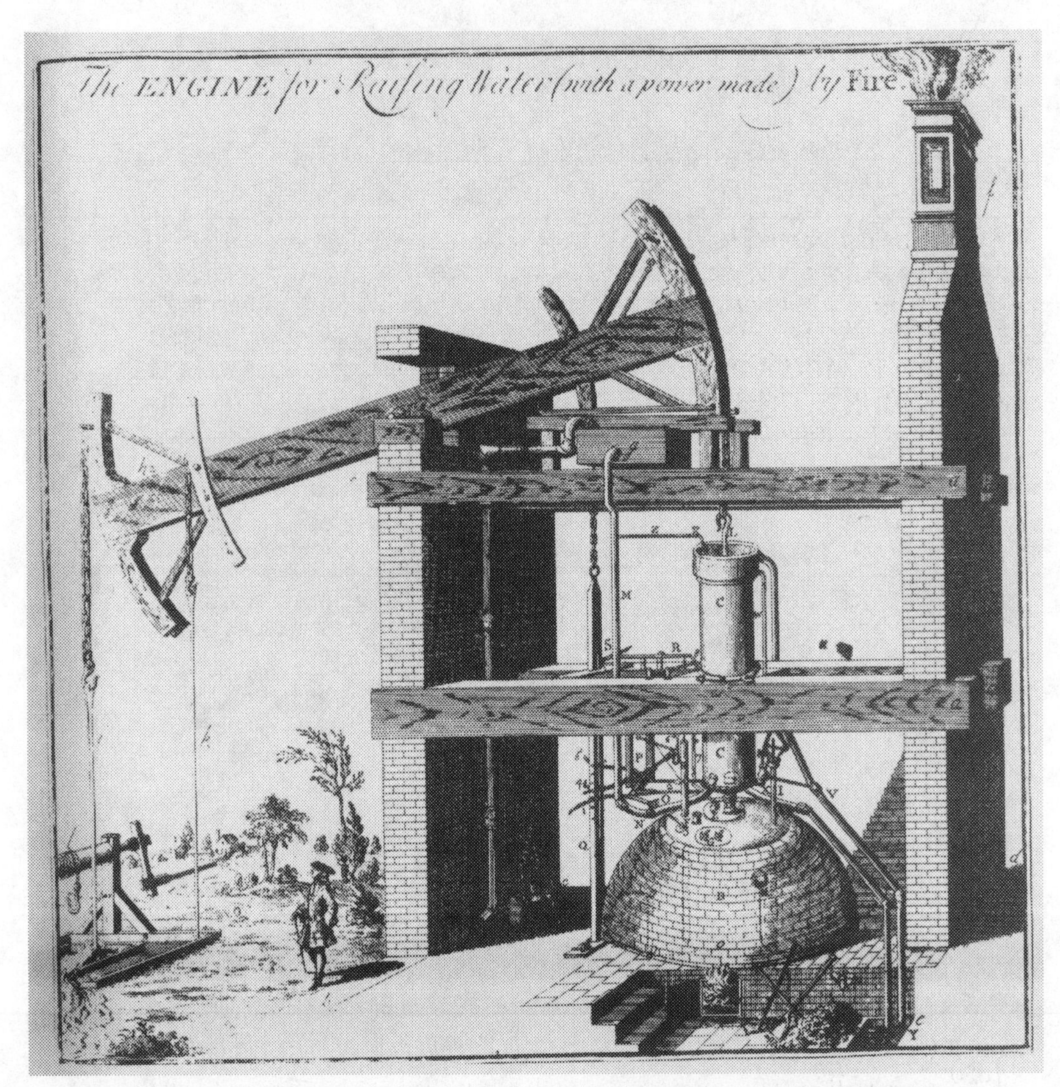


*A Newcomen steam-operated atmospheric engine, used for pumping water out of mines through most of the eighteenth century. From John T. Desaguliers,* Outline of the Experimental and Natural Philosophy, *1728.*

CHAPTER THIRTEEN

# The First Industrial Revolution

During the eighteenth century in Britain, the productivity of several industries was dramatically increased by the introduction of new methods of production. In some industries the new methods involved the replacing of hand manipulations and human muscle power by mechanical gadgets driven by waterwheels and steam engines. In other industries the early changes were less mechanical than organizational.

Those technological changes were part of a much larger, general transformation in the economic and social life of Great Britain. To take only one factor, the population of England and Wales more than tripled from 6.2 million in 1750 to 20.8 million in 1850. However, the fraction of the population in farming fell from 50% to 20% in the same period.

With just those bare facts, you can arrive at some interesting conclusions. All you have to assume is that the population in 1850 was as well fed as in 1750; and that changes in food supply and population were due largely to internal factors; i.e., no significant imports or exports of food or people.

1. Since the farm population declined from about 50 of the total to about 20 , there were proportionally two-fifths as many agricultural workers. But if the food supply per capita of the population changed little, that means the output per worker must have increased by about 2.5 times; a real increase in productivity, as well as in total production.
2. The population in cities and towns increased from 3.1 million in 1750 to 16.6 million in 1850, an increase of five times in 100 years. Just imagine the pressures on home building and the provision of social services, no matter how rudimentary they may have been.
3. You might conclude that a lot of new towns and cities grew up; or at least that many villages turned into towns and towns into cities. For a hundred years, Manchester was doubling its population every 25 years. In 1750 it was 18,000, and in 1850, 300,000: sixteen times the population in a century!
4. If you have no reason to think that urban folk suddenly started to outbreed their rural cousins, then there must have been large scale immigration from the countryside into the towns.
5. Finally, all those extra people in the towns and cities had to have some way to earn their living. Well, of course, they worked in the newly built factories. But, think of the dislocations (at least in the early years) imposed on peasants in the process of becoming factory "hands."

The totality of the technological, economic, demographic, and social changes were so spectacular that they are called *the Industrial Revolution*.

In this chapter we'll look at some of the major technological and economic transformations in eighteenth-century Britain, and then try to unravel some of the causes of change. Along the way we'll keep an eye out for some of the social consequences of the Industrial Revolution.

## *Agricultural Improvements*

There are several ways to double a country's agricultural production. One way is to double the amount of land devoted to agriculture. Yet, by 1700, much of the arable land in England was already being cultivated.

In fact, that achievement in earlier years set the stage for agricultural improvement in the eighteenth century. New arable land comes from two main sources: (a) cutting down trees, and (b) draining swamps. By 1700, large areas of Britain's forests were already gone for fuel and ship building. As we'll see later, there was a serious shortage of wood in England by 1700. Also, much of the fen country of eastern England had already been drained by 1700. Particularly during the seventeenth century, Dutch engineers had been brought in to supervise the work.

With relatively little new arable land to be cultivated, a doubling of output would require a doubling of the productivity of the land: two stalks of grain where once there was one; two cows for one, etc. Of course, it could also require a doubling of agricultural labor. Indeed, that essentially happened in Ireland, where the population increased from 1.25 million in 1600 to 2.5 million in 1700. But in England, a doubling of the land's productivity was accompanied by a doubling in labor productivity as well.

How was that achieved?

The increase in the productivity of English agriculture resulted more from organizational changes than from the introduction of new implements. Of course, it is not any the less technological for that. The main change in farm organization was the enclosure of previously open fields. Traditionally, farm villagers had shared the commons for grazing their animals and had planted their crops either in common or on specified plots in the fields of the district. A somewhat different arrangement, found in some regions, was that each farm family would cultivate a particular parcel of land held on a long-term lease.

Gradually, through the course of the eighteenth century, those traditional systems gave way to **enclosure**. That is, the rights of peasants to the commons were abrogated, which required acts of Parliament (beginning in 1709). The lord of the manor took control of the enclosed land and had it farmed field by field instead of plot by plot. As a result, tenant farmers were turned into farm laborers, with a much reduced capacity to influence decisions about how the land would be farmed. And perhaps that began the process of getting farm "hands" ready to become factory "hands" several generations farther along.

With the decisions now being made by the landowner instead of the tenants, innovation was much easier to achieve. (Traditionally, farmers have been very conservative, and committees find it notoriously difficult to reach unanimity. So, committees of farmers...) On the enclosed land of southern England, progressive landowners began to introduce innovations, many of which came from the low countries across the North Sea.

Curiously, some of these innovations are the same as those we identified for the agricultural revolution of the middle ages (Chapter 5), including such things as crop rotation, and horse-drawn plows. All that means is that the medieval innovations were not as widely dispersed as we might have expected. Actually, we probably should say that the medieval changes spread about as far as they had to in order to satisfy the requirements of the time. Overproduction of food is not a common phenomenon—you don't do the extra work if you can't dispose of the extra produce; and the medieval economy was not organized to expand more than it did.

However, there were some new twists in the old techniques. Root crops (such as turnips) were added to the traditional grains and legumes, and put into a four-year (or four-field) cycle. The new crops helped to reduce the rate of depletion of the soil and provided fodder for meat animals. And while horses continued to replace oxen as draft animals, they began to pull much improved plows, made entirely of iron. These plows, with coulter, share, and mold board wrought into a single implement, were much more efficient than their predecessors. The Rotherham plow of 1730 was one of the earliest of these new plows.

In addition, there was a gradual introduction of entirely new farm implements. The horse-drawn, mechanical seed drill of Jethro Tull (1700) provided a much less wasteful method of seeding. With the seeds planted in straight lines, the crops could be cultivated with horse-drawn hoes. Mechanical threshing equipment was in experimental use in the 1780s.

No single innovation, by itself, could more than double agricultural productivity, but in combination, gradually through the century, the improvements did occur. These were not without dislocations—there is some evidence of a deterioration in the diet of the rural poor after 1750, and, occasionally, bad weather produced crop failures. On the whole, however, the first half of the eighteenth century saw many bumper years, providing providential support for a growing population.

## *Manufacture of Textiles*

In northern climes, the second necessity after food is clothing. In Britain, the traditional clothing material was wool, another agricultural product. Indeed, a lot of the economic history of medieval and renaissance times is devoted to the relative incomes landowners could draw from grain or wool, and how they adjusted production of the two to meet market and price fluctuations.

Traditional "homespun" is cloth made by farm families using relatively primitive spinning and weaving devices. Even as early as medieval times, townsfolk (merchants and artisans) were not satisfied with homespun. In the centuries following 1200, a rather elaborate clothing industry grew up in some towns of Britain (as on the continent). Merchants bought raw wool in the countryside, put it out for carding and spinning to women in their homes, collected the yarn, and sold it to the weavers in town. After weaving, the cloth was successively passed through the hands of sizers, fullers, and dyers, drapers, seamstresses, and haberdashers before being worn at Mass on Easter Sunday. A number of these tradesmen formed guilds to restrict competition within each trade and thus maintain prices.

Because the damp British climate is ideal for sheep, the British textile industry became strong even in medieval times. It was a major supplier to the export market. Many craft shops were about the size of an extended family: a master, a couple of journeymen, and several apprentices. However, some shops became small factories employing, for example, scores of weavers, who walked to work every day to operate the owners' looms.

In the seventeenth century, some enterprising merchants found the regulations of the guilds too restrictive and moved their operations into the countryside. All the spinning and weaving was done by putting it out to farm families. There the spinsters spun, and the farmer wove in his spare time. At the same time, other innovators developed finer techniques in order to make lighter, less scratchy woolen goods. These helped to improve Britain's balance of trade, because the lighter fabrics found a ready market in countries with milder climates than Britain's.

In the eighteenth century, these lighter fabrics began to face increasing competition from cotton goods imported from India: muslins and calicoes. Some English merchants tried to offset that competition by importing raw cotton and fabricating it, but they remained very marginal until after 1770.

Around 1710, cotton output was about two million yards. It took 40 years to double. In the next 40 years, cotton output quadrupled to 16 million yards by 1790; and doubled again in the final decade of the eighteenth century.

Following 1770, cotton became a major industry and an important part of British trade. Slave merchants took on their shameful cargo in Africa and transported the poor victims to the West Indies and the United States, where they slaved on the cotton plantations. Then ships carried raw cotton to be manufactured in England, after which cheap British cotton clothing was shipped to Africa to trade for more slaves.

The acceleration in cotton production after 1770 was founded on inventions in cotton-spinning machinery. This was important because the spinning operation was slow and laborious. As manufacturers tried to increase output, they found that spinning was a bottleneck in the process.

As early as 1738, John Wyatt and Lewis Paul obtained a patent for a spinning machine operated by a waterwheel. Wyatt (1700–1766) came from a family of land surveyors and craftsmen near Birmingham; Paul (d. 1759) belonged to a family of Huguenot immigrants. Their machine used sets of rollers operating at different speeds to grasp and stretch the cotton rovings, which were then twisted and wound onto spindles. Although used in a factory of 50 operators in the 1740s, the equipment was not a commercial success.

*The spinning jenny.*

In 1765 James Hargreaves (d. 1778) invented the spinning jenny with eight spindles, on similar principles, though the wheel was operated by hand. The jenny soon began to appear in cottages and helped rural spinsters to meet the expanding demands of merchants. About the same time, Richard Arkwright (1732–1792), a well-to-do barber, probably with the help of clockmaker John Kay (d. 1764), developed a modification of the Wyatt machine, designed to be operated by a waterwheel. Arkwright built a cotton spinning mill near Derby in 1771. Within twenty years there were 150 cotton mills in the English midlands—doubling in number every three years—surely a revolutionary development. One of Arkwright's later plants was operated by a waterwheel 9 m in diameter, which drove 3,000 spindles. Spinsters left their cottages to become mill hands.

The mechanization of spinning broke the bottleneck. Soon, the pressure for innovation switched to the weaving process. Although Edmund Cartwright (1743–1823) invented a power loom in 1785, it had enough mechanical difficulties that it did not replace the hand loom for some decades. The number of hand-loom weavers increased tremendously between 1780 and 1820, reaching a high of a quarter million. By 1825, the power loom had been sufficiently improved that one boy, working two looms, could produce as much cotton cloth as 15 cottage weavers.

From 1813 the number of power looms in England increased exponentially, with a doubling period of less than four years, reaching 100,000 installed looms by 1833. They could produce as much cloth as 750,000 hand workers. By 1850 there were 250,000 power looms, and only 40,000 underemployed hand-loom weavers.

The tremendous increase in cotton production in England was made possible by these great increases in productivity. That's because the price per unit dropped drastically, and Britain could undersell the traditional suppliers of cotton.

Woolen production also increased in England during this time, though it was outstripped in export quantity by cotton around 1800. Cotton production exceeded woolen for several reasons. First, the supply of raw cotton was much more elastic. Being a plant, its production could be increased as fast as new land was brought into cultivation. All they needed were enough black slaves to tend and harvest the crop. Sheep production, by contrast is more inelastic.

Second, the demand for cotton was also much more elastic. English peasants could now afford the cost of underwear. Large overseas markets in warmer climates (including the plantation slaves) could be tapped as long as prices were below those of the Indian calicoes.

Finally, the cotton fiber, being tougher and more uniform, was more amenable to machine spinning than wool. Indeed, machine spinning really "made" cotton, because handspun cotton was not strong enough to form the warp threads on a loom. Until the advent of mechanically spun cotton, the warp was made of linen (a more expensive material).

The production of cotton spurred expansions in other branches of the industry, including bleaching, dyeing, and other finishing processes. The penetration of textile manufacturing into the economy of Britain in the years after 1770 was much like that of the automobile industry in the North American heartland after 1950.

## *Steam Engines*

For all but the last 300 years of human history, our energy sources have been almost exclusively renewable: human and animal muscles, water, wind, and wood. Pyramids, Gothic cathedrals, voyages of exploration and conquest were all accomplished with such resources, without much digging into the stockpile of the earth's mineral energy. A couple of minor exceptions include sulfur in gunpowder and some coal for heating.

In renaissance times there were a few schemes for using the force of hot air, but they were never practical. (Long before, the Greeks had some temple gadgets operated by steam pressure, masterpieces of ingenuity, but not industrial-strength machinery.) In England, in the middle of the seventeenth century, some patents were granted for intriguing sounding machines that were to run on coal, but few details were given, and there's no evidence any of them amounted to anything.

### *Savery's Steam Pump*

The first significant steam-operated machine was introduced in 1698 by a military engineer, Thomas Savery (1650–1716). Possibly Savery based his design on a earlier machine that was demonstrated in London in the 1660s, but not further exploited. Savery's steam-operated device was not so much an engine as a pump to clear mines of water. His achievement consisted mainly in spreading knowledge of the device and having it put into practice as the "Miner's friend."

Savery's pumping vessel was located about 10 m above the surface of the water and connected to it with a pipe controlled by a one-way valve. Another pipe and valve connected the vessel to a boiler. When the valve from the boiler was opened, steam entered the vessel and condensed there, lowering the pressure in the vessel when the steam valve was closed. That allowed atmospheric pressure to push water up into the vessel. A third pipe ran straight up from near the bottom of the vessel. At the next admission of steam, the pressure forced the water in the vessel up the pipe to about 10 m above the vessel. Closing the valve in that pipe, and then the steam valve, allowed the cycle to be repeated.

A Savery pump was installed in Kensington in 1712. Using about 50 kg of coal a day, the pump developed a power of about 0.7 kW. The pump cycled about four times a minute, delivering 200 L/min of water.

In 1702, Savery described his pump in a pamphlet with a long title:

> *The Miner's Friend, or, an Engine to raise Water by Fire, Described, And of the Manner of Fixing it in Mines, With an Account of the several other Uses it is applicable unto; and an Answer to the Objections made against it.*

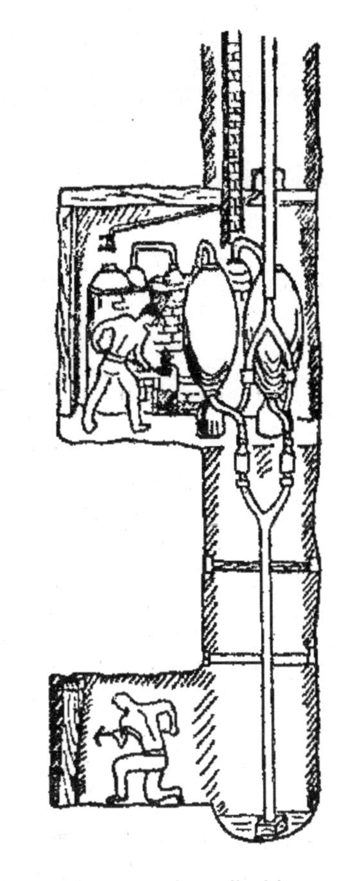

*Savery's pump installed in a mine.*

The objections to Savery's engine were serious. Because the vessel had to be located within 10 m of the surface of the water to be pumped, its boiler (and attending fire) had to be located within the shaft of mines deeper than that. Other pumps would be installed at 20 m intervals to raise the water by stages. But fires in coal mines are dangerous. In addition, the vessel operated at a pressure of about three atmospheres. Metal fabrication at the time was insufficient to guarantee the integrity of pressure vessels. The miner's friend could be a wolf in sheep's clothing. However, it was usefully employed in water supply systems above ground. Savery engines continued to be built in small quantities throughout the century, despite the advantages of the engine designed by Thomas Newcomen early in the century.

### *Newcomen's Atmospheric Engine*

Thomas Newcomen (1663–1729) was an ironmonger and blacksmith in Devonshire. With a local plumber, John Calley (or Cawley), Newcomen experimented on a different kind of engine during the first decade of the eight-

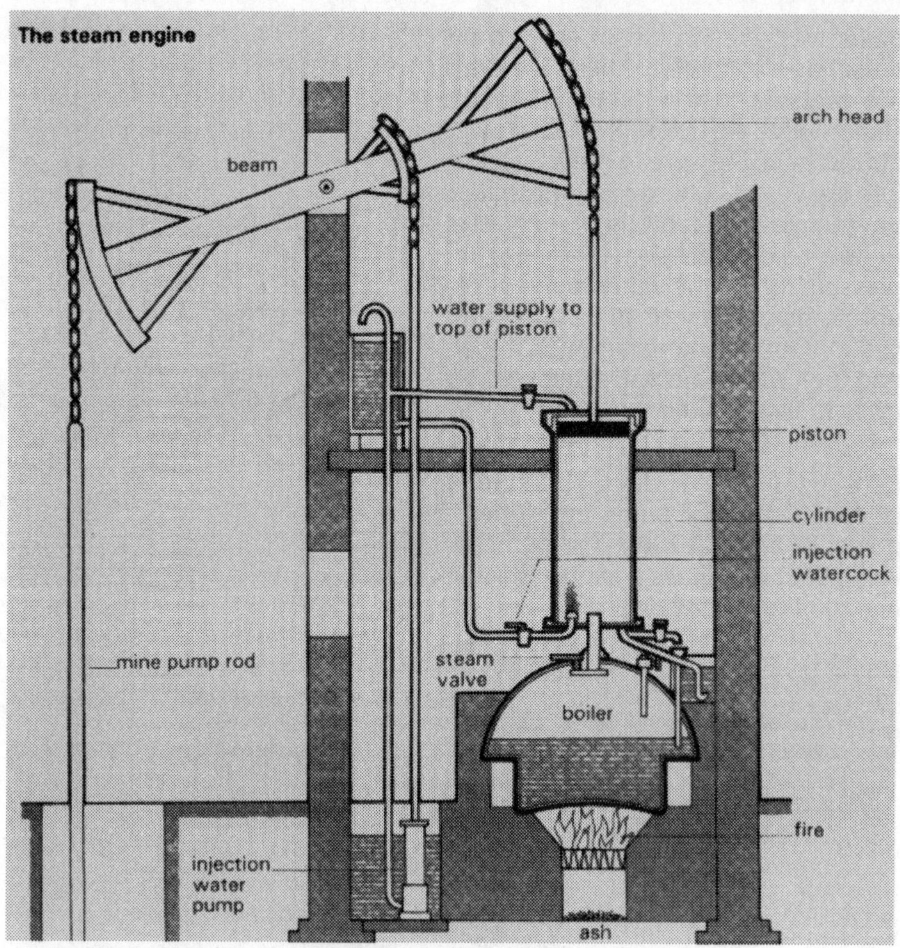


*Schematic diagram of a Newcomen steam engine.*

eenth century. The first successful installation was in Staffordshire in 1712. By 1720, the engine was being widely used and continually being improved by the work of engineers and natural philosophers.

What Newcomen produced should properly be called a steam-operated atmospheric engine. A piston in a cylinder was connected to a pivoted beam above, whose other end connected to a long rod that operated conventional lift pumps. In the resting position the piston was at the top of its travel. Then the operator opened a valve connecting the cylinder to a steam boiler. Once the cylinder was full of steam, the operator closed the steam valve and briefly opened a second valve to admit a jet of cold water. The condensing of the steam lowered the pressure in the cylinder, and atmospheric pressure forced the piston down, raising the pump rods on the other end of the beam. A valve at the bottom of the cylinder allowed the water to leave the cylinder. When the operator opened the steam valve again, atmospheric pressure returned to the cylinder and the weight of the pump rods forced the piston up in the cylinder. Before long, Newcomen made the action of the valves automatic by attaching them to rods worked by the motion of the beam.

Notice that at no time did pressure exceed atmospheric (100 kPa). The power stroke was provided by the pressure of the outside atmosphere pushing the piston against the reduced pressure within the cylinder. For the return stroke, the force of gravity on the pumping gear levered the piston up.

Newcomen's engine was greatly superior to Savery's. In particular, it could be installed at the minehead, with no need to have fire in the mine. However, it was an extremely inefficient engine. In the early years, the mechanical work output was only 0.34% of the energy available in the burning coal. Early engines had a power of only a couple of kilowatts.

About mid-century, engineer John Smeaton (see pp. 248 - 249 in this chapter) performed an extensive series of design tests to improve the Newcomen engine. Armed with his results, Smeaton constructed a gigantic engine, with a cylinder 1.8 m in diameter (weighing 6 t) and a piston stroke of 2.9 m. This engine was supplied by three steam boilers, each 4.6 m in diameter. Its output was about 60 kW, with an efficiency of about 1.0%.

By 1770, about 500 steam engines were in operation in Britain. Most of them were used to pump water out of mines. Occasionally, a Newcomen engine was used to pump water into a reservoir, whence it flowed over a waterwheel to drive machinery. Near the end of the century, steam engines were beginning to drive machinery directly, using a crank to convert the reciprocating action to rotary motion.

*James Watt (1736–1819)*

Watt was born near the mouth of the Clyde River about 30 km from Glasgow. His father was a small merchant who lost his trade through speculation. In 1755, James spent a year in London learning the craft of instrument making. Back in Scotland, he found employment repairing instruments at the University of Glasgow.

*James Watt.*

In 1764, when Watt was asked to repair a model Newcomen engine, he was struck by its inefficient operation. He soon realized there was significant heat loss as the cylinder alternately cooled and heated in each cycle. The first steam admitted to the cylinder always condensed on the cold surface. Only after the steam had warmed the cylinder again could it remain in the vapor state. Not only was this wasteful of heat, it also slowed the rate of operation.

*Watt's steam engine.*

After some months puzzling over various remedies, Watt one day realized in a flash that he needed to have two separate vessels, which could each be maintained at a constant temperature—one for expansion and the other for condensation. If the cylinder was kept hot, the steam would not condense prematurely and the condensation could be managed by opening a valve to a second vessel kept always cold. No longer would cold water need to be injected into the cylinder. Watt soon designed a system in which fluids flowed in jackets around the two vessels, steam for the cylinder, and cold water for the condenser. He also attached a pump to the condenser to remove the condensed steam and keep the pressure down.

As Watt experimented with this design, he soon ran out of capital. He took employment in land surveying and civil engineering, where he invented some useful measuring devices. Although Watt obtained a patent for his steam engine design in 1769, little progress was made for several years. In 1775, Watt and Matthew Boulton (1728–1809), an industrialist in Birmingham, formed the firm of Boulton and Watt to produce the new improved Watt engines. These engines operated at about 4% efficiency, quadrupling the best Smeaton had achieved. (If these sound like tiny efficiencies, realize that because of thermodynamic limitations, it took another hundred years to achieve a further quadrupling of efficiency. And 16% is about the limit for this type of engine.)

Soon, the Watt engines were in demand for pumping water out of the mines in Cornwall. So confident was Watt of his improvements that he charged a royalty on the use of his engines as one-third of the savings in fuel. By 1783, only one Newcomen engine remained in the Cornish mines. The Watt engines also began to drive the wheels of industry. By 1800, Boulton and Watt had sold over 500 engines, with an average output of about 4 kW each.

*Later Developments*

The British government had been granting patents on inventions since 1617. They gave inventors monopoly rights for a period of fourteen years. By those terms the Watt patent would have expired in 1783. However, Boulton and Watt managed to get the patent extended in 1775 for a further 25 years. Competitors rankled under this restriction, and during the 1790s, Boulton and Watt contested several infringements in the courts. After 1800 there was a burst of activity in improving the design of steam engines.

Since early engines suffered from poor workmanship, explosions were minimized by operating them mostly at atmospheric pressure. However, improvements in iron fabrication after 1770 removed the need for such a restriction. Operating at higher pressures meant higher temperatures for the working fluid, and increased efficiency.

Experimenting with high pressure engines began about 1781. Soon, engineers also began using multiple cylinders in their engines: after high pressure steam had been used in one cylinder, it was vented at lower pressure (not yet condensed) to serve a second cylinder. Eventually, the efficiency gained this way made the separate condenser less necessary and it was mostly omitted.

Thus, while James Watt's fame rests mainly on the addition of the separate condenser (not "as every school child knows" for inventing the steam engine), that innovation was really only a brief (though possibly necessary) phase in the development of steam power.

Steam engines in locomotives transformed the landscape in nineteenth century Britain, Europe, and North America (see later in this chapter and in Chapter 14). Yet the first burst of factory production in the latter eighteenth century was powered mostly by waterwheels. The steam engines of Savery, Newcomen, and Watt were used mainly for pumping, for mine drainage, for water supplies to towns, and for raising water to drive a wheel. Even as late as 1850, almost 20% of the textile machinery in England was still operated by waterwheels.

Nonetheless, steam technology was well established by 1800, ready to support the burgeoning industries of the nineteenth century. Though not yet numerous by 1800, steam engines were already being used in a wide variety of factory settings. They also contributed to the industrial revolution in another way: they were built of iron and ran on coal.

## *Manufacture of Iron*

Iron was regularly produced in blast furnaces from the fifteenth century onward (see Chapter 7). With blasts of air forced into the furnace by waterwheels, the iron founders were able to achieve the 1500° C needed to melt iron. The resulting cast (or pig) iron had then to be worked in a forge to reduce its carbon content from about 4% to 1% or less for wrought iron. Smiths performed this task by repeated heating and hammering of chunks of metal to burn out the carbon.

From 1500 to 1700, iron founders enlarged their furnaces and began to operate them continuously for weeks at a time. By these measures they were able to increase the output from about 1 t per day to 2 t, and to reduce the amount of charcoal needed to keep the furnace going.

Nonetheless, with increasing warfare and the greater use of iron cannon both at sea and on land, forests were being seriously depleted to supply the charcoal. Besides, charcoal for smelting iron wasn't the only use for wood. It was also used for building homes and ships, and as fuel for heating. Although coal had been used for heating, its noxious fumes caused its use to be severely limited. As England grew wealthier during the seventeenth century, more wood was needed to fuel the ovens of glassmakers and brewers.

By 1700, the shortage of wood was becoming severe. Coal could not be used for iron smelting because its impurities made the iron brittle. Coke can be made from coal in essentially the same way charcoal is made from wood (that is, by heating out of contact with air). However, for making iron, some of the same damaging impurities remain in the coke. In 1709, Abraham Darby (1677–1717) of Coalbrookdale in Shropshire was fortunate in finding a high-grade coal from which he could produce metallurgical coke.

The use of coke in iron smelting increased slowly during succeeding years. The total production of pig iron in England took 35 years to double from 20 kt in 1745 to 40 kt in 1780. However, from then until 1840, the exponential increase in pig iron production saw a doubling every 11 years—almost 2 Mt in 1840!

This increase in the supply of iron came from other innovations besides the use of coke for smelting. It also depended on an increased supply of coal and ore from the mines, which was assisted by steam pumping engines. In the 1780s, steam engines drove improved blast furnaces to increase production. And techniques for burning the carbon out of cast iron on a larger scale were also introduced.

However, increased production also depends on increased demand. You don't produce more and more iron unless there's a secure market for your enlarged production. In addition to the increase in the traditional demands (like cannon), whole new uses for iron developed during the eighteenth century in Britain. The steam engines for pumping the mines and blasting the furnaces took as much as 10 t each. In addition, agricultural uses began to require more iron—for horseshoes, axes, and plows—so that Bairoch estimated the agricultural demand around 1760 at "between 30 and 50% of the total demand for iron."

*Steam engine installed in a coal mine. England, ca 1790. Anonymous painting.*

Before long, iron became the material of choice for new machinery in the textile industry. John Wilkinson (1728–1808), another famous ironmaster, was closely associated with the firm of Boulton and Watt. He supplied iron cylinders for their improved engines and also used their engine himself. In addition to making cannon and swords, Wilkinson also built iron bridges and boats, and was buried in an iron coffin.

So the eighteenth century saw the transition from the use of wood as the major fuel and building material to coal (often as coke) for fuel and to iron for building. Resource demands shifted from renewable plant materials to non-renewable minerals. It started first in Britain, soon spread to Europe and North America, and is continuing today in the rest of the world, with consequences some of us may die regretting (Chapter 22).

## *Transportation*

The increased economic activity in Britain during the eighteenth century would not have been possible without increasing means for transporting raw materials to manufacturers and finished goods to consumers. Since no place in Britain is more than about 130 km from the sea, a great deal of local transportation has traditionally been by coastal shipping. As well, of course, Britain is ideally located for overseas trade.

*On Water*

Before 1500, most European trade was carried by ships of Venice or of cities on the Baltic Sea belonging to the Hanseatic League. Then, overseas expansions (Chapter 8) gave an impetus to shipping among the nations of the Atlantic seaboard. After 1600, England began to develop from being merely a convenient source of raw materials for continental merchants into a maritime power to be reckoned with. Her successful defeat of the Spanish Armada encouraged English authorities to support a growing merchant fleet. By 1670, much of London's overseas trade was required to be carried in English ships. And growing populations in overseas colonies provided further support for expanding the mercantile fleet.

Between 1630 and 1730, the tonnage of shipping registered at English ports tripled to reach about three hundred and fifty thousand tonnes. By 1775, it had reached about six hundred thousand tonnes. Of that, only about half was used in overseas trade; the remainder was used in the coastal trade "within" England herself. For example, a 1775 estimate puts the number of ships carrying coal (mainly from Newcastle to London) at eighteen hundred, averaging perhaps 100 t each.

This coastal traffic enabled a level of trade that would have been very difficult over land, particularly with bulky goods. As long as it was a matter of shipping from one coastal port to another, bad weather was the major obstacle. But, if either terminus was far inland, the cost of wagon cartage over tracks full of hazards was almost prohibitive. In 1775, an iron firm in Shropshire found that 130 km on a barge down the Severn and 500 km on a ship around Wales, was cheaper for getting pig iron to Chester than directly overland for less than 100 km.

Indeed, before the roads were greatly improved, barges on inland waterways were the best way to transport bulk goods to sea ports. About 1000 km of river ways were naturally navigable by shallow draft barges. These include such rivers as the Severn and the Thames. Between 1600 and 1760, about another 1000 km of waterways were added, mostly by deepening and extending rivers or by bypassing difficult sections. During this period, the English often imported or consulted engineers from the continent, particularly Dutch engineers, who had much hydraulic experience.

Then, between 1760 and 1800, English engineers directed the building of another 1000 km of canals. These were most ambitious, being mostly new waterways, designed to make a network for barge transport throughout central England.

Two of these new canals were built right into coal mines, so the coal could be transported directly to industrial centers at Liverpool and Manchester. For one of these, the engineer James Brindley (1717–1772) designed an aqueduct to carry the canal across a river valley. Yet, Brindley's chief accomplishments were in organizing the vast enterprise: arranging the capital and land rights, supplying and directing the labor. These canals were the earliest large-scale

*The canal bridge at Barton. Engraving 1784.*

projects built with public capital. As such, they provided a training ground for the engineers, bankers, and lawyers who would use their skills to great benefit in the following era of railroads. Before the canal system was superseded by the railways, about 1860, the inland waterway exceeded 6000 km in length.

*On Land*

Travel on land over short distances was needed for exchanging goods between town and countryside. Trips of 10 km by ox-cart or pack-horse required little more than a well-trodden path, though it could become impassable in wet weather. Until the eighteenth century in England, the provision of roads and bridges was entirely in the hands of local parishes.

One special case was the provisioning of London. To feed its growing population, during the last half of the eighteenth century, animals were driven each fall to the Smithfield market from a broad hinterland: a hundred thousand cattle, nearly a million sheep, and scores of thousands of geese and turkeys. The roadway for such traffic needed to be well drained, but not paved.

As trade expanded, civil engineers such as John Metcalf (1717–1810) and Thomas Telford (1757–1834) designed and superintended turnpike roads on a foundation of larger stones, covered with layers of smaller stones, finished with fine gravel.

Until such roads formed an adequate network, long distance travel was often slow and arduous. While post riders could carry mail from London to Edinburgh in under four days, a coach or carriage still took two weeks as late as 1750. Gradually, as turnpike corporations were established after 1700, roads were constructed or improved in stretches spanning several parishes, but it took many years to create a widespread system. By 1785, one could travel by stage-coach from London to Edinburgh in about five days.

Of course, a further advantage of improved roads was to shorten the time for a wagon to get to the nearest canal or seaport. So, while land transport could not be avoided, it was the waterways that sustained the increased traffic in goods until well after 1800.

Then, it was not roads, but *railroads* that carried the burdens of the expanding industrial revolution.

### *Railroads*

The terms "railroad" and "engine" are so closely associated in our minds, we may be surprised that the rails and the engines had separate existences until after 1800. Rails, first of wood, later of iron, were used at mineheads in Germany as early as 1500. Introduced into England after 1600, these rail ways supported horse-drawn carts for moving coal in the vicinity of a mine, or for carrying materials around within an iron works. By 1800, there were several hundred kilometers of such railroads in various parts of England.

The Newcomen and Watt engines of the period were stationary engines of great bulk. One could hardly imagine them installed on a self-propelled wheeled platform—a **locomotive**. When Richard Trevithick (1771–1833) introduced high-pressure steam in the late 1790s, he so reduced the size and weight of engines that he designed and built steam-driven road carriages. Such vehicles were not well suited to the conditions of the roads of the time.

Very soon, Trevithick used his locomotive on a railway and gave a demonstration on a line in Wales in 1804, pulling 10 tons of ore and 70 passengers at a speed of 8 km/h. From then until 1830, various schemes were tried of supplying railways for customers who supplied their own carriages and traction, mostly by horses.

The conclusive marriage of steam and rail was largely the work of George Stephenson (1781–1848). Son of a fireman on a colliery engine near Newcastle, Stephenson grew up in a technological environment. He assisted his father for a while, and attended night school in his late teens in order to understand engines better. After some experience as an instrument maker,

*George Stephenson's "Rocket."*

Stephenson became a consulting engineer and from about 1813 began to promote steam locomotion on railways.

The first fully mechanical rail line connected Liverpool and Manchester. In 1829, Stephenson's "Rocket" sustained the trials set by the line's directors, who then gave him the contract for the eight locomotives with which the line opened the following year.

Although the promoters of early railroads could depend for some experience on the earlier canal building, they still had to show they could make a profit. That didn't take long, as passenger traffic proved to be an unexpected boon. Soon some lines were being projected as much for passengers as for freight, e.g., to seaside resorts. By 1850, the length of rail lines exceeded that of canals in England. After that, the railroad boom sustained an industrial growth far beyond the canals' capacity to cope. Besides the railroads being important consumers of coal and iron production, rails and engines soon became an important part of British exports.

## *"Causes" of the Industrial Revolution*

So far, we've seen some of the main technical changes in a broad sweep of English industry during the eighteenth century. But we haven't examined why those changes happened in England rather than in another country, in the eighteenth century rather than at another time, nor why they should be considered revolutionary.

None of these are simple questions, nor have they received widely agreed upon answers. There are many opinions among the numerous histo-

*George Stephenson driving "Locomotion No. 1" at opening of the Stockton and Darlington Railway, Sep 27, 1825. First passenger-carrying railway open to the public.*

rians and economists who have written on the period and too few hard data to test them. So far, the most they can agree upon is that many factors were involved and that they interacted in a complex way. Rather than rehearse the variety of opinions, I'll spin a web of interactions among those factors, without wishing to claim any more for it than that it should carry at least superficial plausibility.

To dispose of one question quickly: we can call the period revolutionary for two compelling reasons: (1) the pace of change, the acceleration, was exceptional; (2) the extent of the change was so significant that no one afterward could conceive of returning to the previous situation.

To get deeper than that, it will be useful to have a model in mind. For most of human history, we humans have organized ourselves in ways intended to preserve ourselves, our families, our cultures. Our responses to challenges have most often been to try to restore the status quo. Human systems, like natural ones, tend to be self-regulating, using negative feedback. In a given setting a human group has a population suitable to the available food. If more children are born than are needed to maintain a steady population, the amount of food per person declines. When that affects health, there may be a rise in the death rate, or a fall in the birth rate, with the result that the population will return to something close to its previous number. It is this *decline* in food per person with *rise* in numbers that represents a **negative feedback**.

Of course, the equilibrium situation just described assumes a fixed resource base. Occasionally, on a world-wide scale, new technology has expanded the resource base. The agricultural revolution allowed the world population to increase from about five million at 5000 BC to about 200 million at about AD 500. Developments in medieval times allowed a 50% increase in 300 years, from 240 million in 900 to 360 million in 1200. These agricultural innovations allowed periods of relatively rapid expansion, while growth was usually slower at other times.

Despite these occasional spurts of population growth, the vast majority of people still spent their lives tilling the soil. After the innovations, negative feedback simply set the equilibrium population at a new higher level. So, throughout most of the world, 80% to 90% lived rurally until well after 1800. But in England, by 1800, rural population had already fallen to 70%, and by 1825, to 55%.

In England, during the eighteenth century, negative feedback loops were replaced by positive ones in a variety of economic sectors.

*Economic Factors*

As wood shortages developed, there was pressure to exploit coal for heating. The deeper the coal mines got, the greater the need to pump water out. The Newcomen engine met that need and increased the demand for coal to run the engines, as well as for coke to smelt the iron to make the engines. This is one of the primary positive feedback loops in the economy of eighteenth century England.

Overseas commercial activity increased significantly after 1660. London, a fine seaport at the center of a large agricultural hinterland, was a natural leader in that commerce. Between 1650 and 1700, London grew from 7% of the population of England and Wales to almost 10%; with a population of more than half a million, London was the largest city in Europe. To feed that population required the agricultural surplus from most of southern England. Every fall, large droves of cattle, sheep, geese descended on the slaughterhouses of Smithfield. Their requirements for roadways were different from those of wheeled coaches, both of which required regular improvement if London was to be adequately supplied. The better the roads got, the easier was the movement of people, the more the population of London increased, the greater the need for agricultural improvement, the greater the need for iron plowshares and horseshoes—and you can see how these loops of positive feedback can be added to the first one.

To use classical economic terms like supply and demand, we should consider the availability of raw materials and the size of the markets for the products derived from them. As for raw materials, England was well endowed with the iron and coal that fueled the early industrial revolution. Before that, the major English export had been wool, which went some way toward providing the early commercial foundations of the port of London. Later, when cotton became important, British ships brought bales of raw cotton from British colonies (or former colonies) to the spinning mills of Lancashire. All in all, England was well fixed on the supply side of the economic equation.

On the demand side, there are a number of relevant factors. Consider the home market first. Although England had only about a quarter of the population of France, it was a larger region of free trade, because there were internal customs barriers within France. This, coupled with the greater ease of transportation around England, meant that manufactured goods could compete in price for homemade ones sooner in England than in France (all other factors being equal).

All other factors were not equal, and they benefited English industrialization even more. For example, the increasing mechanization of production did not depend merely on a growing population. Even if more people provide a greater demand for goods, they can also provide a greater supply in the sense of more hands to turn out the goods. There are too many examples historically (Ireland) and at present (India) of growing populations *not* mechanizing, to make us pause over this. But the situation may be different with a population that is becoming better off—with more disposable income.

Here, any analysis of the eighteenth-century English scene gets complex. Peasants who were driven off their land by enclosures were not better off. But craft workers in London, whose wages started looking better because food prices were declining, might find an extra shilling in the box at the end of a month. They could buy a Wedgwood creamware jug.

There must have been enough money in the hands of those who *would* buy mass produced goods to justify the mass production. These products aren't fancy enough for aristocrats, but they're dearer than poverty stricken peasants could afford. So, there must have been sufficient demand from yeoman farmers not displaced by enclosures and from the classes of small merchants and craft workers to justify the expenditure on capital plant involved in mechanizing production.

Now, initial capital investments in innovation were not necessarily very large. In many cases, the funds came from the profits of a successful enterprise like an iron smelter. Not a gigantic blast furnace, just a modest little plant, employing a family and a few apprentices. In some cases, an innovator needed more capital than that. Richard Arkwright obtained support from Jedidiah Strutt, who'd done well in supplying an improved hand-operated machine for the manufacture of stockings.

Perhaps the home market by itself was not big enough to support early innovations in manufacturing. But add in colonists overseas: on the eve of the American Revolution, the English colonies in North America had a population of three million, close to half that of England and Wales. Their demand for traditional hand-manufactured goods from England may well have helped provide the extra profits that enabled innovating manufacturers to improve their productivity by mechanizing their operations.

An increased supply of manufactured goods is of no use unless there are customers to buy them. So, demand is an important factor in driving the economic equation. But increased demand is not sufficient to force an enterprise to make innovations. Often you can meet extra demand by hiring more hands. Indeed, some of that must have been done in mid eighteenth-century England. If the demand from the colonies was large enough, and the labor supply at home small enough, that would force up wages (creating another positive feedback loop), but might also encourage a smart entrepreneur to try to improve the productivity of his operation.

Wanting to improve productivity, and being able to, are not the same thing. It's not merely enough to have money to invest in innovation. You need innovation. Where does that come from?

### *Technical Ingenuity*

Innovations come from inventors, people with new ideas. We've described some of them earlier in this chapter. Many of them were men in industry who tried to find ways to improve their work, like the Darbys in iron manufacturing. Newcomen and Calley were not in the engine business, but they did understand the needs of miners in Devon and Cornwall, and had experience in metal fabrication.

Newcomen was a merchant with enough income to be able to experiment. Whether he was aware of contemporary experiments in the Royal Society with the expansive power of steam is uncertain. The key action of squirting water into the cylinder to create a vacuum may well have been the result of an accident during Newcomen's tests, where he'd been cooling the cylinder externally (and *slowly*).

What is the source of Newcomen's technical ingenuity? Perhaps the simplest answer is that he had it naturally—nothing mysterious—just that some children have mechanical aptitudes and some don't. All that's needed is an opportunity to demonstrate it. In Newcomen's case, he was supplying such items as locks to merchants in his native Dartmouth in 1688. And in 1704 he was paid for repairing the town clock. A practical man with a mechanical flair seeing a need in pumping mines and having the skill, luck, and persistence to develop a working engine.

James Watt (not born until seven years after Newcomen's death) was trained as an instrument maker. He thus belonged to a group devoted to mechanical contrivances. Their numbers were growing in the early eighteenth century primarily in response to the needs of British navigators for compasses, sextants, telescopes, and clocks and watches, too. They represented a pool of talent that could be drawn on to devise and maintain various mechanisms. To adapt themselves to engines and textile machinery, they only needed to be able to think a little larger. At the same time, water mills continued to be widely used and to be built and maintained by millwrights, another pool of technical skill, if perhaps of a lower order of precision.

Since such a range of talents also existed in France at the same time, technical ingenuity, added to economic factors, is still not sufficient to account for the industrial revolution.

*Social Factors*

A cluster of factors that influenced England's mechanization of industrial processes are hardly quantifiable, yet nonetheless real. We may call them attitudes: attitudes of rationalism, of individualism and self reliance, of naturalism as opposed to mysticism. An illustration of the difference between England and France in the mid-eighteenth century can be seen in encyclopedias published in the two countries.

A major project in France was the *Encyclopédie* (1751–1772), a product of the efforts of the brightest, most progressive minds in the country. Accompanying its 17 volumes of text were 11 volumes of plates. Although a number of the plates were technical drawings of mechanisms, many more were artists' renditions of technological settings: a shipyard or a flour mill showing the workers at their tasks. The editors seemed to be taking pains to convince their readers that crafts and technology were valuable human activities. By contrast, the contemporary English *Cyclopædia* (1728, 2nd 1738) by Ephraim Chambers contained only diagrams, as if to show that English readers were already convinced.

Although English society was socially stratified much as the French was, there was rather more social mobility in England. Particularly after 1660, men of commercial wealth could purchase large estates and mingle with the aristocracy more freely than was common in France.

Another feature of significant difference was the institution called **primogeniture**, the practice of confining the inheritance of estates and titles to the eldest surviving son. Since primogeniture was practised in England, it meant that many younger sons had to make their own way in the world. Often well educated, they would go into government service or the church, or as London became more prosperous, into commerce. This practice had the result of blurring the distinction between the worlds of aristocrat and businessman.

In France, where primogeniture was not practised, the children of the aristocracy began to clog Paris after 1700. Each of them had a share of the estate "down on the farm," which was run by a manager in order to keep the young aristocrats in high society. Heavily taxed by absentee landlords, neither managers nor peasants had any great incentive to innovate. In England, at the same time, a number of the dukes and barons who owned large estates took an active interest in their operation. They spent their winters in London, at the House of Lords and in society, and often attended the Thursday meetings of the Royal Society. During the summers they were back on their estates, seeing to their management and acting as local magistrates.

It was these English lords who picked up the new agricultural techniques from Holland, and who led the movement to enclose their fields and dispossess their tenants. The 1700s may be too early to talk of "scientific" agricul-

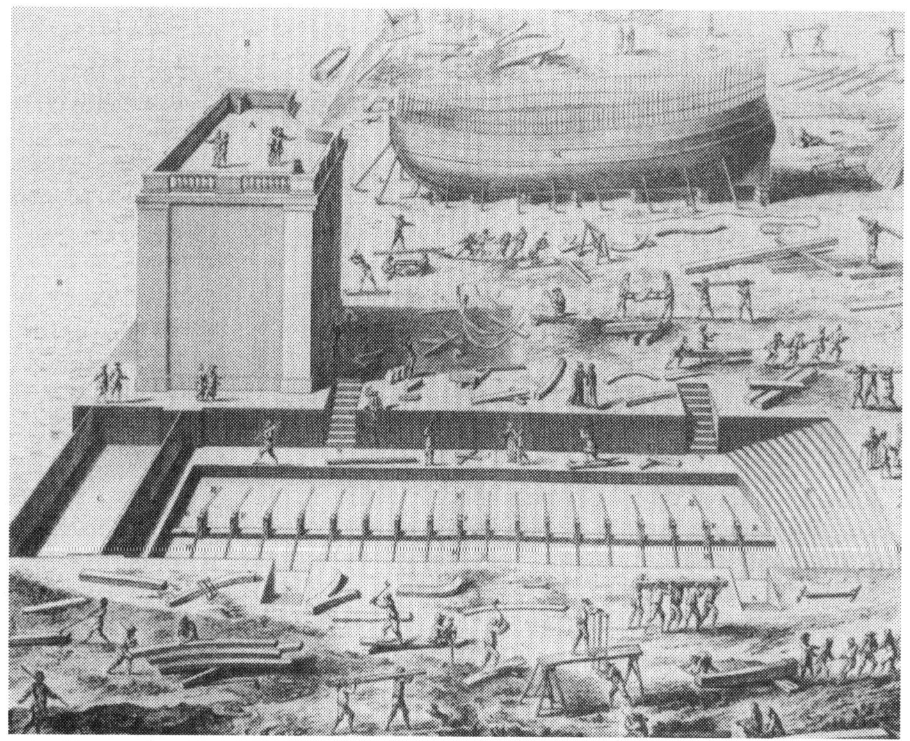

*An illustration of a shipyard, showing a variety of tasks, from the* Encyclopédie *(1751–1772).*

ture, but these lords did learn something about a rational approach to nature from their colleagues in the Royal Society. Treating peasants as units of production may well have been their own idea.

Indeed, attitudes towards workers was one of the features of the old world that modernization had to change. Traditionally, in rather small enterprises, the master and his few workers behaved much like an extended family. Often, they were reluctant to expand if it meant involving strangers. Such attitudes were common on both sides of the English Channel, down to the seventeenth century. They were still common in France 200 years later, but not in England.

Some of this transformation of traditional values has been attributed to the rise of Protestantism. But England was not the only country to undergo the Protestant reformation. Moreover, religious reform was a matter of degree in many places. English Protestants were united in their revulsion against papal authority, yet many early church leaders in England wanted little else to change. In fact, the English civil war of the 1640s was a revolt of gentry and middle-class merchants, individualists both in business and in religion. Their religion often goes by the name of Puritanism, which has too many different definitions to prove very useful in the present context.

In fact, some of the pressure toward economic modernization came from **dissenters**, not so much Puritans as members of groups like Baptists, Quakers, Presbyterians, and Unitarians, who refused to acknowledge the supreme authority of the Church of England. Members of these sects were barred from government and military service, and could not even earn degrees at Oxford or Cambridge.

These dissenting groups operated their own schools, which tended to include "modern" subjects like history, geography, and mathematics in their curriculum, instead of a major concentration on grammar and classical languages. There is ample evidence that members of these dissenting groups were prominent among the new industrialists in the English midlands of the eighteenth century.

So, rational as opposed to traditional business practices emerged sooner in England than in France, and had *some* connection with some kinds of Protestant attitudes. That's a far cry from "Protestantism causes capitalism causes industrialization," but is more in keeping with a feeling that we should expect historical causation to be more subtle than blunt.

There's one other contribution of Protestantism to English industrialization that should not be missed. It has to do with the immigration of French Protestants, called Huguenots, into England toward the end of the seventeenth century. For almost a hundred years, there had been an armed truce between Catholics and Protestants in France; a declaration of limited toleration of Protestants was called the Edict of Nantes (1598). Throughout the seventeenth century, French Catholic leaders urged the king to revoke the edict. Even before Louis XIV did so in 1685, the Huguenots were emigrating to avoid persecution, whether legal or illegal. Then, streams of escapees became rivers. By 1700, some 400,000 Huguenots had sought refuge in Protestant European countries and in America.

Many of the Huguenots were skilled craft workers in textiles, glass making, and watchmaking. Lewis Paul, who participated in early spinning machine inventions was the son of Huguenot immigrants to England. This was a windfall contribution to English technical ingenuity. Immigrants from other countries had brought their technical skills to England earlier: German miners in the sixteenth century and Dutch engineers in the early to mid-seventeenth century. Of course, imported skills cannot carry the whole load; they must be well received by the natives of their adopted country if they are to be effectively deployed.

One final token of English technical ingenuity in the eighteenth century may sound far-fetched. It is offered merely as a possible straw in the wind, with an unlikely enough starting place. One of Galileo's principal exponents in England was John Wilkins (1614–1672). During the period of Cromwell's rule, Wilkins was the warden of Wadham College at Oxford. One of the graduates of Wadham during that time was Charles Morton (1627–1698), who became the master of a dissenting academy in 1662. One of the more famous of that school's pupils was Daniel Defoe (1659–1731), author of *Robinson Crusoe* (1719). Many of the innovators after 1750 in England must have read that book in their youth. Among features such as rugged individualism, *Robinson Crusoe* contains some stirring passages about its hero's skill and ingenuity:

> So I went to work; and here I must needs observe, that as reason is the substance and original of the mathematicks, so by stating and squaring every thing by reason, and by making the most rational judgment of things, every man may be in time master of every mechanick art. I had never handled a tool in my life, and yet in time, by labour, application, and contrivance, I found at last that I wanted nothing but I could have made it, especially if I had had tools; however, I made abundance of things, even without tools, and some with no more tools than an adze and a hatchet, which perhaps were never made that way before, and that with infinite labour.

A conjecture that Galileo influenced James Watt and others by such a route would be impossible to prove—or disprove. Yet there can be little doubt that the rational spirit exhibited by Galileo was similar to the spirit that inspired the inventors of the English industrial revolution.

## *Interactions between Science and Technology*

Although there may be no direct links from scientific discoveries to the early innovations of steam engines and textile machinery, it is nonetheless true that the scientific revolution preceded the industrial revolution. Here we should examine briefly some of the possible influences of science and technology on each other.

Although Francis Bacon had promoted the idea that scientific research would benefit industry, he had not been clear about how to achieve it. And when members of the early Royal Society undertook their "natural history of trades," there were no dramatic benefits to industry. Nor should one expect a study like that to revolutionize the trades; the most it might be expected to do would be to encourage them to be more rational. Even there, little was achieved in the seventeenth century.

Neither can one readily see how Newton's laws of motion would have any immediate application in the design of machinery. But perhaps we shouldn't try to apply a modern perspective to how science might influence technology. Consider some examples of what was happening in the first half of the eighteenth century.

### *Popularizing Natural Philosophy*

The success of Newton's science demonstrated the power of mathematical and experimental analysis. At Cambridge, the mathematical subtleties of Newton's *Principia* soon became the basis for the dreaded mathematical tripos examination. But Newton's work also prompted a large number of popularizations by his followers. A regular stream of expositions on "natural and experimental philosophy" issued from the presses of London. In addition, the authors of those works, and others, began giving lectures in the major urban centers of England—often traveling with a cart load of demonstration apparatus. For, instead of mathematics, these books and lectures used practical machinery to illustrate and explain Newton's physical ideas.

An early prominent lecturer in experimental philosophy was John T. Desaguliers (1683–1744), born in France of Huguenot parents, who brought him to England in 1685. After being educated at Oxford and ordained in the Church of England, Desaguliers moved to London and was made fellow of the Royal Society in 1714. He gave frequent courses of lectures on the new science throughout the rest of his life, as well as being employed as a demonstrator by the Royal Society, as chaplain and consultant on engineering projects to the Duke of Chandos, and as consultant to water works companies in London. In this latter capacity, Desaguliers became acquainted with the principles of waterwheels and steam engines. He seems to have heeded his own advice to philosophers who wished to undertake practical constructions:

> ...to have a compleat Theory, the Undertaker must understand Bricklayer's Work, Mason's Work, Mill-wright's Work, Smith's Work, and Carpenter's Work; the Strength, Duration, and Coherence of Bodies; and must be able to draw not only a general Scheme of the whole Machine, but of every particular Part; and small parts must be drawn by a larger Scale, in order to be fully examined before any thing is begun.

Four printed volumes of Desaguliers's lectures appeared between 1717 and 1744. His detailed description of the Newcomen engine, to illustrate scientific principles, was an early source of information about that engine. Desaguliers was only one of scores of authors and lecturers who were making experimental science and its technological connections available to audiences throughout England. (See engraving from one of his works on page 222.)

Such popularizing could have two major influences. As merchants and gentry became familiar with mechanical ideas, they would more readily invest spare cash in mechanical innovations. In addition, the excitement of these new fields could entice bright youths into considering them as serious occupations.

*John Smeaton (1724–1792)*

Son of an attorney, Smeaton was educated at the Leeds grammar school. Although in his youth he had made mechanical models, he began to work in his father's office at age 16. Three years later he deserted his father's profession and apprenticed to an instrument maker. By 1750 Smeaton had set up his own shop and soon became a prominent engineer. Elected to the Royal Society in 1753, he won its Copley Medal in 1759 for *An Experimental Enquiry concerning the natural Powers of Water and Wind to turn Mills, and other Machines depending on a Circular Motion*. In this research Smeaton used a model waterwheel to determine the conditions by which the greatest power could be obtained from a flowing stream.

In a later research, Smeaton made an exhaustive analysis of the workings of the Newcomen engine. From it, he confidently drew results on which to base the design of an engine ten times more powerful and at least twice as efficient as earlier designs. In both of these researches, Smeaton made a systematic study of the various relevant factors:

> His practice was to adjust the engine to good working order, and then after making a careful observation of its performance in that state, some one circumstance was altered, in quantity or proportion, and then the effect of the engine was tried under such change; all the other circumstances except the one which was the object of the experiment, being kept as nearly as possible unchanged. (J. Farey, 1827)

Although purists will refuse to call such research scientific (it not being an investigation of *nature*), they might allow that its procedures are of the same kind as Galileo applied to motion and Newton to optics.

Smeaton was a rational engineer. Faced with the contract for the Eddystone lighthouse in 1755, he tested samples and proportions of sand and lime until he found a mortar that would set even though submerged periodically by the tides. The rational attitude that made science successful in the seventeenth century contributed to the success of engineering projects in the eighteenth century.

*Smeaton's Eddystone Lighthouse.*

*Science, Clocks, and Navigation*

In the fields of astronomy and navigation, there was a continuing interdependence. The chance discovery of the telescopic effect of two lenses in Holland (technology) enabled Galileo to make great discoveries in the heavens (science). That "revolution" in astronomy was followed by decades of improvement in telescopes and in the accuracy of star charts. The Greenwich Observatory was founded in 1675 to produce star charts. Sea captains needed such charts if they were to achieve navigational improvements to benefit English overseas trade (shorten travel times, reduce insurance costs, etc.).

Accurate navigation also required precise time measurements in order to determine longitude. The basis for accurate clocks was found in scientific research, beginning with Galileo's interest in pendulums. Near the end of his life, he designed a clock where the regularity of pendulum motion could control the rate of the clock's escapement. Galileo never built that clock. But a few years later, in 1657, the Dutch scientist Christiaan Huygens did build a pendulum controlled clock he'd designed. His weight-driven pendulum clock was accurate to within a couple of minutes a day—instead of being out by as much as half an hour a day—a tremendous increase in accuracy. Immediately, people saw that a pendulum clock might be made accurate enough to be used in navigation. Only problem was, a pendulum clock—like the grandfather clock that stands in the hallway—wouldn't work well on board a ship that's pitching and tossing.

That problem was overcome by using spring driven clocks. The falling weight was replaced by a heavy coiled spring. And the pendulum escapement was replaced by a balance wheel controlled by another finer spring. It was done about 1680 by Robert Hooke—the curator of experiments for the Royal Society in London. Then, for the next 80 years or so, English clockmakers worked steadily to produce a really reliable and rugged clock. At this time of great British overseas expansion, the British government was willing to spend money to get a reliable way to determine longitude at sea.

In 1714, the British Parliament offered a prize of twenty thousand pounds to anyone who could provide a means of determining longitude to within half a degree (30 nautical miles). It took 50 years before anyone won the prize. People quickly realized that the only good solution to the problem of longitude would be to have an accurate clock. A really accurate clock: to win the prize, a clock had to be right to within a couple of minutes after sailing all the way to the West Indies and back! That meant gaining or losing no more than about a second a day. A hundred times better than Huygens' clock!

To get that accurate they needed very accurately cut gear wheels; and they had to compensate for the effect of temperature, which causes the pendulum or the balance spring to expand and contract. Huygens, the scientist, wasn't convinced that temperature changes would alter the length of a pendulum. Craftsmen, unhampered by theory, knew better. They had direct experience of metals contracting as they cooled. So, in the 1720s, a number of clockmakers devised pendulums and balance springs that compensated for temperature changes. Here, technology ran ahead of science.

However, science did help to improve clockmaking. For example, a number of mathematicians worked out the geometry of properly meshing gears—what one of them called "the form to give to the teeth of wheels so that they mesh as well as possible without jumps or friction."

But above all, to win the longitude prize, they needed superb craftsmanship. The eventual winner of the twenty thousand pounds—equivalent to some millions of dollars of our money—was a clockmaker named John Harrison (1693–1776). A carpenter's son, Harrison spent most of his adult life devising clocks to win the prize. His ingenuity and craftsmanship were unequalled. One of his early masterpieces had wooden gears and weighed more than 30 kilograms—definitely bigger than a bread box. Finally, in 1764, Harrison produced a brass clock only five inches in diameter, with an accuracy that far surpassed the requirements for the prize. In five months the clock lost only 15 seconds, a mere tenth of a second a day.

*John Harrison's fourth and fifth timepieces, which won him the British Parilament's prize.*

Within a hundred years, British craftsmen had improved the accuracy of clocks by a factor of a thousand! That's the kind of ingenuity that helped Britain make the Industrial Revolution.

*The Lunar Society*

In London, the Royal Society provided opportunities for scientists to communicate their ideas with one another and with interested laymen. Because noblemen could join without demonstrating any scientific accomplishments, the Royal Society was to some degree a high class social club, albeit with scientific interests. You could expect the fellows to be interested in priority disputes between Newton and Leibniz, or in aids to navigation, but by 1710, they had lost their earlier enthusiasm for crafts and trades.

Out in the provinces however, the quickening pace of manufactures after 1750 brought men of science and of industry together to explore their mutual interests. In Birmingham, Matthew Boulton (1728–1809) inherited a metal fabrication business from his father. In cooperation with Erasmus Darwin (1731–1802), a local physician, inventor, and poet, Boulton hosted meetings of men interested in matters technical and scientific. Called the Lunar Society because they met near the full moon so visitors to Boulton's country estate could see their way home afterwards, the number of regular members never exceeded a dozen. However, about a dozen others were closely associated with the "lunatics." The inner group included the dissenting preacher, educator and chemist Joseph Priestley (1733–1804) (who, by the way, was brother-in-law of the iron-master John Wilkinson), chemist James Keir (1735–1820), who was a manager in Boulton's employ, gunsmith Samuel Galton (1753–1832), clockmaker John Whitehurst (1713–1788), James Watt (1736–1819), Josiah Wedgwood (1730–1795), the famous potter, and Richard Lovell Edgeworth (1744–1817), author and inventor who won prizes awarded by the Royal Society of Arts.

Mainly an informal club starting about 1765, the Lunar Society held regular meetings from 1775 to about 1800. The subject matter of the meetings is perhaps less important than the associations they provided among men who were primarily industrialists with others having scientific and inventive talents. In fact, a fine distinction between science and technology was not appropriate in the eighteenth-century context. Although none of the men listed above was of the nobility, all but Galton were fellows of the Royal Society. And many of their friends were also fellows: Benjamin Franklin (1706–1790),

*Erasmus Darwin, one of the founders of the Lunar Society, and the paternal grandfather of Charles Darwin. Charles Darwin's maternal grandfather was Josiah Wedgwood one of the other founders of the Lunar Society.*

a frequent visitor from America; Richard Kirwan (1733–1812), Irish chemist; and Joseph Black (1728–1799), Scots chemist. Most of these fellows of the Royal Society were enrolled on account of a worthy contribution to scientific knowledge. Both Franklin and Kirwan were awarded the Copley medal of the Royal Society, and Wedgwood invented a new optical pyrometer.

These activities might well be considered an early kind of industrial science, by which I mean "using the methods of scientific investigation in seeking to understand and improve industrial processes." Although probably not essential to the first phase of the industrial revolution, such industrial science would eventually preserve the momentum of continued industrialization.

The Lunar Society had two other important consequences. It was the model for dozens of other such associations, frequently more formally organized, in the burgeoning cities of the industrial midlands. The Manchester Literary and Philosophical Society was founded in 1781. Although, as usual, there were a number of medical members, "the great majority of the members were either engaged or interested in the extension of Science and Art to manufacturing purposes." When Erasmus Darwin moved away from the Birmingham area, he continued his association with like-minded folk by founding the Derby Philosophical Society in 1784 (Derby is between Birmingham and Manchester). Similar societies were formed in Bath, Bristol, Exeter, Newcastle, Northampton, Norwich, and Plymouth, among others. One of the major contributions of these provincial societies was to provide technical education. Their members encouraged the revising or founding of educational institutions appropriate to the new industrial age.

Finally, the "lunatics" contributed to the scientific gene pool of the nineteenth century. Erasmus Darwin shared grandparenthood with Josiah Wedgwood of Charles Darwin (Chapter 16), and with Samuel Galton of Sir Francis Galton (1822–1911), anthropologist and biometrician.

## Conclusion

In 1700, with a population near six million, Great Britain (including Scotland, Wales, and England) was a reasonably self-supporting island economy with a significant overseas empire. About four million lived on the land, supplying the agricultural needs of the native population and providing a sizeable export trade in wool and woolen fabrics. The port of London was the terminus of trade with both Europe and the colonies, as well as for the re-export of goods between them.

In the following hundred years, the value of trade increased regularly, except during the war years, 1756–1763, after which a spurt restored the trend—reaching in 1800 about four times the level of 1700. Though the American colonies had been "lost," they continued to engage in valuable trade. Cotton was replacing wool as the major export, and soon British cottons were clothing the world. With a population of almost 11 million in 1800, no more than about six million were rural; urban population had more than doubled.

Britain was becoming the industrial workshop of the world. With adequate resources, entrepreneurs and innovators met or created market demands that could be supplied cheaply by machines that replaced human muscular forces and human finger manipulations.

Why Britain? Although France and Holland had many of the same features and opportunities as Britain, they did not give birth to the mechanization of industry. Why not?

A facile answer, only half joking, is that Holland was too small and France was too big. Between 1700 and 1800, the Dutch population remained constant at two million. Leaders in agriculture, commerce, and industry from about 1590 to 1670, the Dutch had made their whole country into a garden and had nowhere to expand locally. With their small population, they overextended themselves colonially and then had to raise taxes to support too large a naval establishment. With little room to expand, some Dutch innovators emigrated—some engineers to Britain, Christiaan Huygens to Paris.

The Dutch lacked the "critical mass" to engage in the kind of expansion required by the industrial revolution. And, in a sense, the Dutch had already done so well they didn't "need" to industrialize. France, on the other hand, had too much "mass" (inertia) to be the first to mechanize. French population increased from 22 million in 1700 to 29 million in 1800. This was largely a rural population, with little increase in urbanization over the century.

Like the Dutch, the French exported talent to Britain, too, in the form of the Huguenots, whose skills and enterprise contributed to the Industrial Revolution. And throughout the eighteenth century, the French economy tended to be operated on traditional family lines.

Also, the French colonies in North America were in the harsher northern climate, making for a small population and little demand on French industry.

Besides their major export was a luxury item—beaver pelts, rather than the more democratic cotton of the southern states.

Throughout most of the eighteenth century, France endured *social* inertia too, with a strong centralized monarchy and a large aristocracy living on the backs of the peasants. Whenever French science became a matter of practical importance, it was done under government auspices. French engineering education advanced significantly after 1730, but it was mostly for the sake of military enterprises. The *Académie des Sciences* (French equivalent of the Royal Society) consisted of scientists appointed and paid by ministers of the crown.

French industrial innovations were more likely to be in the luxury trade, where identical mass-produced products are the opposite of what is most desirable. And whenever the purchaser is spending unearned money, cost is never a consideration. On the other hand, it was notorious among the English that French peasants couldn't afford even homemade leather shoes.

Britain was favored in her resource base and in the ingenuity of her scientists, engineers, and artisans.

Can we say that the industrial revolution was caused by the scientific revolution? Only, I think, in a muted, but nonetheless positive, way. By contrast with the French, the British clearly had absorbed attitudes of a rational approach to nature and industry that enabled them to replace traditional values with those of progress, capitalism, and competition. This rational attitude had been explicit in the writings of many of the scientists of the seventeenth century.

The whole tenor of British life after 1660 tended toward individualism—in politics, business, industry, science, and religion. Many of these attitudes are so intertwined that we really should hesitate to tease out cause/effect relations. So, we'll not say that the scientific revolution caused the industrial revolution in the sense that one could, in 1700, have guessed that cotton would be a leading economic sector by 1800.

But we will say that attitudes essential to the mechanization of industry were consistent with the attitudes engendered by the new science. The search for causes is hazardous. For example, some historians want to attribute the expansion of scientific activity after 1650 to the quickening pace of the British economy. But the evidence to support that case is as shaky as for the reverse case.

Finally, in technology as in science, there are few if any rules to explain the emergence of individual geniuses. Let us consider them as "sports," mutations from the normal range of human talents. But let us also suppose that they occur at random, with no reason for there to be more or fewer in one generation than another. Then, by analogy with the biological principle of natural selection (Chapter 16), which innovations get adopted in a community depend on the nature of the environment in which they occur. The innovation, the discovery, or the invention is the product of the mutant genius.

Only if the intellectual or economic milieu is able to incorporate the innovation will it have any chance of success.

There is a good *scientific* example for this analysis—the motion of the earth around the sun. It was proposed in antiquity but rejected. The idea cropped up from time to time during the middle ages, and was always rejected. The environment was not suitable for the innovation. Then from the time of Copernicus, a handful of mutants pushed the idea in a climate of opinion that was gradually changing, until by about 1700 it was very widely accepted.

If we apply this analysis to the steam engine, we'll understand more than those historians who keep wondering why the Greeks didn't have steam engines. Although some parts of the utilization of steam could have been developed at any time, such feeble mutants as we know about fell on an entirely unreceptive economy. What would the Romans have done with a steam engine, when they didn't even seem to need waterwheels?

Now, I believe that the innovation was not as simple as all that, anyway, and that knowing about atmospheric pressure might very well have helped Savery and Newcomen. So, the local context of the Newcomen mutation influenced the innovation. (This is an intellectual "inheritance of acquired characteristics," which we reject in the biological realm.) However, the circumstantial pressure of wood shortages and availability of coal provided an environment in which Newcomen would recognize the value of the accidental leak in his cylinder, and entrepreneurs would be willing to give him contracts to build his engine.

It is in the context of such an analysis that I believe we should view the influence of attitudes in the period of the industrial revolution. Whether those attitudes were crucial to the innovators (the mutants) is not the issue; without rational economic attitudes, there would have been too few members of the general public whose acceptance was needed to *adopt* the innovation.

In the years 1700 to 1750, there was a higher concentration of such rational attitudes in England than anywhere else. Of course, attitudes alone did not make the industrial revolution: a whole *cluster* of inter-related factors must take the credit or blame.

*Early locomotives of the Great Western Railroad, England, 1837-55.*

CHAPTER FOURTEEN

# Spreading Industrial Technology

Historians will argue endlessly over which decade saw the start of the industrial revolution in Britain. But they can all agree that it was in full swing by 1800. Once this first-ever, large-scale mechanization of production facilities was under way in Britain, what should happen in other countries? Although we could not have expected to predict the *first* industrial revolution, we might suppose we could guess some of the next developments.

What will you guess? Will manufacturers in other countries rush to emulate the British? Take textiles for instance.

British machine-made textiles with perhaps one-twentieth the labor costs per unit of output will be priced well below the handmade products. With old-fashioned, non-competitive production methods, textile makers in other countries should install mule spinners and mechanical looms. And start scrambling for supplies of raw cotton and for markets for their fabrics. But the British textile firms have contracts with the major cotton growers and are already supplying the main markets.

So, while European countries and the United States had the advantage of a British model to emulate, they faced her formidable trade competition. From 1800 to 1815, European nations under Napoleon also suffered under a British blockade of Atlantic trade routes. On top of everything else, though the Europeans didn't have to invent the machinery, they did have to arrange for the capital to buy it and the labor to operate it. Both of these factors depend on local conditions, economic, political, and social. Maybe history is just too chancy for us to make successful guesses.

Let's look first at the European continent, particularly at the two regions with populations sufficient to mount challenges to British supremacy, namely, France and Germany.

## *European Society from 1800 to 1870*

The early years of the nineteenth century were dominated by the policies of Napoleon. While his armies ranged over Europe, there was too much disruption to allow industrial innovation. Although Napoleon controlled western and central Europe, Britain continued to dominate the seven seas. By the time of Napoleon's final defeat in 1815, European industry was another generation farther behind the British.

On the positive side, Napoleonic reforms eliminated customs barriers within France and reduced the number of German principalities from more than 300 to about 40, of which Prussia was the largest, with about half the population. In France, the aristocratic domains were given to their peasant tenants. Feudal serfdom was gradually eliminated in Germany, too, but more slowly.

In both regions, the move to the towns was considerably slower than in Britain. In England and Wales, half the population was urban by 1851; that did not occur in Germany until 1891, nor in France until 1931. With relatively larger rural populations on generally smaller holdings than in Britain, agricultural innovation occurred more slowly on the continent. However, although 75% of the French lived in rural areas in 1851, only 53% of the population was employed in agriculture. That means that trades and manufactures tended to be rurally dispersed in France for a much longer time than in Britain. This kept enterprises small, making capital accumulation and innovation more difficult. At the same time, it made cheap labor available at prices that could often encourage the continuation of hand manufacturing.

The continental attitude to labor was not as "modern" and unfeeling as in Britain. Social traditions in France and Germany encouraged group cooperation and responsibility, as opposed to the individualism that supported money-making practices in Britain. Entrepreneurs were more likely to be hated than envied. As a result, there was more government intervention in industry on the continent than there had been in Britain.

Government activities in Europe took several forms. In order to justify any additional textile production, for example, import duties were imposed on the British products. Behind these tariff walls, local manufacturers could compete with imports cheaper than they could produce. In some cases, governments opened manufactories or provided capital for them. Since France and Germany are not islands, their internal transportation facilities were extremely important for advancing trade. As governments saw more need for roads, canals, and, later, railways, their investment in those enterprises assisted the process of industrialization.

Germany is a somewhat special case in that no unified German nation existed until 1871. The area called Germany can be identified as such before then (though rather inexactly) by referring to regions that are primarily German speaking. The political unification under the King of Prussia was not a single step, but for our purposes its most important predecessor was the establishment of a German customs union in 1834. That move, initiated by Prussia, signalled the importance of trade matters to German governments. It gave manufacturers freer access to local markets, in default of foreign markets in which they could not as yet compete.

For her part, Britain tried to protect her industrial dominance by prohibiting the export of machinery and machine makers. These bans were never totally effective and were gradually lifted; the last, on textile machinery, was ended in 1841.

*"By Industry We Thrive: 'Progress' Our Motto" An inspiring engraved poster extolling the virtues of industry. Smithsonian Institution.*

Through the first half of the nineteenth century, French industry was generally ahead of German industry. But the latter was expanding at a greater rate and began to surpass the French after 1860. Then, after their stunning loss in the Franco-Prussian war (1870–71), the French ceded their prime industrial north-east to the new German Empire. With the two national economies having been roughly equal in 1865, that transfer gave the German economy a significant lead over the French. Thereafter, German industry outpaced the French at an increasing rate.

## *The United States of America*

The former British colonies along the eastern seaboard of North America had a population of 2.5 million at the time of their revolutionary war. Yet they looked across the Appalachian mountains to a potential nation with an area equal to all of Europe. During the nineteenth century, Americans proceeded to fill that nation, increasing in population more than 12 times in 100 years, reaching 76 million in 1900. With all that territory, urbanization was rather gradual, reaching 50% about 1920.

For most of the time from 1775 to 1900, Americans had access to cheap agricultural land. Thus, wages in industry had to be high enough to keep workers from being tempted to move west to homestead. Indeed, there's a paradox here: if wages got higher, workers could save sooner to pay for a plot of land in the new territories. The turnover in labor must have been considerable. Really skilled labor might be kept at home with sufficiently high wages. Hence, there was a desire among American businessmen to make industrial processes mechanical as quickly as possible. Pay high wages to skilled personnel to design and build machines that reduced the need for unskilled labor.

As far as traditional attitudes were concerned, the challenges of a new land tended to encourage a spirit of innovation and adventure. The Americans largely lacked the social blocks to industrial expansion that impeded the French and the Germans. Besides, being of British stock, they were more likely to share British attitudes than continental ones.

### *Textiles in Europe and America*

The French textile industry developed initially in a very different way from the British. Before the 1790s, the greatest profits were to be made in fine silks and linens for aristocratic customers. And thus it was in the luxury trade that manufacturers had sufficient capital to introduce mechanization. One clear sign of this process is the invention in France of a device for creating elaborate patterns in weaving by a special mechanism. This machine, known as a Jacquard, after its final developer, J.M. Jacquard (1752–1834), used a sequence of punched cards to control the draw cords that lifted particular sets of threads in the warp for each pass of the weft. Put into use about 1804 on hand looms, the Jacquard was later adapted to power looms.

The full mechanization of the French textile industry took many years. And it required imported British workers to transfer their skills to their local pupils. By the 1830s, the French cotton industry had reached about a quarter of the size of the British, which by then had seven times the output of 1800. Also in the 1830s, the American cotton industry was about the same size as the French, while the German was only about 15% of the French.

In the following decades, continental cotton manufacturing expanded somewhat faster than in Britain, so that in 1873, Germany and France produced about 35% as much as Britain. By then, the French output had fallen to about two-thirds of the German.

In the United States, cotton manufacturing became concentrated in the New England states. Despite having raw cotton produced within their own nation, the Americans needed a protective tariff to be able to compete with British cottons. With their expanding population and economy, American cotton manufacturing reached half the British rate by 1870, and surpassed it before 1900.

In the first decade of the twentieth century, worldwide consumption of cotton was roughly a hundred times greater than it had been a century earlier, while world population did not so much as double. We should realize that

*A handloom with Jacquard action.*

by 1900, the use of cotton had gone far beyond dresses, shirts, and underwear. It also included sheeting for bedclothes, ticking for mattresses, chintz for curtains, gauze for bandages, canvas for sails and tents, blue denim for overalls, etc. The cheapness of cotton made it useful everywhere. Cotton was the "plastic" of the nineteenth century.

The distribution of the consumption of raw cotton by manufacturing nations in the early 1900s was roughly as follows:

| | |
|---|---|
| United States | 30 % |
| Great Britain | 25 |
| Germany | 10 |
| Russia | 8 |
| France | 7 |
| Rest of Europe | 12 |
| Rest of World | 8 |

Although Russia (with a population 3.5 times that of France) was outproducing France in quantity by 1900, her output of fine cottons was relatively small.

*Watch factory work room. Waltham, Massachusetts. Machine tools at each work station are attached to conveyor belts powered by water turbine engines. Smithsonian Institution, Collection of Business Americana.*

## *Powering Industry*

In 1840, when railroads were still in their infancy, Britain had a total of 460 MW of power in steam engines, mostly stationary engines driving pumps in mining, bellows and forges in metallurgy, and mules and looms in textiles. Yet, despite the predominance of steam, a third of woolen machinery and an eighth of cotton machinery were still being driven by waterwheels. On the continent, as you might expect, the proportion of water-driven equipment stayed higher longer. This was not only because of technical and economic backwardness, but also because France and Germany had many fast streams to exploit and had smaller enterprises, which could get by with waterwheels operating at less than 10 kW.

In France, the use of hydraulic energy was given a boost by an important technological innovation, the water turbine. In its simplest form, a reaction turbine has hollow vanes supplied with water from a vertical pipe. The vanes move in reaction to jets of water streaming out through holes in the vanes. The principle is very much like the rotating lawn sprinkler. With careful design, a reaction turbine can be a very powerful package. Following some theoretical work in the eighteenth century, Bénoit Fourneyron (1802–1867) produced, in 1837, a water-driven turbine weighing 18 kg, 30 cm in diameter, delivering 45 kW at 2300 revolutions per minute, with an efficiency of 80%. (The maximum possible efficiency of an *overshot* waterwheel is 75%.)

In less than a decade, Fourneyron turbines were also being used to advantage in the textile mills of Massachusetts. These turbines reduced the importance of steam power in industry in France and the United States as long as coal was scarce and expensive.

Of course, some steam engines had been exported from Britain even as early as the 1720s. Yet they did not appear in great numbers outside Britain until well after 1800. By then, high-pressure engines were seen to be superior to the atmospheric engines of Newcomen and Watt. In the United States, Oliver Evans (1755–1819) obtained a patent for a high pressure engine in 1804. Before 1840, British numerical superiority in steam power was surpassed by the Americans, with almost 40% of their engines being used for transportation, especially on the larger rivers of the central United States. By comparison, the French and Germans in 1840 had steam power capacities of 15% and 7% of the British, respectively. In following decades, particularly with the rapid growth of railroads, the Germans and French increased their steam power plant more rapidly than the British. Nevertheless, just before 1900, when Britain had passed 1000 MW of steam capacity, German capacity was 58% of the British, French was 43%, and the American was 132% of the British.

## *Coal and Iron*

Naturally, steam engines that run on coal will be installed more rapidly where coal is cheaper. That depends partly on where the coal comes from: imported coal is likely to be more expensive, if only because of transportation costs. In France and Germany, coal deposits relatively near the surface are scarcer than in Britain. In addition, French and German forests and woodlots were not being depleted as rapidly as those in Britain. As a result, and also because this industry continued to be smaller and more widely dispersed on the continent than in Britain, charcoal was used there for smelting iron considerably longer than in Britain.

The deeper coal deposits in northern France and western Germany did not begin to be exploited until after 1830. And, in their smelters, they did not use coke more than charcoal until after 1850.

In 1865, Britain was still producing about 50% of the world's pig iron, with the United States and Germany each contributing about 10%. For the rest of the century, Britain's output increased at less than 2% per year, while the American increased at almost 10% (surpassing Britain just before 1890). Germany's output increased at a little over 6% per year, and surpassed Britain just after 1900. In 1907, the United States was producing 43% of the world's pig iron and Germany 21%. Britain's share had dropped below 17%.

Clearly, by 1900, if not before, Britain's dominance in heavy industry had passed to the United States and Germany. We cannot understand the political events of the next quarter century without realizing the importance of this economic fact. From the time of Germany's political unification in 1871, her

*Abraham Lincoln travelling to Gettysburg, November 1863.*

leaders played a conscious game of technological and economic catch up. By 1900, the Germans began to think they'd won, which made a great war thinkable, if not ultimately, winnable. Meanwhile, in that lusty nation across the Atlantic, the Americans just kept on exploiting the resources of a whole continent.

## *Canals and Railroads*

As in Britain, canals led to early improvement in the transport of goods in continental countries. Yet like their experience in other industries, the French and Germans lagged behind the British in canal construction. In 1800, each of them had about 1000 km of inland waterway, which the British had possessed 100 years earlier. Then, in the next 50 years, both France and Germany constructed about the same 2500 km of canal the British had built during the eighteenth century. After 1850, although barge traffic on the Rhine continued to be important, the subsequent history of transportation in France and Germany takes us to the railroad.

With her head start, Britain had built more than 10,000 km of rail lines by 1850. By then, the Germans had 55% of Britain's lines, and the French less than 30%. In the following twenty years, the French outbuilt the other two, adding 13,900 km of line, while the British added 13,700 km, and the Germans 11,600 km. The French system was almost as large as the German in 1869.

But then, after the Franco-Prussian war, the German victors surged ahead. By the early 1900s, Germany with two-thirds of the area of France, had a rail network 20% larger, which by then was almost 60% larger than the British.

This greater density of rail installation was essential to the development of Germany's continental economy to the level where it could challenge the British, particularly in industries where the materials to be transported are heavy and bulky. Just imagine trying to supply a steel mill with a train of pack horses!

In the United States, transportation started from a very different base than continental Europe at the same time. Colonial Americans had stuck pretty close to the coast and used the Atlantic for communications back to the old country. Communications among colonies developed slowly. After 1776, there were more reasons for the colonies to communicate with one another. While coastal shipping remained important, inland exploitation required other modes of transportation.

Turnpike construction began in Pennsylvania in 1792. By 1810, 300 turnpike companies had been chartered in New York and New England. One of the most striking of the early roads was the Cumberland Road or National Pike. Starting in Cumberland, Maryland, it reached 210 km to Wheeling, West Virginia by 1818, and was extended westward through Columbus and Springfield, Ohio, past Indianapolis and into Illinois by 1838. Spanning a distance of almost a thousand kilometers, the National Pike averaged 20 m in width. Traffic was heavy, with passengers carried in Concord coaches and goods in Conestoga freighters drawn by six or eight horses.

Water transport to the west was facilitated by the 584 km Erie Canal, completed in 1825. This opened up the Great Lakes and their vast hinterland to eastern traffic. Other canals linked navigable rivers to create an important water route for the goods of American industry. By 1850, some 6000 km of canals had been built.

Traveling on the rivers and canals of America were steamboats and tugs, beginning with Robert Fulton's *Clermont* in 1807. Although Fulton (1765–1815) had been preceded in trials by Oliver Evans and John Fitch, his was the first commercially successful steamboat. The steamboat was particularly important on the Mississippi and Ohio Rivers as it provided sufficient power to allow boats to travel upstream.

*Robert Fulton's Clermont, the first commerically successful steamboat. On the Hudson river in New York State, around 1807.*

Railroad construction began in America almost as soon as in Britain. More than 1700 km of line had been built by 1835. States east of the Mississippi were covered with very nearly 50,000 km of rail lines by 1860. After the Civil War, the railroad won the west, as we've seen in countless old movies. In 1869, the Union Pacific, out of Omaha, linked up with the Central Pacific, out of San Francisco, almost 3000 km away. In 1886, the Canadian Pacific Railway completed a single line from tide-water to tide-water, Montreal to Vancouver, a distance of almost 4700 km.

In North America, the railroad provided the transcontinental bonds that tied nations together. The engineers and laborers built their ribbons of steel across hostile and desolate territory in order to link new settlements on the west coast with the established economies of the east. After the link was made, it enabled the territory in between to be gradually brought under control. California was the thirty-first state (of 50) to join the union; British Columbia was the sixth province (of 10) to join the dominion.

By the early 1900s, the United States had 381,000 km of railway. With its vast territory that represented a *coverage* of a mere 0.5 km of rail line for each 10 square km of land area. However, it also provided a transportation *opportunity* to its citizens, with 5 km of rail per thousand of population. For comparison purposes, the same values are given in the table for countries in Europe at the same time.

*The ceremony of driving the last spike in the completion of the Canadian Pacific Railway line providing rail service across Canada from "sea to sea," i.e., Montreal to Vancouver, 1886.*

| | Coverage | Opportunity |
|---|---|---|
| United States | 0.5 | 5.0 |
| Germany | 1.6 | 1.4 |
| British Isles | 1.2 | 0.9 |
| France | 0.9 | 1.2 |
| Italy | 0.6 | 0.5 |
| Russia | 0.1 | 0.6 |

These indices provide a very rough indication of economic capacity as far as transportation facilities are concerned. They show, for example, that relative both to area and population, German rail lines were ahead of both Britain and France. Russia with its large area (Siberia not included in the index) had low coverage, but relative to population provided an opportunity slightly greater than in Italy.

## *Conclusion*

During the nineteenth century, other countries began to respond to the challenge of industrialization that Britain had provided. On the continent, any serious response was delayed until after 1815, and traditional attitudes in France and Germany meant that the initial response was slow and awkward. By 1850, both countries had industrial programs well under way. In both countries, industrialization was more a matter of governmental policy than it had been in Britain in the eighteenth century.

By 1900, the German industrial plant had surpassed the British in some (but not all) fields. In the latter half of the nineteenth century, some British enterprises were a hundred years old, often still using time-tested methods, and even century-old steam engines. German entrepreneurs were "new men" with driving needs to succeed ("Number 2 tries harder"). They were supported by a general and technical educational system far ahead of the British, and by investment bankers who were more interested in the domestic scene than was true in Britain. The British had been leading the way for most of a century because of their attitude of *laissez-faire*; now, when government intervention would have helped, they were not prepared to move quickly enough to sustain their lead. To the Germans it looked as if the British had "shot their bolt," and now it was Germany's turn to lead.

American industrial enterprise, on the other hand, was what you might expect from British colonists landing on a virgin coast. With a whole continent to fill, the population of the United States in 1900 was close to double the roughly 40 million each in Britain, France, and Germany. Even without export markets, the Americans had a growing home market to provide all the demand that grasping merchants could wish for.

We've concentrated in this chapter on the same industrial factors that revolutionized the British economy: textiles, steam, iron, and transportation. Europeans and Americans did use British models to develop those factors of their economy. However, after 1850, new industries like steel, chemicals, and electricity helped to support the German and American challenge to British economic supremacy.

*John A. Roebling's railroad bridge across the Niagara River (1855).*

*The great lens of the Paris Academy of Sciences (1782). To produce intense heat for chemical experiments.*

## CHAPTER FIFTEEN

# Chemistry and Atoms

Broadly speaking, chemistry is the science of materials. Early chemical investigators were puzzled about what things were made of and how substances combined to form different substances.

Even in ancient times, artisans knew quite a lot about the properties of materials and how they behaved. Cooking, potting, the smelting of metals—all involve chemical processes. The alchemists mentioned earlier (Chapters 4 and 6) followed and extended these crafts, and also tried to make a consistent theory for chemical processes.

Although we're inclined to consider alchemical theories to be rather juvenile, they do demonstrate that men were speculating about the nature of materials and their transformations. And throughout the long history of alchemy (down to the seventeenth century), they added progressively more knowledge about the behavior of substances.

By 1650, developments in other sciences were encouraging men interested in materials to match their theories of the structure of matter more closely with their experience. The science of chemistry grew gradually out of alchemical knowledge and the principles of the "experimental philosophy," such as was promoted by Francis Bacon (Chapter 11).

Because the materials in the world around us are so complex, chemists took more than a century to develop a theory of chemical elements sufficient to fit their experience. Yet, even with limited theories, they continued to add to their understanding of chemical reactions. As we'll see, this was an international effort, with Scottish, English, French, and German participants. In the 1780s, a major contribution came from the French chemist Antoine Lavoisier, with a list of elements based on his oxygen theory of combustion. After 1800, chemists really got down to business. They extended and corrected Lavoisier's work; and John Dalton invented the atomic theory of matter. In the 1850s, chemists began to examine the chemical composition of animal and vegetable matter, and founded organic chemistry, with its structures of carbon chain molecules. By the 1870s, Dmitri Mendeleev's periodic table of the elements put inorganic chemistry firmly on the path still essentially being followed today.

## *Chemistry Emerges from Alchemy*

The gradual transition from the craft of alchemy to the science of chemistry involved generalizing from particular experiences, and finding a vocabulary in which to express those generalizations. For example, vapors that come from heating sulfur and heating greenstone are both what we'd called sulfur dioxide, but to the alchemists they were respectively "spirits of sulfur" and "spirits of vitriol." Different terms because they arose from different materials or processes. And, by calling them "spirits," the alchemists hardly distinguished those vapors from souls, ghosts, and other spirits. They lacked not only a precise theory of spirits, but of states of matter in general.

Indeed, Aristotle's elements, earth, water, air, and fire, were as much an attempt to understand states of matter as the nature of substances, but did little more than give them names. Until after 1700, chemists were inclined to consider different vapors as modifications of air, rather than as distinctly different materials. This traditional identification continued long after Johann Baptista van Helmont (1577–1644) introduced the term "gas." (Van Helmont's term, "gas," is a Dutch transliteration of the Greek word, χαος, the "void," from which we get the word "chaos" in English.) Van Helmont's work added to our knowledge of gases, but was still limited by identifying the gases from their sources. Van Helmont had such names as wild gas (perhaps carbon dioxide), windy gas for the air, fat gas for methane, dry gas, smoky gas, and so on. Occasionally, he did recognize that two very different sources could produce the same gas.

## *Early Chemical Industry*

From ancient times, small amounts of sulfuric acid had been used for bleaching fabrics. As the output of cloth in Britain increased significantly during the eighteenth century, the demand for acid also increased. Frequently, the cottagers in northern England used the lactic acid from sour milk as a bleaching agent. They soaked their cloths in sour milk and then laid them out in the fields to take advantage of the bleaching action of the sun as well.

That was a time-consuming operation and could be improved by the use of a stronger acid. Before long, the process of manufacturing sulfuric acid was industrialized in a series of steps until its price had fallen to one-hundredth of its former value.

When the fumes from burning sulfur are dissolved in water, sulfurous acid is formed. To obtain sulfuric acid, the sulfur fumes must be further oxidized. Traditionally, the oxidizing agent was provided by the burning of a small quantity of saltpeter (potassium nitrate) along with the sulfur. Near the end of the seventeenth century, the French chemist, Nicolas Lemery (1645–1715), described the production of a few liters of sulfuric acid by burning a couple of kilograms of sulfur and a tenth as much saltpeter on an earthenware stand in the center of a large pot of water. The absorption of the vapors into the water requires a large surface of water.

*Bleaching cloth with sour milk.*

This process was enlarged by a London quack-doctor, Joshua Ward (1685–1761). He used enormous glass vessels with enough water in the bottom to produce about 15 L of sulfuric acid in each operation. Later, John Roebuck (1718–1794) expanded the quantities much more. Roebuck, a physician and chemist, was James Watt's first patron in the 1760s. In 1746, Roebuck industrialized Ward's process by constructing a cubical lead chamber about 2 m on a side. He put water on the floor of the chamber to a depth of a few centimeters, and then had the burning mixture of sulfur and saltpeter run into the chamber in an iron wagon on rails.

Such large-scale production dropped the price of sulfuric acid sufficiently that it became widely used in the textile industry.

The industrial production of sulfuric acid did not depend on any new developments in chemical theory. Nonetheless, an increasing dependence of industry on chemical processes did gradually encourage chemists to pay more attention to ways in which they could understand and improve those processes.

## *A New Theory of Combustion*

In the early 1700s, new theories and new experiences began to lead to a deeper understanding of chemical processes. Two particular developments would eventually be combined and transformed by Lavoisier to produce his new system of chemistry. Those developments were a new theory of combustion and the discovery of new gases.

*Georg Ernst Stahl.*

In Germany, the physician Georg Ernst Stahl (1660–1734) introduced a new theory of combustion. Following the interests of his teacher J.J. Becher (1635–1682), Stahl began to concentrate on an explanation of the phenomena associated with burning. Rather than simply dismissing it as the work of fire particles, Stahl wanted to account for the fact that metals gained in weight when they were burned. That is, the "ash" of a metal like lead (called a *calx*) weighed more than the original metal.

Stahl proposed that a "fire principle," which he called *phlogiston*, was expelled from the metal when it was heated. Of course, that meant that for him the metal was a compound of the calx and phlogiston. Although the air did not enter into the reaction in Stahl's theory, it did have a role to play. For example, when a candle is burned in a jar and extinguishes after a while, Stahl explained that the air was absorbing the phlogiston until it became saturated and could absorb no more.

Since a metal could be recovered from its calx by roasting with charcoal, Stahl proposed that charcoal (carbon) was particularly rich in phlogiston. Later chemists would occasionally even identify carbon as actually being phlogiston.

Stahl's phlogiston theory seemed adequate to explain some of the burning puzzles that bothered chemists. However, they had some trouble making the theory consistent. Burning wood seemed to lose weight, while burning metals gained weight. Sometimes they said phlogiston has negative weight; at other times they suggested that other principles must be involved.

## *Collecting Gases*

While Stahl and others were working out the details of the phlogiston theory of combustion, English experimenters were gaining more experience with gases. For a long time, experimenters had rather neglected gases. That was because they couldn't see the gases and had no way to collect and contain them. These are not vapors, like the ones alchemists condensed by distilling, but substances called gases that can only be condensed at very low temperatures—far below zero. In some chemical reactions a gas is produced, which just goes off into the surrounding air. In other reactions, one of the gases in the air combines with a solid to produce a new substance. Early chemists didn't even realize those things were happening.

Even if they did suspect something, investigators had no way to distinguish one gas from another. Their theory told them they were still dealing with air, but that somehow it had changed. It was equivalent to thinking of

*A pneumatic trough belonging to Joseph Priestley.*

changing water by dissolving different substances in it. For example, blue vitriol turns water blue. Sugar makes water sweet, and so on. So, if you burned sulfur, you'd get sulfurous air. And if you burned phosphorus, you'd get phosphorous air. Since sulfur occurs free in nature, it was a common substance with the alchemists. Phosphorus was not discovered until 1670. It was one of the substances left behind when chemists distilled human urine. God only knows why they did that!

*Stephen Hales (1677–1761)*

Chemists' knowledge of gases began to improve after 1727, when Hales, an English clergyman, described an apparatus for collecting gases. Suppose you have a chemical reaction that you think gives off a gas—perhaps you've seen bubbling when heating a couple of substances together. To follow Hales's technique, you'd enclose your reaction in a container with a hose or tube running out of the top. Then you'd lead the tube into the mouth of a bottle of water sitting upside down in a trough of water. As the gas comes out of the reaction, it goes through the tube, bubbles up into the bottle, and displaces the water in the bottle. Soon, you have a bottle full of gas. This apparatus is called the pneumatic trough, and it's still used today in school chemistry labs.

Hales was sure plants needed air to grow, and suspected that, somehow, air was incorporated into the substance of a plant. So he heated plant material and collected the gas that was given off. Seeing that a gas was given off confirmed Hales' suspicion, but he didn't go any farther. He never questioned whether that gas was ordinary air or something else.

## *Discovery of New Gases*

### *Joseph Black (1728–1799)*

The clear identification of gases other than air began about 1750 with the work of Black, a Scottish chemistry professor. He identified what we call carbon dioxide, but he didn't actually collect any of it. A brief description of his procedure shows an interesting aspect of science: how can we say we've identified something, even though we can't see it or collect it?

*Joseph Black.*

Black was investigating some of the properties of alkaline substances. He knew from earlier work that alkalis are neutralized by acids. When they're mixed, acids and alkalis combine to form salts, which are neutral, neither acidic nor alkaline. For example, the alkali, sodium carbonate, combines with the acid, hydrochloric acid, to produce sodium chloride, ordinary table salt. While that process is going on though, there's a lot of bubbling; that is, gas is also produced. Black called this gas fixed air, air that had been fixed in (or attached to) the alkali. So, sodium carbonate plus hydrochloric acid gives salt and fixed air.

Then, instead of mixing the sodium carbonate with acid, Black heated it. The alkali changed its form, becoming what we call sodium oxide, but in Black's time it was simply called soda. When Black added hydrochloric acid to the soda (or sodium oxide), he again got salt (sodium chloride), but without any bubbling; no fixed air was given off. Black decided that the previous heating had driven off the fixed air from the sodium carbonate; that, in fact, sodium carbonate consisted of a combination of soda and fixed air. He felt fixed air was not merely mixed up with the soda. Because soda and sodium carbonate have some different properties, he saw the sodium carbonate as an intimate "compound" of soda and fixed air, needing a lot of heat to separate them. By repeating the experiments and carefully weighing the various solid compounds involved in the two reactions, Black showed that both methods produced the same amount of salt.

Black decided that meant the same amount of fixed air came out of the sodium carbonate, whether it was driven off by heat or by the reaction with acid. Let's put numbers to it. Black took two equal portions of sodium carbonate (say a hundred grams) and treated each by a different process. First, he poured hydrochloric acid on one 100 g portion of sodium carbonate until there was no more bubbling. The reaction of the alkali and the acid produced salt. When Black dried the salt he found it weighed 109 g. Remember that 109. Then Black strongly heated the other 100 g portion of sodium carbonate until all he had left was soda, 58 g of soda. The other 42 g must have been fixed air that'd been driven off. And when Black poured hydrochloric acid on the 58 g of soda, he got—again—that same 109 g of salt. Everything was consistent. Even though he might have used Hales's pneumatic trough, Black did these experiments without ever collecting the fixed air. He was confident his results were correct because he'd carefully weighed all the solid compounds involved in his reactions. He didn't have to weigh the acid, and he couldn't weigh the fixed air. He assumed that mass was conserved.

Black published the results of his experiments in 1756. During the next twenty years, other investigators in Britain collected other gases and identified them. Henry Cavendish (1731–1810) found hydrogen in 1765, though he called it "inflammable air."

*Joseph Priestley (1733–1804)*

*Joseph Priestley.*

Priestley isolated oxygen in 1774, though he called it "dephlogisticated air" or "respirable air," meaning the part of the air from which we and animals get support for our lives when we breathe.

Born of humble parents in Yorkshire, Priestley was a leading activist in religious dissent, a Unitarian clergyman, theologian, teacher, and publicist. He turned to chemical investigations in the early 1770s and collected various gases. He used mercury in his pneumatic trough instead of water, because gases are much less soluble in mercury. Priestley produced gases by heating substances or by pouring acids on solids. Working consciously within the paradigm of the phlogiston theory, he concluded from some experiments that ordinary air had several constituents:

> There are, I believe, very few maxims in philosophy that have laid firmer hold upon the mind, than that air, meaning always atmospherical air (free from various foreign matters, which were always supposed to be dissolved and intermixed with it) *is* a simple elementary substance, indestructible and unalterable, at least as much so as water is supposed to be. In the course of my inquiries, I was soon satisfied that atmospheric air is not an unalterable thing; for that the phlogiston with which it becomes loaded from bodies burning in it, and animals breathing it, and

*Priestley's home being destroyed by an mob angry because of Priestley's republican sympathies. Priestley eventually moved to the United States.*

> various other chemical processes, so far alters and depraves it, as to render it altogether unfit for inflammation, respiration, and other purposes to which it is subservient; and I had discovered that agitation in water, the process of vegetation, and probably other natural processes, by taking out the superfluous phlogiston, restore it to its original purity. But I own I had no idea of the possibility of going any farther in this way, and thereby procuring air purer than the best common air. I might, indeed, have naturally imagined that such would be air that should contain less phlogiston than the air of the atmosphere; but I had no idea that such a composition was possible.

Priestly went on to describe how, in 1774, he had placed a calx of mercury at the surface of mercury in a tube and heated it by concentrating the sun's rays with a lens. He tested the gas that was evolved and found that it supported the vigorous burning of a candle. Initially, he thought the gas was related to an acid vapor he'd previously worked with. Only about a year later, after other experiments, did he decide that he had actually produced the most respirable part of the air. He found, for example, that a mouse enclosed in this gas lived twice as long (about half an hour) as in ordinary air, and, in fact, revived when released. Priestley called his new gas *dephlogisticated air*, because he assumed that it was phlogiston that fouled ordinary air, and this new gas was eminently pure.

Priestley's religious and political activities, and his sympathy with the principles of the French revolution, led to his house near Birmingham being wrecked by an angry mob in 1791. In 1794, he emigrated to America, where he died, still a proponent of the phlogiston theory.

## *Antoine Laurent Lavoisier (1743–1794)*

Priestley got his respirable air by heating mercuric oxide. But it wasn't called an oxide at that time. Priestley called the oxide, a calx. He said the calx absorbed phlogiston from the air, and the resulting air was thereby dephlogisticated. We now say the heat separated the oxygen from the mercury in the mercuric oxide. Getting from dephlogisticated air and calxes to oxygen and oxides was the work of Lavoisier.

*Antoine Lavoisier and his wife and collaborator, Marie-Anne Pierrette Paulze.*

Lavoisier belonged to a wealthy family in Paris and engaged in numerous public service activities. In addition, he bought a tax collection franchise, for which involvement he became a victim of the Terror in the second phase of the French Revolution. Lavoisier devoted much time to science. He received a good education and undertook his chemical investigations with an analytical turn of mind, more like a physicist than like the majority of chemists who came from a medical or pharmaceutical tradition; that is, he was more interested in quantitative results, whereas the others concentrated on qualities.

In 1772, Lavoisier found that the product of burning phosphorus (phosphoric acid vapor) weighed considerably more than the original phosphorus he'd started with. However, when the phosphorus was burned in a sealed container, there was no change in weight; but when the seal was broken air rushed in. That convinced Lavoisier that the reaction involved a combination

*Lavoisier being arrested by revolutionaries. Being a tax collector, he was associated with the monarchy. He was later guillotined.*

of phosphorus with some part of the air. He assumed that mass was conserved.

In 1775, following the work of Priestley and others, Lavoisier performed a classic experiment with mercury enclosed with air in a container where he could measure the level of the mercury. Using a lens, Lavoisier heated the surface of the mercury until some of it calcined (that is, produced mercury calx, or "oxidized," as we would say). The volume of air in the vessel decreased by a measured amount. Then, at a lower temperature, the application of heat caused the mercuric calx to become mercury again. And exactly the same volume of gas was evolved as had originally been absorbed. As Priestley had done, Lavoisier soon showed that the part of the air involved was the "eminently respirable" part.

In the early 1780s, Lavoisier decided that Priestley's dephlogisticated air should be called **oxygen**, a term he coined from Greek words meaning "acid former." This was because the products of burning substances like sulfur and phosphorus formed acids when dissolved in water. And thus, with Lavoisier, we arrive at the theory of combustion we're all familiar with. Instead of a burning substance giving off phlogiston into the air, we have the substance taking up oxygen from the air. The old calxes that yielded metals when roasted with charcoal turn out to be oxides.

*The Composition of Water*

To see the contrast between Lavoisier's new oxygen theory and the old phlogiston theory, consider a specific case, one that had an interesting technological spin-off. It involves the composition of water. Originally, water had been considered by Aristotle to be one of the four elements. Then, after Henry Cavendish had discovered flammable air, he found that when he burned it, the only product was water. So, did Cavendish say water is a compound, containing inflammable air? He did not!

Cavendish knew from Priestley that the atmosphere contained dephlogisticated air. So, he said that flammable air was a compound of water and phlogiston and that Priestley's gas was really dephlogisticated water, that is,

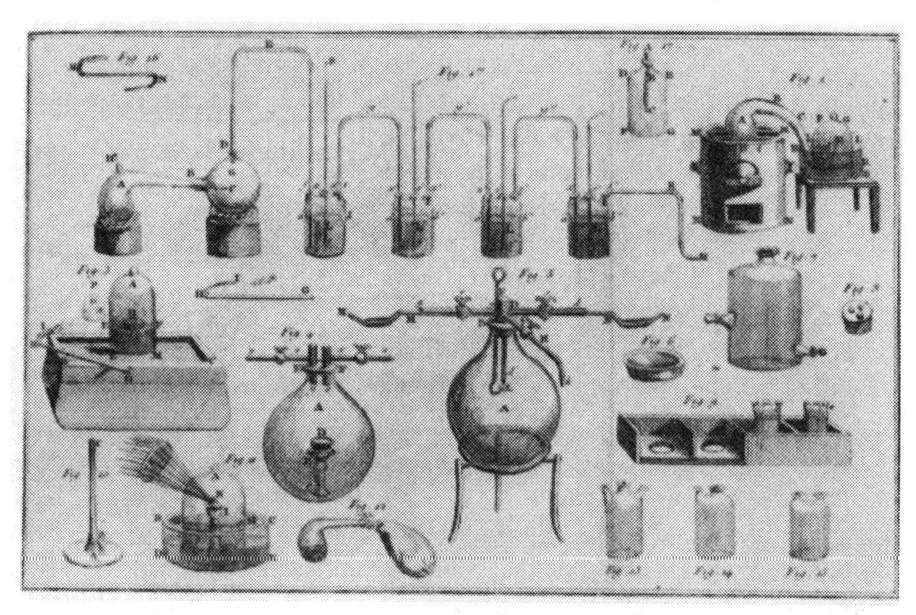

*Apparatus used by Lavoisier. From Lavoisier's* Traité Élémentaire de Chimie *(1789). This illustration, like others in the book, was drawn by Mme. Lavoisier.*

water with the phlogiston removed. So, Cavendish said, the two gases we call hydrogen and oxygen were really two forms of water, one rich in phlogiston, the other poor in phlogiston. When the two gases combined, it was just phlogiston-rich water combining with phlogiston-poor water. The surplus phlogiston in the one cancelled the deficit in the other, and water was left behind. This way, Cavendish could still consider water to be an element.

Lavoisier, with his new oxygen theory, described Cavendish's experiment of combining the two gases the same way we do. Hydrogen and oxygen combine to form water, which makes water a compound. Yet, still only in theory. How could Lavoisier find evidence to support his idea that water was a compound and not an element?

Because of Cavendish's interpretation, the synthesis of water from oxygen and hydrogen could not be considered conclusive. Lavoisier would have to analyze water, that is, break it down into oxygen and hydrogen. In fact, the actual production of both hydrogen and oxygen in gaseous form did not occur until after 1800, using electrolysis. Lavoisier figured out a way to extract oxygen from water to form an oxide with iron, and release hydrogen. As long as he could establish that the product with iron was a genuine oxide (i.e., identical with the calx of iron he could produce by burning iron in oxygen), the test he devised would be an adequate confirmation that water is a compound of hydrogen and oxygen.

Lavoisier put a measured amount of distilled water into a vessel where he could boil it. The steam from the vessel passed along a glass tube containing thin strips of iron. The glass tube was laid across a hot furnace and kept just at low red heat. The steam then passed through a cooling apparatus and on to a container where gases or condensates were collected. Lavoisier described the results in his *Traité Élémentaire de Chimie* (Paris 1789), as translated by Robert Kerr and published in Edinburgh in 1790:

> ...we obtain 416 cubical inches, or 15 grs. of inflammable gas, thirteen times lighter than atmospheric air. By examining the water which has been distilled [during the experiment], it is found to have lost 100 grs. and the 274 grs. of iron confined in the tube are found to have acquired 85 grs. additional weight, and its magnitude is considerably augmented. The iron is now hardly at all attractable by the magnet; it dissolves in acids without effervescence; and, in short, it is converted into a black oxyd, precisely similar to that which has been burnt in oxygen gas.

Once Lavoisier got hydrogen out of water, he had a new technique for manufacturing hydrogen. Previously he had to get it from pouring acid over a metal like zinc. But acid was in demand for making explosives. Now, the government found a way to take advantage of Lavoisier's method of getting hydrogen out of water. Ballooning had begun about 1783 (with the Montgolfier brothers, though they used hot air), and the French saw uses for balloons in military reconnaissance.

So Lavoisier devised a field kit for manufacturing hydrogen on the spot. All the soldiers needed was a boiler and fire, and tubing to run the steam through a red hot gun barrel, with the hydrogen emerging at the other end to inflate the balloon.

*Lavoisier's List of Elements*

During the 1780s, Lavoisier, along with a number of colleagues, undertook a reformulation of chemical theory, which they frankly considered to be revolutionary. Their oxygen theory was revolutionary enough (essentially turning the phlogiston theory upside down); their approach to *elements* was equally revolutionary. Gone forever were the four elements of Aristotle. Now Lavoisier provided a growing list of elementary substances (one which further work would extend) and a precise operational definition :

> ...if by the term *elements* we mean to express those simple and indivisible atoms of which matter is composed, it is extremely probable we know nothing at all about them; but, if we apply the term *elements*, or *principles of bodies* to express our idea of the last point which analysis is capable of reaching, we must admit as elements all the substances into which we are capable, by any means, to reduce bodies by decomposition. Not that we are entitled to affirm that these substances we consider simple may not be compounded of two, or ever of a greater number of principles; but since these principles cannot be separated, or rather since we have not till now discovered a way of separating them, they act with regard to us as simple substances, and we ought never to suppose

## TABLE OF SIMPLE SUBSTANCES.

Simple fubftances belonging to all the kingdoms of nature, which may be confidered as the elements of bodies.

| *New Names.* | *Correfpondent old Names.* |
|---|---|
| Light | Light. |
| Caloric | Heat.<br>Principle or element of heat.<br>Fire. Igneous fluid.<br>Matter of fire and of heat. |
| Oxygen | Dephlogifticated air.<br>Empyreal air.<br>Vital air, or<br>Bafe of vital air. |
| Azote | Phlogifticated air or gas.<br>Mephitis, or its bafe. |
| Hydrogen | Inflammable air or gas,<br>or the bafe of inflammable air. |

Oxydable and Acidifiable fimple Subftances not Metallic.

| *New Names.* | *Correfpondent old names.* |
|---|---|
| Sulphur<br>Phofphorus<br>Charcoal | The fame names. |
| Muriatic radical<br>Fluoric radical<br>Boracic radical | Still unknown. |

Oxydable and Acidifiable fimple Metallic Bodies.

| *New Names.* | *Correfpondent Old Names.* |
|---|---|
| Antimony | Regulus of Antimony. |
| Arfenic | Regulus of Arfenic. |
| Bifmuth | Regulus of Bifmuth. |
| Cobalt | Regulus of Cobalt. |
| Copper | Regulus of Copper. |
| Gold | Regulus of Gold. |
| Iron | Regulus of Iron. |
| Lead | Regulus of Lead. |
| Manganefe | Regulus of Manganefe. |
| Mercury | Regulus of Mercury. |
| Molybdena | Regulus of Molybdena. |
| Nickel | Regulus of Nickel. |
| Platina | Regulus of Platina. |
| Silver | Regulus of Silver. |
| Tin | Regulus of Tin. |
| Tungftein | Regulus of Tungftein. |
| Zinc | Regulus of Zinc. |

Salifiable fimple Earthy Subftances.

| *New Names.* | *Correfpondent old Names.* |
|---|---|
| Lime | Chalk, calcareous earth.<br>Quicklime. |
| Magnefia | Magnefia, bafe of Epfom falt.<br>Calcined or cauftic magnefia. |
| Barytes | Barytes, or heavy earth. |
| Argill | Clay, earth of alum. |
| Silex | Siliceous or vitrifiable earth. |

*Lavoisier's Table of Simple Substances.*

> them compounded until experiment and observation has proved them to be so.

Lavoisier's "Table of simple substances" shows that he has admitted his inability to decompose the "muriatic" and "fluoric" radicals (to the chlorine and fluorine of the next generation). Also, notice that he has included "light" and "caloric," which would later be identified as forms of energy instead of substances. The final five substances are actually oxides, soon to be reduced to calcium, magnesium, barium, aluminum, and silicon.

Lavoisier concluded his description of the table with the remark that he had omitted potash and soda

> because they are evidently compound substances, though we are ignorant as yet what are the elements *they are composed of.*

Lavoisier's suspicions were soon confirmed by Humphrey Davy in England.

## *Humphrey Davy (1778–1829)*

Humphrey Davy was born in Cornwall in southwestern England, in 1778. In his early 20s, he became a chemistry professor in London. Davy was knighted in 1812, served as President of the Royal Society from 1820 to 1827, and died when he was just past 50.

*Humphrey Davy*

After training himself by reading and experimenting, Davy became, in turn, an apothecary's assistant, a medical research chemist in Bristol, and, in 1800, moved to London to take charge of chemical research at the newly founded Royal Institution.

The Royal Institution had been founded in 1799 by Count Rumford, an American, loyal to George the third, who had to leave the colonies after independence. Among other activities in his colorful career, Rumford married Lavoisier's widow.

The Royal Institution soon became a center of chemical research and popular scientific lecturing. Just at the time Davy started at the Royal Institution, the voltaic electric battery was invented (Chapter 17). Soon it was found that the electric battery could be used for decomposing compounds into their elements. Two English investigators used a battery to decompose water—showing that it consisted of hydrogen and oxygen. They confirmed Lavoisier's theory that water was a compound. With electricity, both gases could be easily collected. This process of decomposition by electricity is called **electrolysis**.

### *New Elements*

Soon, Davy used electrolysis to decompose other substances that had previously been considered to be elements. He began with the conviction that the number of elementary substances had to be small. (If he or Aristotle should visit us today and learn there are about 100 elements, they'd both react in horror: "You may have 100 somethings, but you daren't call them *elements*!" They'd be overjoyed by our electrons, protons, and neutrons, and even more by quarks: "Now they're *really* elements.") Davy considered the voltaic battery an ideal analytical tool, because its strength could be increased indefinitely—just add more plates or cells. With such power he thought he could break substances down to their few ultimate elementary constituents.

Unfortunately for his deep-seated convictions, all Davy did was add to Lavoisier's list. With his new method of electrolysis Davy confirmed Lavoisier's earlier suspicions. In 1807–1808, Davy isolated sodium and potassium, and also calcium, magnesium, barium, and strontium. Lavoisier would have been pleased, even though Davy wasn't.

### *A New Theory of Acids*

Although Davy's electro-analysis of those substances confirmed Lavoisier's hunch, in other work he disconfirmed Lavoisier's theory of acids. Lavoisier had named oxygen the "acid-former," from the Greek word *oxy*, for sharp. He thought every acid contained oxygen, and that it was the oxygen in a substance that made it an acid. Davy found out differently. He did it by a careful analysis of what we call hydrochloric acid—in 1800 it was called marine acid. Marine acid had been first made about 1650 by pouring sulfuric acid over sea salt. Then, in the 1770s, a Swedish chemist had extracted a greenish gas from marine acid. He called it "dephlogisticated marine acid air." Obviously not an element. Later, Lavoisier's followers renamed this green gas "oxidized marine acid," showing that they thought it contained oxygen, and therefore was a compound, still not an element. Using electrolysis, Davy showed that the green gas was an element.

About 1808, Davy began to search for the oxygen in marine acid. He tried every analytical technique he could think of. They all failed. That marine acid gas yielded no further components. Eventually, in 1810, Davy was forced to conclude that it was an element, which he named chlorine.

Davy also showed that marine acid is a compound of hydrogen and chlorine (i.e., hydrochloric acid). In fact, later investigators would show that it's hydrogen, not oxygen, that makes a compound acidic. So, the name oxygen reminds us of Lavoisier's outmoded theory of acids. As Shakespeare's Juliet said, "What's in a name?" The answer in chemistry is sometimes "nothing—that's just what it's called," and sometimes "a lot" (that is, it reflects a theory of chemistry, whether dated or current). Salt's just a name, but if we call it sodium chloride, that tells us what it's made of. And the name of a compound like carbon dioxide not only tells us it contains the elements carbon and oxygen, but also that it contains twice as many atoms of oxygen as of carbon.

Although Lavoisier had talked atoms, John Dalton made a proper theory of atoms.

## *John Dalton (1766–1844)*

John Dalton was born in northern England of humble Quaker parents. He received a good elementary education and became a schoolmaster in his early twenties. He spent most of his life in Manchester teaching and making scientific investigations. Dalton's methodical nature is shown by the fact, that for fifty-seven years, until the day of his death, he kept a careful daily record of the local temperature, rainfall, and barometric pressure. Despite Dalton's humble circumstances, he was eventually honored with membership in the major scientific societies of Europe.

*John Dalton.*

Dalton's major idea was that matter was composed of tiny particles called atoms; that all the atoms of any one element are alike; that elements differ from one another because their atoms have different weights; and that compounds consist of small clusters of atoms. He called these clusters of atoms, **molecules**.

By 1800, chemists had realized that compounds had a constant composition. That is, the elements forming a compound are always found in the same proportion by weight. For example, in sodium chloride, 40 percent of its weight is sodium, and 60 percent is chlorine. And it doesn't matter where the salt comes from, as long as it's pure. It can be extracted from the sea, anywhere in the world. Or it can be produced in the reaction of sodium carbonate with hydrochloric acid. Later, Davy combined sodium and chlorine directly, by putting sodium in a chlorine atmosphere. Still the same proportions: 40% sodium, 60% chlorine.

Dalton's atomic theory easily explained that constant proportion: every molecule of sodium chloride is composed of one atom of sodium and one atom of chlorine. And since the atoms of sodium all weigh the same, and also the chlorines, their proportions in sodium chloride must be constant.

Although the atomic theory is the foundation of chemistry, Dalton came at the problem through physics. As Dalton made his meteorological investigations, he tried to understand the effects on atmospheric pressure of the

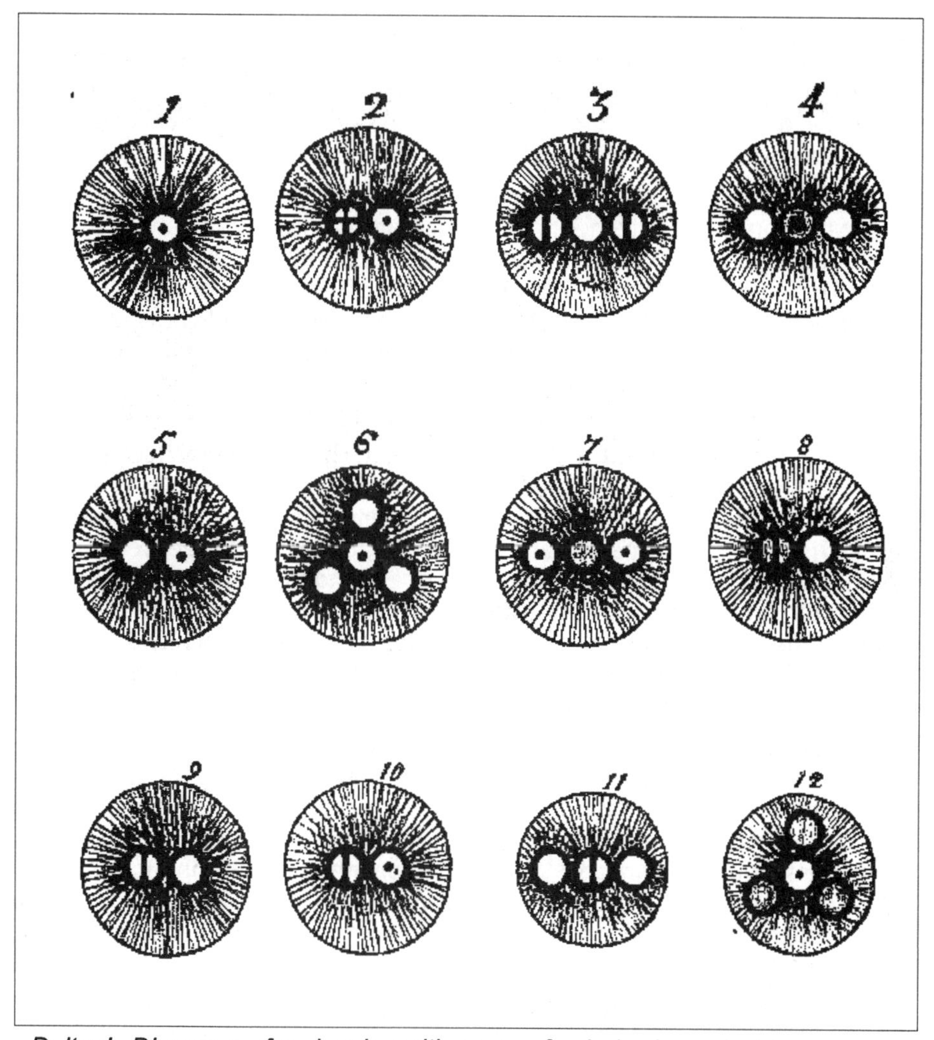


*Dalton's Diagrams of molecules with areas of caloric shown inside.*

presence of water vapor (which the climate of Manchester had in abundance). That led him to the reverse problem of the effects of the pressures of various gases on the surface of water in a closed container. Although earlier, chemists had realized that the atmosphere contained both oxygen and nitrogen, they hadn't been sure how the two gases were combined, that is, whether they were in chemical combination or were merely mixed mechanically.

Dalton found that each enclosed gas exerted its pressure individually, independent of the presence of any other gas. That convinced him that the atmosphere was simply a mixture of its gases, that they weren't combined chemically. Then, as Dalton examined the matter more closely, he was led to the notion that the various gases could be distinguished from one another simply by the weight of their atoms. Somehow, in his mechanical picture, he imagined that identical oxygen "atoms" (we'd say molecules of $O_2$) could only repel other atoms of their own kind—as if they couldn't even "see" the atoms of other gases.

In fact, on the kinetic theory of gases, Dalton's pressure relations are evidence that the molecules of gases neither attract nor repel; rather they behave like so many tiny, hard balls that bounce off one another without loss of energy when they collide. Nonetheless, Dalton's picture had enormously significant consequences for chemistry. By assuming that in chemical reactions substances acted only via their constituent atoms, Dalton could explain the fact that compounds contained their components in fixed proportions.

Then, Dalton derived another important chemical principle from his atomic theory. It's called the law of multiple proportions. To see how it works consider this analogy. Imagine atoms to be skaters in a large arena. Start with a bunch of carbon atoms skating slowly around all linked together, hand to hand to hand. Suddenly, the gates burst open, and many pairs of oxygen atoms enter, skating vigorously. As they jostle about, the carbons and oxygens intermingle. Soon, carbon dioxide molecules are formed, each with one carbon linked to two oxygens. When all the carbon has been oxidized, extra pairs of oxygen atoms that didn't find a carbon continue to skate around.

Now, suppose that in a neighboring arena, there's another bunch of carbon atoms skating around, but this time the gate opens only a crack. Only a few pairs of oxygen atoms can manage to squeeze in. They still find it more fun to skate with carbons than with each other, but there aren't enough pairs of oxygen for all the carbons. So, we end up with each carbon atom skating around with only one oxygen atom—carbon *mon*oxide molecules. So, carbon atoms can form molecules with either one oxygen atom at a time, or two, depending on how much oxygen there is.

From this kind of analysis using his theory of atoms, Dalton predicted that the weight of oxygen combining with 100 g of carbon in one compound, carbon dioxide, would be double that in the other, carbon monoxide. And that was something chemists could check—even if they didn't believe in atoms. They did check, and found that Dalton was right. Not only for the two combinations of carbon and oxygen, but in all other cases where elements can combine in more than one proportion (that is, in *multiple* proportions) to form different compounds of the same elements. Because of his atomic theory, Dalton predicted that these proportions would be ratios of small whole numbers—like 2 to 1, or 4 to 3. Simple arithmetic with atoms; not complicated fractions.

When we explain the law of multiple proportions using atoms, it seems almost trivial. But you have to realize that chemists don't actually work with atoms; they work with substances. When they found that substances conformed to the law Dalton had derived from his atomic theory, you might suppose they'd see it as a confirmation of the theory. After all, that's the way science works. When a deduction from a theory is confirmed, that tends to support the theory. Yet, throughout the nineteenth century, many chemists refused to believe that atoms actually existed. But that didn't keep them from using laws they could confirm, no matter what silly assumptions they'd originated from. Of course, the atomic assumption isn't silly, and Dalton's laws made chemists pay attention to the relative weights of the elements.

By about 1810, chemists had enough principles so they could confidently analyze a wide range of substances. They could measure atomic weights—or as they called them, equivalent weights. In the 1820s, the acknowledged leader in chemistry was Jons Jakob Berzelius (1779–1848), a Swedish professor who became an doctor of medicine at age 22. Berzelius purified and analyzed over 2000 compounds. He analyzed every mineral he could find among the rocks of Sweden. He also calculated the atomic weights of many elements. Berzelius published his results and trained chemists. In this way, he made chemistry into a profession.

## *Organic Chemistry*

Soon, chemists were analyzing other substances besides minerals—substances like wood and leather. They called the compounds in minerals **inorganic** to distinguish them from the organic compounds that make up living organisms. Chemists eventually found that almost all organic compounds contain carbon so that **organic chemistry** came to mean compounds containing carbon. They include, for example, the compounds of carbon and hydrogen, called the hydrocarbons.

Now, if you work on the compounds in living organisms, you're really doing what today we call biochemistry (see Chapter 21). In the 1800s, organic chemists spent most of their time on substances that were once alive, but aren't now, like coal, which is concentrated plant material hundreds of millions of years old. Nowadays, organic chemists work mainly with hydrocarbons found in petroleum, the remains of ancient marine organisms.

For a while, some chemists supposed that organic compounds must involve some "vital force" and hesitated to assume that organic compounds would follow the same rules as inorganic ones. As they conducted more analyses, however, they found no evidence of any vital force. Since organic molecules are generally considerably larger (i.e., more atoms) than inorganic ones, chemists needed to take greater care to determine precisely what relative numbers of atoms of carbon, hydrogen, and oxygen they contained. In the 1820s, chemists were initially confused to find that two quite different organic compounds had the same empirical formula (e.g., both ethyl alcohol and dimethyl ether have the formula $C_2H_6O$). They called such compounds *isomers*, and eventually realized that the atoms must be arranged differently in various isomers.

The simplest hydrocarbon is the gas methane. Methane was originally called marsh gas, because it's one of the main products of decaying vegetation found in swampy areas. It's also the major gaseous component of coal. As chemists analyzed coal, they found other substances in it besides methane—various liquid and tarry materials. One of the most important light liquids was benzene. When chemists analyzed benzene, they found it contained equivalent amounts of carbon and hydrogen. Those who believed in atoms were able to determine that there must be six atoms each of carbon and hydrogen in benzene.

*The Dream of Kekulé (1829–1896)*

Most hydrocarbons have more hydrogen atoms than carbons, twice as many or more. Benzene, with equal numbers of carbon and hydrogen atoms, was peculiar. How could the atoms be arranged in the benzene molecule? In the 1850s, organic chemists were getting the idea that atoms were strung out linearly in carbon compounds. Friedrich Kekulé, one of the chemists who could believe in atoms, began to consider the detailed structure of organic molecules. This interest might have come from his early training as an architect. In 1858, Friedrich Kekulé published a paper in which he asserted that carbon possessed four of what he called "affinity units" (what we'd call valence bonds):

*Friedrich August Kekulé von Stradonitz.*

> The simplest case...of the linking together of two carbon atoms is that one affinity unit of one atom is bound to one of the other. Of the 2 x 4 affinity units of the two carbon atoms, two are thus used to hold both atoms themselves together; there still remain six extra which can be bound by the atoms of other elements.

With this idea Kekulé provided an adequate structure for the many organic compounds that are built up on a carbon-chain backbone. But the six carbons in benzene should be expected to have 14 affinity units available for attachment to hydrogens (as in the hydrocarbon hexane). Somehow or other, eight affinity units were missing. Kekulé puzzled over the structure of benzene for some years.

Then in 1865, as Kekulé reported it, he fell into a reverie before the fire, and awoke with the idea of the six carbons in benzene twisting around to form a circle. Whether he saw six snakes in a ring, mouth to tail, or six monkeys holding hands, Kekulé's dream provided the clue to the solution of the benzene problem. He immediately realized that to use up all the affinity units, three of the carbon-carbon bonds would have to be doubled.

Before long, chemists were describing the shape of the carbon atom in three dimensions. They pictured the carbon atom as a solid with four triangu-

*The structure of the benzene molecule. Three of the bonds between carbon atoms are double bonds.*

lar faces: a tetrahedron. That gave the simplest hydrocarbon, methane, a structure of one carbon atom with one hydrogen atom at each of the four corners.

During the middle years of the nineteenth century, organic chemists began to create actual organic compounds from simpler components: organic synthesis. This encouraged people to believe that life was nothing more than a series of complicated chemical reactions, rather than some special kind of living force—which they called the vital force back then. Organic compounds could now be made in test-tubes from lifeless materials, without the intervention of living organisms—except as one wag put it, "You mean chemists aren't living organisms?"

## *Applied Synthetic Structural Chemistry*

For thousands of years fabrics had been dyed with natural coloring materials, like indigo, which comes from the leaves of a tropical plant. By the 1850s, chemists had found a substance, called aniline, both in the dye indigo and in the tar that comes from coal. From that, they eventually figured out how to manufacture synthetic indigo, the first of many important synthetic compounds. Aniline is a relatively simple modification of benzene. In aniline, there's an ammonia molecule replacing one of the six hydrogens attached to the six carbons in benzene. Aniline was an important compound, because it became a building block in the manufacture of many commercial chemicals.

From the 1830s, Germany was the center of chemical research. German universities turned out research chemists by the score. One of these, August Wilhelm von Hofmann (1818–1892), moved to London in 1845, to become the head of the newly founded Royal College of Chemistry. Hofmann concentrated his research on organic compounds containing nitrogen. In 1856, his pupil, William Perkin (1838–1907), discovered a derivative of aniline that provided a purplish dye, mauve. Perkin soon set up a plant to produce the dye and by the early 1870s was wealthy enough to retire.

*William Perkin at age 14.*

Although the British had early success in the manufacture of synthetic organic compounds, the fundamental strength of the German chemical establishment soon came to the fore. By the 1880s, German chemical output surpassed all others and, by 1900, was supplying 90% of the market.

This German pre-eminence in chemical manufacture applied to explosives as well as to dyestuffs and contributed to the extension of industrial competition to the political realm, with the resulting outbreak of war in 1914.

## *The Periodic Table of the Elements*

One of the reasons chemists had resisted Dalton's idea of atoms was that atomic weights didn't tell them anything about the chemical properties of the elements. For example, why elements of quite different atomic weights had similar chemical properties, and elements of similar atomic weights often had very different properties. Then along came the periodic table—an arrangement of the elements in order of increasing atomic weight, broken into rows and columns so that elements with similar properties fall in the same column. Seeing the elements arranged this way helped chemists to account for many of the similarities in the behavior of certain sets of elements. Let's see how chemists arrived at this arrangement.

Back to the alkalis. Earlier, we saw how Davy discovered sodium and potassium, elements with many similarities: they're both soft metals with similar densities and melting points. And they make similar looking compounds, like sodium chloride and potassium chloride. By 1861, chemists had isolated three more similar metals: lithium, cesium, and rubidium. So, chemists called the five the alkali metals. Although the alkali metals all have similar properties, they have very different atomic weights. In fact, their atomic weights range all the way from 7 to 133. Another chemical family was labelled the halogens, which means "salt-former." The halogens include fluorine, chlorine, bromine, and iodine, with atomic weights ranging from 19 to 127. Now, it happens that the intervals in weight between the various alkali metals are very similar to the intervals between the halogens. Chemists in the 1860s began to think this couldn't be just a coincidence.

*Julius Lothar Meyer.*

Finally, two of them constructed a periodic table by placing the elements in a table of rows and columns in the order of increasing atomic weight. The table is "periodic" in the sense that similar properties recur periodically as one runs through a list of the elements arranged in order of increasing atomic weight. The two chemists were Dmitri Mendeleev (1834–1907) in Russia and Julius Lothar Meyer (1830–1895) in Germany. An example of simultaneous independent invention.

*Dmitri Mendeleev.*

They made their tables of rows and columns so that as you go across each row you move from family to family—lithium, beryllium, boron, carbon, nitrogen, oxygen, fluorine—and then start a new row with sodium, which puts all the members of each family into a column of their own. When you put sodium and potassium under lithium, each in its own row; then chlorine and bromine come under fluorine, and so on. (Actually, Mendeleev's first published table had rows and columns interchanged.)

From about 1830, occasional chemists had attempted to make a rational classification of the elements. However, most chemists paid little attention to them until the late 1860s. At that time, both Meyer and Mendeleev were composing textbooks in chemistry. In order to clarify their subjects, they sought a form in which to express the relations among the elements. Both men created similar tables, but Mendeleev's was published first and was bolder in its speculation about the nature of missing elements. The leaving of gaps for as yet undiscovered elements was, in fact, the most striking feature of the periodic tables of both men.

Mendeleev, for example, left a space in his table for an element in the same family as aluminum, assigning to it an atomic weight of about 68. He

**TABELLE II**

| REIHEN | GRUPPE I. — $R^2O$ | GRUPPE II. — $RO$ | GRUPPE III. — $R^2O^3$ | GRUPPE IV. $RH^4$ $RO^2$ | GRUPPE V. $RH^3$ $R^2O^5$ | GRUPPE VI. $RH^2$ $RO^3$ | GRUPPE VII. $RH$ $R^2O^7$ | GRUPPE VIII. — $RO^4$ |
|---|---|---|---|---|---|---|---|---|
| 1 | H = 1 | | | | | | | |
| 2 | Li = 7 | Be = 9,4 | B = 11 | C = 12 | N = 14 | O = 16 | F = 19 | |
| 3 | Na = 23 | Mg = 24 | Al = 27,3 | Si = 28 | P = 31 | S = 32 | Cl = 35,5 | |
| 4 | K = 39 | Ca = 40 | — = 44 | Ti = 48 | V = 51 | Cr = 52 | Mn = 55 | Fe = 56, Co = 59, Ni = 59, Cu = 63. |
| 5 | (Cu = 63) | Zn = 65 | — = 68 | — = 72 | As = 75 | Se = 78 | Br = 80 | |
| 6 | Rb = 85 | Sr = 87 | ?Yt = 88 | Zr = 90 | Nb = 94 | Mo = 96 | — = 100 | Ru = 104, Rh = 104, Pd = 106, Ag = 108. |
| 7 | (Ag = 108) | Cd = 112 | In = 113 | Sn = 118 | Sb = 122 | Te = 125 | J = 127 | |
| 8 | Cs = 133 | Ba = 137 | ?Di = 138 | ?Ce = 140 | — | — | — | — — — — |
| 9 | (—) | — | — | — | — | — | — | |
| 10 | — | — | ?Er = 178 | ?La = 180 | Ta = 182 | W = 184 | — | Os = 195, Ir = 197, Pt = 198, Au = 199. |
| 11 | (Au = 199) | Hg = 200 | Tl = 204 | Pb = 207 | Bi = 208 | — | — | |
| 12 | — | — | — | Th = 231 | — | U = 240 | — | — — — — |

*Mendeleev's Periodic Table. Note the empty spaces for undiscovered elements.*

called this element eka-aluminum (*eka* being the Sanskrit prefix for one, presumably meaning "one of the same family"). Mendeleev even suggested the form of the element's compounds with other elements and its density (essentially by interpolating from the values of the known neighboring elements). When another chemist reported the new element gallium in 1874, Mendeleev quickly showed that its properties matched those of his eka-aluminum. This and other successes soon gave chemists confidence in the value of the periodic table.

Chemists could now use the periodic table to explain chemical properties. Later, physicists explained the periodic table (Chapter 19).

## *Conclusion*

Over a couple of thousand years of alchemy and chemistry, experience and theory interacted back and forth. From Aristotle's four elements and the alchemists' spirits, chemists eventually passed through Stahl's phlogiston theory and a variety of airs, to Lavoisier's new set of elements and the oxygen theory of combustion.

For Lavoisier, an element was any substance that could not be reduced to simpler components. He recognized that this depended on the analytical techniques chemists could use. When Davy used the new technique of electrolysis, he was able to break the alkaline compounds into their constituent elements, confirming what Lavoisier already had suspected. Davy started a trend, so that Mendeleev knew about 60 elements, double the number Lavoisier had known 80 years before.

John Dalton established the atomic theory. He assumed that each element had its own atomic weight, and that chemists could distinguish elements by their atomic weights. That helped chemists to analyze compounds correctly, but it gave them a lot of headaches, because determining atomic weights wasn't easy, and atomic weights didn't tell them anything about chemical properties. Nonetheless, Berzelius analyzed many compounds and improved our knowledge of atomic weights.

Soon, Berzelius and others switched their attention from the composition of inorganic rocks to organic compounds, especially the ones found in coal. With the work of Kekulé in the mid 1860s, chemists were able to describe the structures of organic molecules.

Finally, Mendeleev and Meyer created the periodic table. That gave chemists a way to see how the chemical properties of the elements were related to one another. It was an exciting time for chemists.

## The Modern Periodic Table

| 1A | 2A | 3B | 4B | 5B | 6B | 7B | | 8 | | 1B | 2B | 3A | 4A | 5A | 6A | 7A | 0 |
|---|---|---|---|---|---|---|---|---|---|---|---|---|---|---|---|---|---|
| 1<br>**H**<br>1.0 | | | | | | | | | | | | | | | | | 2<br>**He**<br>4.0 |
| 3<br>**Li**<br>6.9 | 4<br>**Be**<br>9.0 | | | | | | | | | | | 5<br>**B**<br>10.8 | 6<br>**C**<br>12.0 | 7<br>**N**<br>14.0 | 8<br>**O**<br>16.0 | 9<br>**F**<br>19.0 | 10<br>**Ne**<br>20.2 |
| 11<br>**Na**<br>23.0 | 12<br>**Mg**<br>24.3 | | | | | | | | | | | 13<br>**Al**<br>27.0 | 14<br>**Si**<br>28.0 | 15<br>**P**<br>31.0 | 16<br>**S**<br>32.1 | 17<br>**Cl**<br>35.5 | 18<br>**Ar**<br>39.9 |
| 19<br>**K**<br>39.0 | 20<br>**Ca**<br>40.1 | 21<br>**Sc**<br>45.0 | 22<br>**Ti**<br>47.9 | 23<br>**V**<br>50.9 | 24<br>**Cr**<br>52.0 | 25<br>**Mn**<br>54.9 | 26<br>**Fe**<br>55.8 | 27<br>**Co**<br>58.9 | 28<br>**Ni**<br>58.7 | 29<br>**Cu**<br>63.5 | 30<br>**Zn**<br>65.4 | 31<br>**Ga**<br>69.7 | 32<br>**Ge**<br>72.6 | 33<br>**As**<br>74.9 | 34<br>**Se**<br>79.0 | 35<br>**Br**<br>79.9 | 36<br>**Kr**<br>83.8 |
| 37<br>**Rb**<br>85.5 | 38<br>**Sr**<br>87.6 | 39<br>**Y**<br>88.9 | 40<br>**Zr**<br>91.2 | 41<br>**Nb**<br>92.9 | 42<br>**Mo**<br>95.9 | 43<br>**Tc**<br>(98) | 44<br>**Ru**<br>101.1 | 45<br>**Rh**<br>102.9 | 46<br>**Pd**<br>106.4 | 47<br>**Ag**<br>107.9 | 48<br>**Cd**<br>112.4 | 49<br>**In**<br>114.8 | 50<br>**Sn**<br>118.7 | 51<br>**Sb**<br>121.7 | 52<br>**Te**<br>127.6 | 53<br>**I**<br>126.9 | 54<br>**Xe**<br>131.3 |
| 55<br>**Cs**<br>132.9 | 56<br>**Ba**<br>137.3 | 57<br>**La**<br>138.9 | 72<br>**Hf**<br>178.5 | 73<br>**Ta**<br>180.9 | 74<br>**W**<br>183.8 | 75<br>**Re**<br>186.2 | 76<br>**Os**<br>190.2 | 77<br>**Ir**<br>192.2 | 78<br>**Pt**<br>195.1 | 79<br>**Au**<br>197.0 | 80<br>**Hg**<br>200.7 | 81<br>**Tl**<br>204.4 | 82<br>**Pb**<br>207.2 | 83<br>**Bi**<br>209.0 | 84<br>**Po**<br>(209) | 85<br>**At**<br>(210) | 86<br>**Rn**<br>(222) |
| 87<br>**Fr**<br>(223) | 88<br>**Ra**<br>226.0 | 89<br>**Ac**<br>227.0 | 104<br>**Unq**<br>(261) | 105<br>**Unp**<br>(262) | 106<br>**Unh**<br>(263) | 107<br>? | | | | | | | | | | | |
| | | follow<br>**La** | 58<br>**Ce**<br>140.1 | 59<br>**Pr**<br>140.9 | 60<br>**Nd**<br>144.2 | 61<br>**Pm**<br>(145) | 62<br>**Sm**<br>150.4 | 63<br>**Eu**<br>152.0 | 64<br>**Gd**<br>157.2 | 65<br>**Tb**<br>158.9 | 66<br>**Dy**<br>162.5 | 67<br>**Ho**<br>164.9 | 68<br>**Er**<br>167.3 | 69<br>**Tm**<br>168.9 | 70<br>**Yb**<br>173.0 | 71<br>**Lu**<br>175.0 | |
| | | follow<br>**Ac** | 90<br>**Th**<br>232.0 | 91<br>**Pa**<br>231.0 | 92<br>**U**<br>238.0 | 93<br>**Np**<br>237.0 | 94<br>**Pu**<br>(244) | 95<br>**Am**<br>(243) | 96<br>**Cm**<br>(247) | 97<br>**Bk**<br>(247) | 98<br>**Cf**<br>(251) | 99<br>**Es**<br>(254) | 100<br>**Fm**<br>(257) | 101<br>**Md**<br>(258) | 102<br>**No**<br>(259) | 103<br>**Lr**<br>(260) | |

*The elements are arranged in order of increasing atomic number, the top number in each box. The elements in each column have similar chemical properties, which are identified as groups. The bottom number in each box identifies the average atomic mass for each element, based on the proportions found in nature. Atomic masses in parentheses identify the most common isotope of short-lived and artificial elements.*

*Charles Darwin as a young man.*

## CHAPTER SIXTEEN

# Old Rocks and New Creatures

When you look at the earth and its creatures, you see bewildering variety. Mountains and glaciers, elephants and roses—they're a lot more complicated than the orbit of a planet or the path of a cannonball. How do you make science out of them? In fact, the sciences of geology and biology took longer to establish than astronomy and physics.

### *Geological Theories Up to 1700*

The theories and models of any science have to match the evidence that's claimed to support them. The evidence in geology consists of observations of the earth around us—the oceans and the mountains and the minerals within the earth. Before 1700, theories that tried to account for the structure of the earth were pretty vague. The most popular theory was nearly two thousand years old. Aristotle again! He treated the earth as a kind of living organism. The very hills were alive—changing as the minerals within them grew or changed their form.

Aristotle's way of looking at the earth was imported into medieval Europe, along with the rest of his philosophy. And it continued with little change until about 1700. For example, when miners in Germany in the 1500s found silver deposits associated with those of lead, they thought lead was the seed from which silver grew, that through a process of growth and change lead became tin, then bismuth, and finally silver, if only they could wait long enough. When they came upon a vein of bismuth, the miners would say, "Alas, we've come too soon."

Sometimes miners even found rocks shaped like living organisms—fossils. Today we consider fossils to be the petrified remains of ancient life. But 500 years ago, they thought the rocks had just grown that way. So they had Aristotle's theory and even a little evidence that seemed to support it. Such a theory was really more of a cosmology than a useful theory to influence the way scientists or craftsmen worked. Without a useful theory, men interested in the earth just accumulated experience.

After 1500, the conjunction of mechanical printing and increased mining activity resulted in the publication of a few books on mining and metallurgy. The most famous of these was the work of a German physician, Georg Bauer (1494–1555). He tended patients in an important mining region and spent much time observing and thinking about geology and mining. Among Bauer's several publications was his book *De re metallica* ("On matters metallic"), where his name is Latinized to Georgius Agricola. Here, Bauer described many details of the technology of mining. In other works he carefully described and classified many minerals and the contexts in which they were found. He is acknowledged today as the founder of mineralogy.

Bauer used the term *fossil* in the general sense meaning merely "things that are dug up." Although Leonardo da Vinci and others had suggested that the rocks we call fossils are the remains of ancient creatures, that view was generally rejected until after 1650. In 1669, a Danish physician and anatomist, Neils Stensen (1638–1686), published a detailed comparison of certain fossils and the shelled creatures they resembled. Since he was living in Florence at the time, he is most generally known by the name he used there, Nicolaus Steno. He readily extended his view of fossil shells to other fossils:

> What hath been said of shells, the same is to be said of other parts of animals, and of the animals themselves buried under ground, of which number are the teeth of sea-dogs, the teeth of the fish aquila, fishes' backbones, all sorts of whole fishes, skulls, horns, teeth, thigh-bones and other bones of terrestrial animals.

Steno also proposed a general theory for the formation of the various strata of the earth.

Until Steno's time, the geological theories of Aristotle were still widely accepted. However, as Aristotle's natural philosophy became generally discredited by the work of Galileo and Newton and others, scientists found it easier to doubt his views of the earth too.

Now, scientists are usually uncomfortable without a general theory around which to organize their investigations. As Aristotle declined, geologists cast about for an alternative. They found it ready-made in *Genesis*, particularly in the first chapter, and in the account of The Flood. In the scientific and religious climate of the seventeenth century, they readily transformed their views from the world as a living, pulsing creature, to a mechanism created by God the Eternal Engineer.

The model of the earth that could be pieced together out of *Genesis* seemed to make perfect sense. If you could imagine a series of floods, then each one could deposit a new layer of material—strata! And of course, a gigantic deluge would readily account for finding the fossil remains of sea creatures high up in the mountains. A perfect theory!

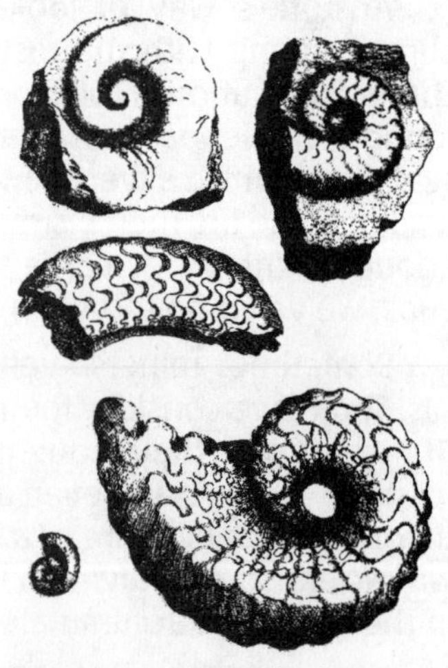

*Drawings of fossils from Leibniz's* History of the Earth.

## Geological Theories in the Eighteenth Century

Once geologists began to fit their observations to the Bible, they were inclined to use the Bible to supply information not found in their observations. For example, the age of the earth. From medieval times, numerous scholars had used the genealogies in the Bible to make an estimate of the age of the earth, assuming six literal days for Creation. They mostly got values in the range of 5000 to 6000 years. The English in the seventeenth century adopted the spurious precision supplied by an Irish archbishop, James Ussher (1581–1656), who dated Creation to 9 a.m. 23 October 4004 BC.

Mining and canal digging during the eighteenth century continued to confirm the stratigraphic structure below the earth's surface. Geologists who attributed these layers to a series of floods came to be called Neptunists, after the god of the sea. Other geologists, called Vulcanists, rejected the flood hypothesis. They claimed that many layers of rock looked more like they'd been melted than deposited by water. So they attributed rock formation to the action of volcanoes. Yet both Neptunists and Vulcanists clung to the idea that the earth was young.

As geologists observed the earth's structure in greater detail, they realized that within a brief span of 6000 years, really profound changes in the earth would have had to occur. Whether the action had been by floods or volcanoes, it must have been catastrophic—that is, a lot of change in a short time. So, both Neptunists and Vulcanists were called catastrophists.

Catastrophic views of the earth's development held sway until after 1800. Geologists worked hard to fit all their findings into the sequence described in the first chapter of *Genesis*.

## Biological Classification

At the same time, naturalists were trying to fit their ideas of living organisms into the *Genesis* account: "Whatever Adam called each creature, that was its name."

However, as European explorers brought home more and more exotic animals and plants from around the world, naturalists had more creatures than Adam had names. Some naturalists felt that just making up new names wasn't good enough. They looked for a way to classify and name the creatures in a logical, systematic way. After all, lions, tigers, and leopards are similar, but different from apes and monkeys. A Swedish botanist, Carl Linnaeus (1707–1778), finally created a scheme that naturalists could adopt.

*Carl Linnaeus in the clothing of a Lapp huntsman.*

Linnaeus was interested in plants from his youth, and while training for medicine at the University of Uppsala, assisted the professor of botany there. When he read a book describing the functions of stamens and pistils in plant propagation, he resolved to use the number and arrangement of stamens as the basis for a classification scheme. He tested this scheme on the plants in the university garden and on trips throughout Sweden. While traveling through Europe after receiving his degree, Linnaeus consulted with other naturalists and found general approval for his scheme. From then on, he enlisted the aid of correspondents from around the world to add specimens to his classification. In 1741, he became professor of botany at Uppsala.

Linnaeus published his naming scheme in a book titled *Systema Naturae* ("The System of Nature"). The first edition, only a dozen pages long, came out in 1735. Before he died in 1778, Linnaeus had produced a twelfth edition of his scheme that was 1500 pages long, including animals classified according the chambers of their hearts.

Linnaeus was aiming at a *natural* system of classification, based essentially on the notion that God had filled organic creation without gaps. The task of taxonomists was to find an arrangement that put species with the greatest similarities closest to one another. Linnaeus realized that his scheme, based on a single feature like stamens or hearts, was only a first step in arriving at the proper arrangement of organisms created by God's plenitude. Later taxonomists added other features to get closer to an ideal that might seem to be always beyond their grasp. Nonetheless, the Linnaean scheme provided the foundation upon which they built.

In particular, Linnaeus was responsible for the identification of species by two names, the **binomial** system. As an example, consider the biological classification for a house cat—let's call her Tabitha. What categories does she fit into? Since Tabitha has fur and suckles her young, she belongs to the class, Mammalia, along with squirrels, horses, whales, and us. The mammal class is subdivided into various orders. The order that Tabitha belongs to is called Carnivora. Its root meaning is meat-eater, but there are meat-eaters in other orders besides the Carnivora.

Biologists don't classify animals by their habits, but by their forms and structures.

Biologists identify the various orders by the particular shapes and arrangements of their teeth, skulls, and feet. An example of the detail involved can be seen in the following quotation from a modern textbook. It describes how the surfaces of teeth (the surfaces a dentist attacks when he's drilling) vary among mammalian orders:

> The crowns and the occlusal surfaces of the cheek teeth vary structurally according to diet. In omnivores like some rodents, some carnivores, some primates, and pigs—the cusps of the molars form separate rounded hillocks that crush and grind food. In herbivores the molars may either have cusps that form ridges, or crescents. The teeth finely section and grind vegetation. In many insectivores, bats, and carnivores, there

*Linnaeus in his study.*

are sectoral teeth. These have blade-like cutting edges that section food by shearing against the edges of their counterparts in the opposing jaw.

The carnivores include wolves, bears, skunks, walruses, and, of course, Tabitha and all the other cats. Biologists put all the cat-like animals into a "family" called the Felidae, from the Latin word for cat. Then they subdivide the family into **genera** (the plural of the word *genus*). Tabitha belongs to the genus *Felis*, while the larger cats, like lions and tigers, belong to the genus *Panthera*. Each genus contains a number of species. Lions, for example are *Panthera leo*. And finally, Tabitha belongs to the species *Felis catus*—along with Garfield, Thomasina, Sylvester, Deuteronomy, Topaz, and your cat. Of course, the species *Felis catus* includes varieties like Siamese, Persian, Manx, and so on.

Every species carries two names: the generic name, *Felis*, and the specific name, *catus*. That's the binomial naming system Linnaeus invented. Using the Latin word for man, he put us into the genus *Homo*. And our species is *Homo sapiens*, wise human.

In summary: Tabitha and her cousins belong to the species *F. catus*, the genus *Felis*, the family Felidae, the order Carnivora, and the class Mammalia.

Linnaeus' classification scheme put order into the world of living things. In the 1700s, this order was seen as the handiwork of God. All the various combinations of characters were simply part of God's design. Yet, however similar neighboring species might be, they were distinct and separate. Certainly not connected by any kind of evolution.

Since naturalists were as willing to accept the *Genesis* account as the geologists, they had no worries about time. They could easily accept that all the earth's creatures had been created on the third, fifth, and sixth days, "all according to their kind," in exactly the same form as they could now be found around the earth. Until they were confronted with the fossil remains of creatures of different forms!

## *Challenges to a Youthful Earth*

By 1800, geologists were making discoveries that began to challenge the traditional picture. In the lower layers of cliff faces, exposed by weathering, they were finding fossil remains of creatures different from any living species. They were different enough that biologists couldn't fit them into any existing species or genera, yet similar enough to belong to classes like fish, amphibians, and reptiles. In low (old) layers, they found invertebrates like clams. Just above that, the only vertebrates they found were species of fish. Higher still they found amphibians as well, like giant salamanders. Even higher, reptiles—like dinosaurs—were added, and so on. Some people said these fossils must have been put there by God—*as* fossils. Others thought the fossils were the remains of creatures that had once been alive. Creatures that were now extinct.

Since geologists didn't find any fossils of modern animals in the deeper layers, biologists began to imagine a series of separate creations. That God had created fish first, then amphibians, then reptiles, and most recently birds and mammals. They argued that each creation had been wiped out by a flood, and then replaced. Along with a higher creation. A series of progressive creations—it was as though God was expanding and improving his design after each flood.

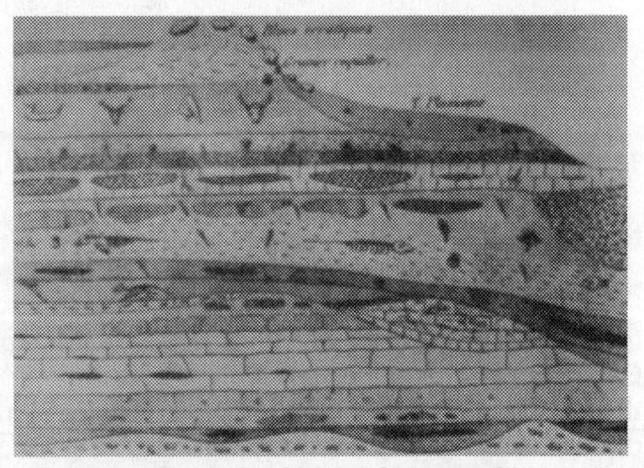

*Georges Cuvier's sketch of strata with imbedded fossils.*

Soon, some geologists began to insist that the earth was older than the biblical 6000 years. They now considered the six "days" of creation to be ages, perhaps ten thousand years each, with humans coming only in the last age of creation, 6000 years ago. By the early 1800s, many geologists and biologists agreed that the earth and its creatures had developed through a series of progressive catastrophes—wrought by the hand of God.

But an alternative to catastrophism was emerging. It began with new views on the formation of the earth itself. The Scottish geologist, James Hutton (1726–1797), proposed that the earth had been formed by a natural process, gradually and slowly, requiring an immense amount of time.

Hutton was born in Edinburgh and was educated for medicine there and in Paris. Back in Scotland, he operated a farm inherited from his father. Once he got it running smoothly, he moved to Edinburgh and lived the life of a scholar on its proceeds.

In 1785, Hutton presented a paper to the recently formed Royal Society of Edinburgh, *Theory of the earth, or an investigation of the laws observable in the composition, dissolution and restoration of land upon the globe.* Hutton refused to speculate on the actions of hypothetical deluges. Instead he resolved to confine his analysis of the earth's structure to processes that could be observed or derived by analogy from observations. While admitting that some rock formations were of aqueous origin, Hutton claimed that many others resulted from pressure and internal heat. He also imagined rocks being worn away by wind and water, or gradually uplifted by the action of heat. Such processes were necessarily slow, unlikely to be able to generate all the features of the earth in a mere 6000 years.

In 1795, Hutton published a detailed elaboration of his paper, *Theory of the Earth.* There he made explicit his sense of the long history of the earth:

> We find no vestige of a beginning,—no prospect of an end.

At the time, the Neptunist school was still ascendent, and Hutton's book found little support.

About 1800, a French biologist, Jean Baptiste de Lamarck (1744–1829), described the development of life as slow, gradual, and natural—matching Hutton's view of the development of the earth.

Lamarck saw that there were many similarities between extinct animals and existing ones. Instead of invoking separate supernatural creations, he proposed that natural forces could account for the changes of animals through time. Lamarck suggested the animals themselves could change, in response to changes in their environment.

Here's what Lamarck wrote about his most famous example, the giraffe:

> We know that this animal, the tallest of mammals, dwells in the interior of Africa, in places where the soil, almost always arid and without grass, obliges it to browse on trees and to strain itself continuously to reach them. This habit sustained for long, has had the result in all members of its race that the forelegs have grown longer than the hind legs and that its neck has become so stretched, that the giraffe, without standing on its hind legs, lifts its head to a height of six metres.

Lamarck's idea is called the inheritance of acquired characteristics. If it were true, then the children of today's weight lifters would grow up with heavier muscles simply because their parents worked out. And all academics would have studious children. They don't!

Lamarck considered these changes to be slow. So he visualized a great age for the earth of perhaps millions of years. But people couldn't accept the idea the earth was that old. And Lamarck's proposal of a *natural* progression of species was unacceptable because it seemed atheistic.

### *Charles Lyell (1797–1875) and Uniformitarianism*

In 1800, Hutton's views were as unacceptable as Lamarck's. But they were revived about 1830, by Lyell, an English geologist. Lyell's opposition to catastrophism was strong and consistent. He claimed that the earth has been formed by the same kinds of processes we see acting today. That is, to explain the deep Niagara gorge, you don't invoke special catastrophes like the rush of waters from Noah's flood. Instead, you suppose that the Niagara River has eroded the ground away, bit by bit, over large stretches of time, as it still does today. Lyell's approach to the history of the earth is called **uniformitarianism**.

Lyell's principle of uniformity made geology scientific. Lyell wrote that if we want to understand the earth, we have to assume

> that all former changes of the organic and inorganic creation [result from an] uninterrupted succession of physical events governed by the laws now in operation.

He added that geology would lack a scientific foundation unless we made that kind of assumption. Although Lyell did accept the action of volcanoes and earthquakes (which he studied), he considered their effects to be relatively local, not *global*, as the catastrophists proposed.

Lyell *started* with a principle of uniformity. Yet he was willing to modify that view when he found evidence to the contrary. For example, we now attribute part of the Niagara River formation to a rather catastrophic event when the glaciers of the last ice age began to melt, about 15–20 thousand years ago. The land around the river took its present shape as the glaciers receded. Of course, it has been gradual erosion that has moved the falls in the Niagara River from Queenston/Lewiston upstream 10 km to its present location.

However, we could only get to a more refined science of geology by starting with Lyell's assumption of uniformity. You just have to be willing to concede non-uniformity when you're faced with facts that require it—like a mountain bursting up out of the sea in less than a month. Yet, today, with a much fuller understanding of geological processes, we consider even catastrophic events to be natural.

Since Lyell attributed the earth's formation to slow processes like erosion or the silting of rivers, he agreed with Hutton and Lamarck that the earth must be very old. But he didn't agree with Lamarck's progression of species. So, as he was writing his *Principles of Geology* in the 1830s, Lyell devoted part of it to criticizing Lamarck. Lyell applied his doctrine of uniformity to life forms the same way he did to the earth. That is, he considered that the processes that acted in the past were the same ones we see acting in the present. Since we don't see species changing now, we can't suppose they changed in the past. Lyell believed there had been one ancient, original creation of life. His one exception to this was his belief that there'd been a separate, more recent creation of humans.

So Lyell's strict application of the principle of uniformity led him to deny any progression in the living world. But even though Lyell couldn't accept that species changed, he did accept that species could disappear. He explained that a species could become extinct by competing for food with a better endowed species—and losing. And in that discussion he used the famous phrase "the struggle for existence."

## *The Malthusian Doctrine*

The idea of the struggle for existence had originated in a very different context. It came in the writings of an English clergyman, Thomas Robert Malthus (1766–1834). Just before 1800, Malthus published a short book titled *An Essay on the Principle of Population as It Affects the Future Improvement of Society*. He wrote the book because he wanted to be more realistic about the future of society than utopian writers of the time. French utopians like Jean Jacques Rousseau were suggesting simple schemes for reforming society and perfecting mankind. Malthus's father was a strong supporter of those schemes. He discussed them with his son Robert, who'd recently graduated in mathematics from Cambridge University. Robert also wanted to promote the happiness of mankind, but he claimed the utopians were ignoring some cold hard facts.

*Rev. Thomas Malthus.*

Malthus said history showed that the common lot of mankind was misery and vice, because people were always living on the edge of starvation. Whenever they found additional food, their numbers increased because fewer starved and more married. Usually, Malthus thought, the population would soon expand beyond the additional food supply, so they'd be back in misery again. Population size was limited by miseries such as famine or war. Whenever their prospects brightened, the poor would beget more children than they ought to. For Malthus, responsible breeding meant you only had as many children as you could afford to feed. But, as a clergyman in 1800, Malthus was opposed to artificial birth control; he considered it a vice to limit population that way.

So, Malthus's *Essay on Population* was a moral tract—how people ought to behave, and what would happen if they didn't. Population, he said, *is* always limited, by one of three means: by misery, vice, or moral restraint (that is, self-control).

Malthus realized that the utopians could answer his argument by claiming that if there are more people we should simply improve the capacity of the land to produce more food—what we call today a technological fix. Malthus dealt with that possibility, using a mathematical analysis. He looked at the way agriculture was being improved in his own time and estimated how much the food supply could be increased in each generation—say 25 years. Suppose, he thought, we produce a million tonnes of food a year now. Malthus considered that at best we might be able to produce two million tonnes 25 years from now, and three million tonnes 50 years from now. That is, an increase of a million tonnes every 25 years. After all, the amount of land for growing food is limited, and you'd have to expect that where one stalk of grain was growing now, you'd have five growing a century from now. Malthus was willing to imagine technological improvement able to do that well: to provide five times as much food in a hundred years.

But, in that same hundred years, Malthus claimed, a population that started out at a million could reach 16 million. It's just a lot easier to increase population than to increase food supply. If every couple raises four children, then the population will double in each 25-year generation: a million now, then 2, 4, 8, 16 million after four doublings in a hundred years.

So, while in theory you might get five times as much food in a hundred years, you'd also get 16 times as many people in the same time. You wouldn't actually expect so great a population increase, because then everybody would end up with only a third as much food as their great-great grandparents had had.

To support his argument, Malthus collected population statistics from all over the world. He travelled across Europe examining census records. And he found ample evidence to suggest that human populations could double in each generation if there were no checks to that growth. Of course, usually there *were* checks—misery and vice. Malthus wasn't claiming that populations did double every generation—only that they would if unchecked. And he claimed the utopians were being wildly optimistic to think that some technological fix would solve the problem.

Malthus's solution was moral restraint. As a result, he opposed increasing welfare payments to the poor, or family allowances, because they'd only provide money for more children, and thus, more misery and vice. He argued that governments should give just enough to keep indigents from dying. Family policy and government policy were the same for Malthus—beget only the number of children your present resources can feed.

In his essay, Malthus briefly applied his principle to the animal world. There, he said, populations had a tendency to even more than double in every generation. Animals breed like rabbits. But animal populations are kept relatively constant by the struggle for existence. Foxes keep the rabbit population in check. And foxes are limited by how many rabbits there are.

Malthus's argument was an important starting point both for Charles Darwin and Alfred Russel Wallace as they worked out their theory of evolution.

## *The Theory of Evolution*

The *Origin of Species* by Charles Darwin was published in London in November 1859. Its impact was so great that a second printing double the size was ordered immediately.

Since Darwin didn't originate the *idea* of evolution, we'll have to examine why his name is so closely attached to the theory. Particularly because the very same theory was conceived independently by another Englishman, Alfred Russel Wallace.

By itself, the word "evolution" simply means that the plants and animals we have today are descended from earlier plants and animals—all the way back to the beginning. That there weren't any supernatural creations along the way—no special interventions by God in the process. Other men had that idea before Darwin and Wallace—men like Lamarck—but they could never explain exactly how one species evolved into a different species. Lamarck's theory of the inheritance of acquired characteristics didn't really work.

When we talk about "the theory of evolution" nowadays, we really mean the theory of evolution *by natural selection*, begun by Darwin and Wallace, and developed by later biologists. This points to the mechanism—the process—of natural selection, as the way to explain how new species could arise from existing species.

Darwin and Wallace both began with the idea of Malthus—that plants and animals usually beget many more offspring than survive to maturity. In fact, the population of any one species tends to remain relatively stable. Since many more are born than survive to maturity, Wallace asked the simple question: Which ones survive and which die early? His answer was

> those that die must be the weakest...while those that prolong their existence can only be the most perfect in health and vigour...It is..."a struggle for existence" in which the weakest and least perfectly organized must always succumb.

Both Wallace and Darwin stressed the importance of variability among the offspring of any species. The offspring vary among themselves and are not exactly identical to their parents. With the weaker dying early, the vigor of the species is maintained generation after generation. No evolution in that. In fact, a couple of other men got that idea before Darwin and Wallace. This is simply natural selection to preserve the species.

But then Darwin and Wallace took a crucial further step, quite independently of each other. They imagined the consequences of a change in external physical conditions; a change in climate makes things wetter, for example. Let's consider a species of rat. Features that gave vigor to successive generations of rats in the old environment may be less useful in the new environment. But, some members of the species might by chance be better prepared—perhaps with a little webbing between their toes. They differ from the normal variety of rat, but not enough to have deprived them of vigor in the old environment. Now, in the new environment, that webbing between their toes may help them to get around better than those without it. Over many generations, with every increase in webbing giving added advantage, a web-toed variety will eventually replace the previous dominant variety. The new variety, which originated by chance, has been naturally selected. Not specially created by God.

Then, as Wallace put it: the continuation of natural selection over a vast amount of time can promote a multitude of variations. So the descendants could depart farther and farther from the original form, ultimately producing a new species—like muskrats—and a new genus, and so on. Darwin provided a neat summary:

> Yearly more are bred than can survive; the smallest grain in the balance, in the long run, must tell on which death shall fall, and which shall survive. Let this work of selection on the one hand and death on the other, go on for thousands of generations. Who will pretend to affirm that it would produce no effect, when we remember what, in a few years, breeders effected in cattle and sheep by this identical principle of selection?

Darwin wrote that in 1844, but didn't publish it for 14 years. Let's turn to Darwin's career and see where these ideas came from.

## *Charles Darwin (1809–1882) and Natural Selection*

Charles Darwin was born in the midlands of England, the son of a well-to-do physician. In his youth, Darwin was fond of tramping through the English countryside and developed an interest in identifying insects and rocks. After trying and rejecting a medical education, he graduated from Cambridge University, and seemed destined for the life of a country parson. Just then, in 1831, an event occurred to change Darwin's life. He was invited to accompany the captain of a British naval vessel on an expedition to chart the coast of South America. Darwin spent five years aboard HMS *Beagle*. While the *Beagle* plied the South

*Charles Darwin.*

American coast, Darwin spent his time ashore observing plants, animals, and land formations. In those five years, he transformed himself into a full-fledged naturalist.

*The H.M.S. Beagle in the Straits of Magellan.*

Darwin's scientific transformation actually began in geology. During the summer of 1831, he had accompanied one of his professors on a geological trip through Wales. Thus, we may not be surprised that among the books Darwin carried aboard the *Beagle* was the just published first volume of Lyell's *Principles of Geology*. (The further two volumes were sent to him along the way as they were published.)

Darwin's early interest in geology was important for the theory of evolution, because it contributed the idea of the long stretches of time needed for the action of natural selection to modify species. One of Darwin's first forays into geological theory on his own was an explanation of the formation of coral reefs and atolls. There was general agreement that they were formed by numerous colonies of single-celled marine animals, which can only live within about ten meters of the surface of the ocean. But how could the shape, depth, and extent of atolls and reefs be explained?

Darwin deduced his theory from seeing off the Atlantic coast of Africa what seemed to be a clear example of an island's gradually subsidence. He argued that the coral skeletons were produced on the tops of submerged mountains in the sea. As the mountains subsided, the corals kept building up their circular reefs, which could from time to time reappear as atolls slightly above the surface of the ocean. Because of the tiny size of the animals and the slowness of the growth process, large numbers of animals and long periods of time would be needed to build the atolls and large reefs naturalists had found (e.g., the 2000 km long Great Barrier Reef off the northeast coast of Australia). Curiously enough, Darwin worked out this theory before he'd ever seen a coral atoll. His theory turned out to require some modification by later naturalists.

At the various stops of the *Beagle* along the coast of South America, Darwin spent time ashore while the ship's crew conducted their hydrographic survey of coastal waters. Besides continuing his geological observations, Darwin also collected many specimens of plants, insects, and birds. After identifying and cataloging them, he sent them back to England for further examination.

After the *Beagle* returned to England, the bird specimens collected by Darwin and others on the Galapagos Islands were classified by John Gould, an expert in ornithology. Darwin had been strongly inclined to think of the finches in the collection merely as varieties of a single species. However, Gould asserted that they actually belonged to more than a dozen different species (in several genera). Almost immediately, Darwin began to consider that these different species had descended from a common stock—descent by modification, as he would eventually call it. Darwin wondered what could have caused the differences among the various species of finch. One of them had a much longer beak and attacked the bark of trees like a woodpecker. Another had a strong, parrot-like beak, which enabled it to eat much larger seeds than ordinary finches. Then Darwin realized that those finches on the Galapagos were similar to, but distinct from, a species of finch on the mainland in Ecuador.

*Darwin's Finches.*

From his geological studies Darwin knew that the islands—of volcanic origin—were much more recent than the mainland. Why should islands with an environment so different from the mainland be populated by birds so similar—and yet so distinctive? Instead of imagining that God had arbitrarily created them that way, Darwin began to look for a natural explanation. He imagined that a few of the mainland finches had invaded the islands—probably blown there by strong winds. Then, over succeeding generations, chance variations had developed naturally. So that new species of finches had arisen with characteristics that made them better able to consume the different kinds of food found on the Galapagos—different from island to island, and also very different from their normal diet on the mainland. This was Darwin's first glimmer of the idea of natural selection.

Then, as Darwin wrote many years later:

> In 1837 I opened my first note-book for facts in relation to the origin of species, about which I had long reflected, and never ceased working on for twenty years.

Soon, he was consulting breeders of domestic animals to see how rapidly new characteristics could be developed. What he learned from pigeon fanciers and others about artificial selection, formed a key step in his argument on natural selection.

By 1838, Darwin was putting his whole argument together—incorporating ideas about gradual change from Lyell's geology, ideas about population and the struggle for existence from his reading of Malthus, plus his own observations of variations. In 1844, Darwin wrote an essay of more than 200 pages outlining his ideas. But he didn't publish it. Instead, he discussed it with friends and collected more and more information to support his argument. Historians feel that Darwin wasn't prepared to face the criticism he expected would greet his bold new idea—the idea that new species arise from a parent stock through chance variation and natural selection.

However, in that same year, 1844, a Scottish journalist named Robert Chambers published a book that proposed the general idea of evolution for the whole universe and everything in it. Chambers's book, *The Vestiges of the Natural History of Creation*, was issued anonymously. Chambers wasn't antireligious, but he was going against scientific orthodoxy. Instead of having God dabbling all the time to create first this and then that, Chambers claimed that God had established two major principles of action—*gravitation* for the larger physical realm and *development* for the organic realm. God had started things off along a path and then let them follow their own natural bent. Chambers had the stars evolving, and the earth—in Lyell's uniformitarian fashion; and he had plants and animals developing progressively, just as Lamarck had done.

*The Vestiges of Creation* was a popular success. It was denounced by scientists, which only made it more interesting to the general public. Although the book was largely unsubstantiated theory, it helped pave the way for Darwin by putting the idea of evolution before the public. As Loren Eiseley has written, Chambers "drew the lightning strokes of criticism upon himself." As a result, people were less shocked when Darwin's *Origin* appeared 15 years later.

Although Darwin seems to have had the kind of intuitive insight into the logic of nature that Galileo and Newton had, he also spent many years investigating minute details. In the 1850s he published four monographs describing the class of marine animals known as barnacles. He also corresponded widely with naturalists and breeders of plants and animals to learn as much as he could about the structure and behavior of organisms.

During this period, Darwin spent his life as a semi-invalid, living in a country house near Down in Sussex, a few miles from London. In 1856 he began—methodically—to write a large book about evolution by natural selection, full of footnotes. If he'd ever finished it, the book might have been ready in about ten years.

## *Evolution Goes Public*

Out of the blue, in June of 1858, Darwin received a letter from an English naturalist working on an island in the south seas east of Borneo. Alfred Russel Wallace, 14 years younger than Darwin, was enclosing an article for Mr. Darwin to send off for publication if he thought it "sufficiently novel and interesting." Novel? Interesting? Darwin was thunderstruck! He immediately wrote a letter to Lyell:

> Your words have come true with a vengeance—that I should be forestalled...I never saw a more striking coincidence. If Wallace had my ear-

> lier essay he could not have made a better short abstract. Even his terms now stand as chapter headings of my book...all my originality, whatever it may amount to, will be smashed.

What should Darwin do? He consulted influential men in London. He'd long been on friendly terms with Lyell, now Sir Charles Lyell, and with Joseph Hooker, a director of the Kew Botanical Gardens. Being familiar with Darwin's 1844 essay, they arranged for an immediate joint publication of Wallace's recent paper and extracts from Darwin's earlier writings. That assured Darwin's priority by showing that his ideas dated back at least 14 years. So assured that we've all heard of Darwin. Who's ever heard of Alfred Russel Wallace?

*Alfred Russel Wallace (1823–1913)*

Of a rather different social class than Darwin, Wallace left grammar school at age 14 and joined his elder brother as a land surveyor. Until 1848, he performed that work, except during a slack period when he taught school for a year. Wallace came late to his interest in nature, but after developing a fascination for beetle collecting, he soon resolved to journey to the Amazon to collect specimens for sale. He published an account of his Amazon travels of 1848–49 in 1853, and then set off for the East Indies, where he remained until 1862.

*Alfred Russel Wallace.*

While in the East Indies, Wallace (like Darwin before him) became increasingly curious about the process by which new species arise. By 1855, he had seen enough evidence to publish an article contending that new species originate in space and time coincident with an already existing related species. Early in 1858, during a bout of fever, Wallace was cogitating about Malthus and the "survival of the fittest." Suddenly there flashed into his mind the idea that if the weaker die earlier than the fitter, then the species would necessarily be improved. Within a couple of days, he was able to write out the 4000-word paper that he sent to Darwin, titled "On the tendency of varieties to depart indefinitely from the original type." Its argument is the one sketched earlier in this chapter.

On his return to England, Wallace continued to support Darwin's theory of evolution, remarking that it was entirely just to judge their contributions by the amount of time each had spent on the theory—one month for Wallace, compared to twenty years for Darwin. Wallace spent the remainder of his life in writing, lecturing, and traveling. In his own time he was generally acknowledged as the co-founder of evolution by natural selection and was awarded the first Darwin medal of the Royal Society when it was struck in 1890.

### *The **Origin of Species** Published*

Now, Darwin's friends urged him to get his book published quickly. He immediately set aside his larger work and composed what he called an "abstract of an essay," derived from his full documentation of the theory of natural selection. Even then, the printed book was 500 pages long. Darwin wrote it in eight months, and the book went on sale at the end of 1859. The full title is *On the origin of species by means of natural selection: or, the preservation of favoured races in the struggle for life.*

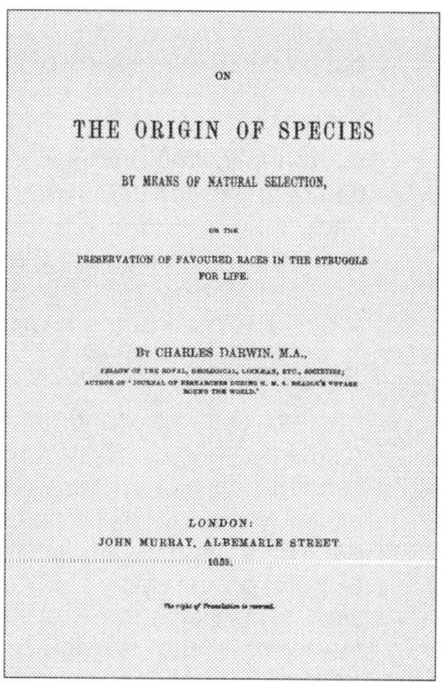

ON

THE ORIGIN OF SPECIES

BY MEANS OF NATURAL SELECTION,

OR THE

PRESERVATION OF FAVOURED RACES IN THE STRUGGLE FOR LIFE.

BY CHARLES DARWIN, M.A.,

FELLOW OF THE ROYAL, GEOLOGICAL, LINNÆAN, ETC., SOCIETIES;
AUTHOR OF 'JOURNAL OF RESEARCHES DURING H. M. S. BEAGLE'S VOYAGE ROUND THE WORLD.'

LONDON:
JOHN MURRAY, ALBEMARLE STREET
1859.

*The right of Translation is reserved.*

*Title page of the first edition of* On the Origin of Species.

Darwin's *Origin of Species* was a masterpiece of scientific reasoning. It was a coherent, consistent argument, in which Darwin marshalled his evidence bearing on the theory of natural selection. His first chapter described the varieties of creatures that can be produced by the artificial selection of stock breeders. In the second chapter he discussed the kinds of variations that can be observed in nature. Darwin titled the next three chapters, "The Struggle for Existence," "Natural Selection," and "The Laws of Variation." These first five chapters contained the core of Darwin's argument. The final paragraph of Chapter Five is worth quoting:

> Whatever the cause may be of each slight difference in the offspring from their parents—and a cause for each must exist—it is the steady accumulation, through natural selection, of such differences, when beneficial to the individual, that give rise to all the more important modifications of structure, by which the innumerable beings on the face of this earth are enabled to struggle with each other, and the best adapted to survive.

Then, in a move unique in the annals of science, Darwin devoted a whole chapter to the difficulties of his theory. He showed that he knew what the problems were and tried to find ways around them. His readers could judge how well he succeeded.

Darwin devoted most of the rest of the book to showing how descent with modification could explain a variety of biological facts, such as, for example, the geographical distribution of organisms.

The final chapter was a résumé of the book:

> As this whole volume is one long argument, it may be convenient to the reader to have the leading facts and inferences briefly recapitulated: That many and grave objections may be advanced against the theory of descent with modification through natural selection, I do not deny. I have

> endeavoured to give them their full force. Nothing at first can be appear more difficult to believe than that the more complex organs and instincts should have been perfected, not by means superior to...human reason, but by the accumulation of innumerable slight variations, each good for the individual possessor. Nevertheless, this difficulty, though appearing to our imagination insuperably great, cannot be considered real if we admit the following propositions; namely,—that gradations in the perfection of any organ or instinct...either do now exist or could have existed, each good of its kind;—that all organs and instincts are, in ever so slight a degree, variable;—and lastly, that there is a struggle for existence leading to the preservation of each profitable deviation of structure or instinct. The truth of these propositions cannot, I think, be disputed.

Finally, there's real poetry in Darwin's concluding paragraph:

> It is interesting to contemplate an entangled bank, clothed with many plants of many kinds, with birds singing on the bushes, with various insects flitting about, and with worms crawling through the damp earth; and to reflect that these elaborately constructed forms, so different from each other, and dependent on each other in so complex a manner, have all been produced by laws acting around us....There is grandeur in this view, of life with its several powers having been originally breathed into a few forms or into one; and that, whilst this planet has gone cycling on according to the fixed law of gravity, from so simple a beginning endless forms most beautiful and most wonderful have been, and are being, evolved.

The *Origin of Species* was an instant sensation. It went through six editions in Darwin's lifetime. Forty thousand copies were sold in England during the 30 years up to 1890, and many more in the United States.

Just because Darwin's *Origin of Species* sold well, doesn't mean it was immediately accepted. Many people read it in order to object to it. "Can you imagine? That Mr. Darwin thinks animals just evolved? Next thing you'll know he'll be saying we're descended from the apes!" Eventually Darwin did say something like that—as my clergyman grandfather wrote on a note inserted in the front of his copy of Darwin's later book *The Descent of Man*. It was a real Victorian scandal!

Public reaction came to a head at the 1860 meeting of the British Association for the Advancement of Science held at Oxford. Darwin didn't attend. But he was ably defended by a young disciple in his mid-thirties, who later came to be called Darwin's bulldog. Thomas Henry Huxley (1825–1895) was a surgeon and a naturalist.

The attack on Darwin and evolution was led publicly by the Bishop of Oxford, Samuel (Soapy Sam) Wilberforce, though he was coached by Darwin's jealous scientific opponents. To a packed house, Wilberforce used his oratorical skills to full effect, playing on Victorian sensibilities by wondering if Darwin and Huxley were ready to claim that women as well as men could be derived from beasts. "Tell us, Mr. Huxley, do you claim your descent from the apes through your grandfather, or your grandmother?"

*Thomas Henry Huxley.*

Huxley rose to reply, fully aware that the bishop had played into his hands. After a brief explanation of Darwin's theory as a legitimate scientific hypothesis, he turned to Wilberforce: "I would never have dreamed of raising in this discussion such a question as My Lord Bishop has asked; but since he asked it, I answer without hesitation that I would not be ashamed to have an ape for my grandfather. What would make me feel shame would be to have to acknowledge as ancestor a man of such great influence and intelligence who, nevertheless uses those high endowments to introduce prejudice and ridicule into a sober scientific debate."

Ladies fainted at the disrespect toward a bishop, while undergraduates cheered for the same reason. The debate over evolution between scientists and religionists went on for a long time. It's still going on.

## *Evolution Unifies Biology*

What contribution did Darwin, Wallace, and the theory of natural selection actually make to science?

As early as 1844, Robert Chambers had written in his *Vestiges of the Natural History of Creation* that as gravitation provides "one comprehensive law" for the inorganic world, so the organic world "rests in like manner on one law"; he called this the law of development. Although premature, Chambers clearly pointed the way to the significance of the theory of evolution as Darwin presented it in his *Origin of Species*. That Darwin felt the same way is suggested by the "grandeur in this view of life" sentence, with which he concluded the *Origin* (see above).

To see all organisms as related by the evolutionary tree (or bush) does for biology what Newton's gravitational theory did for physics. Just as gravity provides a single basis for phenomena as diverse as falling apples and spinning galaxies, so evolution provides a foundation for seeing algae, corals, redwoods; gnats, humans, and whales all as part of a single organic web.

Very soon, the work of taxonomists took on a very different character than it had with Linnaeus and the notion of fixed species along God's great chain of being. Now, a natural scheme of classification would be one that attempted to demonstrate evolutionary relationships. That's not easy to do, because the lines of descent according to natural selection depend on the myriad ways in which environmental changes affect the varieties of a species that happen to be present. Nonetheless, taxonomists now had a much clearer vision of their task than when they were trying to plumb the depths of God's plenitude.

How you look at the organic world strongly affects what questions you ask about it. For example, if you take seriously the evolutionary relation between humans and apes, you can investigate brain functions in apes and relate them to human anatomy and physiology (always being careful, of course, to acknowledge that there must be *some* differences). Thus, the anatomist F.G. Parsons could write in 1911 (*Encyclopedia Britannica* 4: 402) that the brain of a gibbon "apart from all question of comparative anatomy, forms a useful means of demonstrating to a junior class" the main convolutions of human brains.

Apart from the "cosmological" significance of the unitary view of life provided by the theory of evolution, biologists in many fields found themselves asking new questions within that framework. Not only did classification take on a new meaning, but so also did studies in anatomy, physiology, and embryology. Indeed, comparative studies among classes and orders only make sense from an evolutionary point of view. These biological fields flowered in the decades following 1860.

Such contributions to research programs brought many biologists and geologists onto Darwin's side. Charles Lyell, for example, in a later edition of his *Principles of Geology* supported the theory of evolution—though somewhat grudgingly. Although his uniformitarianism had given Darwin the long stretches of time needed for the slow changes wrought by natural selection, Lyell had to overcome his previous objections to progression in the organic world. He just didn't find it easy to change his mind.

### *Physicists Attack Evolution*

In the early 1860s, some physicists mounted a direct challenge to that idea of long stretches of time. Chief among them was William Thomson (1824–1907), Lord Kelvin, the leading British scientist of the nineteenth century. Kelvin wrote articles and gave speeches claiming that the earth was probably less than 50 million years old. Darwin needed hundreds of millions of years. For, if all the small changes had to be compressed into a few million years, we'd expect to see such changes happening around us. Darwin had argued that we don't see them because they happen too slowly for us to observe.

*William Thomson, Lord Kelvin.*

Kelvin's claims gave Darwin a lot of trouble, because they were buttressed by the appearance of mathematical certainty. Kelvin calculated the age of the sun from the length of time a ball of hydrogen that size could continue to burn. That worked out to less than 100 million years. And, of course, the earth couldn't be any older than the sun.

Then Kelvin calculated the age of the earth directly, using the idea of its rate of cooling. Temperature measurements in mines showed that the earth was warmer down below than at the surface. That meant the earth was cooling down from the temperature it had when it was formed (roughly the present surface temperature of the sun). So, Kelvin calculated how long the earth must have taken to cool down, and be at a temperature suitable for life. He got about 30 million years.

These two independent calculations—one for the sun, the other for the earth—tended to support each other. The earth was between 30 and 100 million years old, say 50 million years. Kelvin's arguments almost took poor Darwin's breath away. All he could do was to suggest that changes might have occurred more rapidly in ancient times than he'd supposed.

It happens that both Kelvin's calculations were wildly wrong, because he didn't know what we now know. The sun's heat doesn't come from burning hydrogen, but from a thermonuclear reaction. Nuclear reactions are so much more energetic than burning that they use up the sun's hydrogen very much more slowly than Kelvin calculated. And the earth's internal heat is not just what's left over from when it was formed; radioactive rocks are continually producing more heat within the earth. So, new discoveries since Kelvin have given back ample time for evolution. Recent estimates put the age of the earth at about five billion years—a hundred times what Kelvin allowed. Evolutionists can breathe easily again.

### *Evolution and Genetics*

Another big problem in Darwin's and Wallace's theory of natural selection was that they had only one mechanism where they really needed two. It was all very well to say that the fittest varieties would be selected naturally in the struggle for existence. The mechanism of selection may work all right. But what's the mechanism for creating the varieties in the first place? How do those chance variations arise?

Neither Darwin nor Wallace had a satisfactory answer. Both of them were adamant in rejecting Lamarck's answer: that the variations somehow came from the deliberate effort of creatures to adapt to new environments. And yet, without anything better to put in its place, after kicking Lamarck out through the parlor, Darwin sneaked him back in through the kitchen. In his later writings, Darwin developed a strange theory: that new characteristics were *imposed* on creatures by a changed environment. And then were collected by the blood and deposited in the sperm or egg—to be passed on to the next generation. Really just a disguised form of Lamarck's inheritance of acquired characteristics.

Despite the great value of the theory of evolution in organizing biology, it didn't really explain the source of the variations, which were naturally selected. That explanation requires a correct theory of genetics. Oddly enough, had Darwin or Wallace only known it, the basic theory of genetics was found only five years after Darwin's *Origin* was first published. In central Europe in 1866, an Augustinian priest named Gregor Mendel published a clear statement of the laws of inheritance—firmly supported by experimental results.

Mendel's paper suffered instant obscurity. Biologists noticed his work only 34 years later, in 1900, when they were already on the right track themselves (Chapter 21).

## *Influence of Evolution beyond Science*

Evolution is both a scientific theory and a general idea in western culture. It is important to try to keep these two separate at least to some extent—while realizing that they will inevitably interact.

As a scientific theory, evolution supplied the basis for research programs in fields as diverse as embryology and anthropology. Within science it is subject to the canons of scientific reliability and applied in circumstances where scientists find it helpful.

As a general idea, evolution challenged cherished beliefs that lay at the very core of western European civilization. After the publication of Darwin's *Origin of Species*, theologians, philosophers, and lay people reacted in a variety of ways to the new idea. Either they rejected it outright, or they considered how their view of the world would have to be modified to accommodate it. We must understand that these people reacted to the *idea* and were not signficantly influenced by the continuing modifications and developments of the scientific theory over the century following 1859.

In the middle of the nineteenth century, there were many lay people with interests in natural history and natural theology. They quickly realized that the *Origin of Species* would have a significant impact on those interests, that it could alter their views of order in organic creation and of the relation of God to the world.

This was particularly true of clergymen in England. Their reactions ran the gamut from instantly opposing evolution to sympathetically examining how they could change their teachings to accommodate the idea of evolution. Could they, for example, allow evolution to be the process employed by God in populating the earth? Could they limit the idea of God's continuing activity in the world to the spiritual realm without diminishing their idea of God's omnipotence? In general, could they maintain the major tenets of Christian theology within the context of evolutionary theory? Some could; others had more difficulty. This process went on for a long time.

### *Evolution and Creation*

In recent years, evolution has appeared in news headlines because of strong doubts raised against it by evangelical Christians. Basing themselves essentially on Biblical texts—the Word of God—they have sought equal time for "creation science" in school science classes. Controversies continue to rage from time to time in the media, in court rooms, and at meetings of school boards.

This is a touchy issue, which is more inclined to generate heat than light. At one level the issue is really a matter of cosmology or world-view—what one believes about the nature and structure of the universe. Attitudes about cosmology have implications also in morals and politics, and even in economics, as well as in religion and science. They are largely matters of conviction in which rational argument seems generally to take a back seat to passion and emotion.

At another level, the evolution/creation argument may have a direct impact on scientific enquiry. As long as the distinction between evolution as world-view and as scientific theory is not kept clear, arguments about the former can impact on the latter. Many scientists oppose the creationist attacks on evolution in order to preserve scientific integrity. They are unhappy to see their principles of experimental validation and rationality on trial in civil courts. They feel that the public's appreciation of the capacity of those principles to advance knowledge is being undermined.

On the other hand, the creationists are inclined to consider that science today represents the constituted authority, and that it is running out of control. Thus, the evolution/creation controversy has become a symbol for ongoing struggles about the distribution of power in a democracy: the rights of parents to determine their children's beliefs, and the rights of scientists to control the canons of scientific evidence, among others.

*Social Darwinism*

In the last half of the nineteenth century a number of writers proposed that evolution supported the view that human behavior was, or should be, controlled by the evolutionary idea of the "survival of the fittest." They saw social interaction as a struggle among competing groups and individuals. Since most of these writers came from the more well-to-do classes in England and America, their message was that governments should not intervene in society, but allow the enterprising to succeed and the inefficient to fail. They promoted economic *laissez-faire*, which literally means "permit to act," and may be loosely interpreted as "leave us alone to make out." They claimed that large-scale capital enterprises should be left free to act in their own best interests, independent of their effect on employees or customers. In a free market the fittest would survive.

In the same vein, the struggle for existence was applied to warfare among nations, and could even be used to justify genocide; that the weeding out of "inferior races" would produce a general improvement in human kind.

You should not suppose that these were new ideas engendered by Darwin and the theory of evolution. People simply grasped at evolution to justify behaviors and attitudes they had already developed in their own self-interest. As long as evolution was seen to be a positive addition to our knowledge of the world, its principles were used to strengthen traditional prejudices.

Generally speaking, there is little evidence that Darwin himself countenanced such applications of the theory of evolution. Indeed, many of these ideas derived more from the writings of Herbert Spencer than from the *Origin of Species*. Spencer began to write about the idea of evolution even before

1859, and it was he who coined the phrase "survival of the fittest." Spencer's writings were devoted to philosophical and social ideas, and were not significantly scientific. Darwin's reasoned arguments in the *Origin* merely added a patina of scientific respectibility to those ideas, for which there was very little real support in the detailed theory of natural selection.

Now that the neo-Darwinian genetical theory of natural selection is firmly established (see Chapter 21), we should be wary of attempting to apply it to realms beyond the reach of scientific testability. While evolution and natural selection may provide the basis for fruitful hypotheses or analogies in other realms, such hypotheses need to be tested in any such realm according to the standard canons of scientific evidence.

All too often, imaginative writers will take hold of a poorly understood new idea and try to fit it into their own pet schemes for saving the world. In matters of human behavior and the social good, people often grasp at straws that will help to simplify existence in this complex modern world. They see scientific theories as fair game, because of the generally high regard in which science is held today. Their major error is to attribute general *truth* to scientific principles that possess only a defined validity within the range of experience in which they have been tested. While extrapolation of such principles into other areas is an important scientific activity, no extrapolation can be accepted without being tested. Assertions of validity beyond the range of testing are simply false, no matter how persuasively or loudly they may be proclaimed.

A more recent application of biological theory to human behavior has arisen under the name of sociobiology. It will be discussed in Chapter 21.

## *Conclusion*

From 1500 to the 1850s, our knowledge of the world gradually increased as a result of exploration and digging. Geological layers were identified by the fossils they contained. Linnaeus brought order to the world of organisms by his binomial classification system. Through most of this period, scientists stuck with the theory of creation and existence they'd made to fit the Bible.

Despite Galileo's troubles with the authorities in Rome, Christianity and science were mostly mutually supportive from 1500 until past 1800.

Indeed, there's a certain irony in that interaction between science and religion. The American anthropologist, Loren Eiseley, says that while recent scientists have accused religion of opposing progress by its preconceived beliefs, it's more complicated than that. Eiseley wrote:

> By establishing species as a fixed point for examining the organic world, Linnaeus gave the concept of species a precision and fixity which it did not originally possess. Until the scientific idea of species acquired form and distinctness there could be no dogma of special creation in the modern sense. That only happened when naturalists of the seventeenth

> century began to substitute exactness of definition for the previous vague characterizations of the objects of nature.
>
> As scientific delight and enthusiasm over the naming of new species grew with the expanding world of the voyagers, the conviction of the stability and permanence of the living world increased. Strict definition, so necessary to scientifically accurate analysis, led in the end to the total crystallization of the idea of order. It was Linnaeus, with his proclamation that species were absolutely fixed since the beginning, who intensified the theological trend. Henceforth the church would take the fixity of species for granted. Science, in its desire for classification and order, found itself satisfactorily allied with a Christian dogma whose refinements it had contributed to produce.

New, naturalistic foundations for geology and biology were attempted by Hutton and Lamarck, though they weren't widely influential in their own lifetimes. But Hutton's views, taken up by Lyell, made geology firmly scientific. A scientific theory is one we can adjust to conform to the evidence. A theory delivered by God in the Bible may well be true—it's just not *scientific* because we can't adjust it, even when there's evidence to the contrary.

The scientific foundation for biological relations and changes was laid by Charles Darwin in his book *On the Origin of Species*.

The theory of evolution by natural selection needed the geological doctrine of uniformity to provide the long time-span required for the myriad of steps from fish to horses. The mechanism of natural selection conceived by Darwin and Wallace also required the careful observation of varieties and species in nature—combined with the insight of Malthus' essay *On Population*.

Others before Darwin and Wallace had pieces of the puzzle. But Darwin and Wallace put them all together. They framed the basic argument and supported it with a wealth of detail. Like Newton, they get the lion's share of the credit (and blame too, as it happens), despite the essential contributions of many others.

With the theory of evolution, biology was put on a firm scientific foundation. The theory of natural selection could be tested and modified as required by the evidence.

Recall that science involves a close interaction among speculation, reasoning, and direct evidence (whether from experimentation and analysis as in physics and chemistry, or from observation and classification as in geology and biology, where experiments in the sense of the physical sciences often cannot be done). The period when that conjunction of activities can be found in a field of human investigation marks the time of its becoming scientific in the sense we've defined. Thus, astronomy became scientific with Ptolemy, about 150; physics became scientific with Newton, about 1690; chemistry became scientific with Lavoisier, about 1790; geology became scientific with Lyell, about 1830; biology became scientific with Darwin and Wallace, about 1860.

*Michael Faraday.*

CHAPTER SEVENTEEN

# Currents and Waves

Isaac Newton cast a long shadow over investigations of nature throughout the eighteenth century. For more than a hundred years, scientists mined his two great works, the *Principia* and the *Opticks*, for details to elaborate or correct. More than that, those two books laid down (both explicitly and by example) the procedures that Newton's successors would follow.

The *Principia* was the handbook of techniques in mathematical physics; grand speculations in gravitation were coupled with detailed analyses of relations among physical quantities, tested by precise measurements in laboratory and observatory. Though many of those measurements were made by others, in the *Opticks* Newton showed himself to be as superb an experimenter as he was an analyst.

Scientists followed Newton in a variety of ways. Keen mathematical analysts on the continent (not many in England) examined the *Principia* in great detail, testing his logic as they converted the form of his relations to the style of calculus devised by Leibniz. They elaborated his analytical techniques into a mathematical physics of great power.

On the other hand, experimenters in England and America used the techniques they found in the *Opticks* to put the study of electricity on a firm foundation. Toward the end of the eighteenth century, they had clarified electrical phenomena sufficiently to be able to analyze them in the style already found successful in mechanics, both terrestrial and celestial.

The study of optical phenomena lingered longer within the channels established by Newton. Only at about 1800 did investigators begin to break away from Newton's particle model for light, and use a wave model to make optics as mathematical as mechanics and electricity.

Newton seems to have wished to account for all phenomena by the analysis of motions and forces. Such a program, if successful, would provide a unified view of all physical phenomena. Thus, Newton handled the refraction of light as an interaction between luminous particles and the atoms of matter. He was premature by almost three centuries. Now physicists in the latter stages of the twentieth century have made a grand unification of the forces of nature their major objective.

Actually, the earliest successful unification of physical fields occurred in a very different way—in the principle of energy conservation, partly by adding novel experimental investigations to Newton's analyses, and partly by indulging in wild speculation. However, as a result of the canons of logic and evidence laid down in the seventeenth century, wild speculations were not admitted into the accepted corpus of physics until they'd been thoroughly tamed.

## *Particles or Waves of Light*

Luminous phenomena have intrigued humans for a long time. Our fascination with light may well date all the way back to our primitive ancestors' first consciousness of fire. As for written sources, we can find the prominence of light expressed right at the beginning of *Genesis*:

> In the beginning of Creation, when God made heaven and earth, the earth was without form and void, with darkness over the face of the abyss, and a mighty wind that swept over the surface of the waters. God said, 'Let there be light,' and there was light; and God saw that the light was good, and he separated light from darkness. He called the light day, and the darkness night.

The writer of those lines seems to be putting the creation of light even before that of the material world. First God, then light, then the world. So, what is light? Is it matter or force or something else? The ancient Greeks knew that *sound* was produced by vibrating objects. They figured we hear sounds as vibrations spread out in waves through the air. But light seemed to travel in narrow beams. It didn't spread out. The Greeks said light travels in rays, not waves.

For two thousand years, light was considered to be some kind of material—particles of light streaming out from the sun, for example. After 1800, scientists challenged that particle idea. Thomas Young found evidence that light did travel in waves, that light was much like sound. Except that light didn't need air to travel in. So scientists like James Clerk Maxwell invented a hypothetical medium called the **æther** for light to travel in. For them, light wasn't made of material particles, but was electromagnetic energy spreading out in waves in the æther. They couldn't tell what the æther was made of, but they were sure it wasn't made of atoms.

## *Early Ideas about Light*

Early Greeks thought we see things because the things themselves send out little images which strike our eyes. That was easily disproved. You can't see in a dark room. The sun or a lamp has to be shining. Eventually, mathematicians decided that light from a lamp was reflected from objects to our eyes. In the third century BCE, Euclid worked out the geometry of reflected light rays. He described how light worked in simple cases, without worrying much about what light was.

In the middle ages, Arab and Latin scholars extended Euclid's work. They also dissected vertebrate eyes and began to figure out how our eye's lens and retina help us to see.

During the scientific revolution, Kepler and others established a simple geometry of light rays to explain the refraction of light in lenses. And Galileo used lenses to make telescopes and microscopes.

Then René Descartes undertook to describe the actual nature of light. He wanted to fit everything, including light, into one consistent, mechanical system. Descartes pictured the universe full of matter, of particles of various sizes whirling and colliding. Since particles of air transmitted sound, Descartes invented particles to transmit light. He knew that scientists like Galileo and Mersenne had worked out how vibrating objects created sound waves.

The vibrations of a plucked harp string make the air vibrate, and the vibrations spread out in waves through the air particles. Air is a material medium: it carries sound at about 300 m/s. The distance between the crests of the waves–the **wavelength**–ranges from a few millimeters for high pitched sounds to a few meters for deep bass sounds. Descartes' picture for light was a little like that. For him, light consisted of a particular set of particles that transmitted light by bumping one another along. But he didn't say they were waves. The Dutch scientist, Christiaan Huygens, did say that light was waves. Near the end of his life, about 1690, Huygens published a *Treatise on Light*. He claimed that light had to involve material particles, something like sound. Huygens wrote:

> It is inconceivable to doubt that light consists of the motion of some sort of matter. For, if we consider its production, we see that here on Earth it comes mainly from fire and flame, which are clearly bodies in rapid motion, since they dissolve and melt many other bodies, even the most solid. If we consider its effects, we see that when light is collected, it has the property of burning as fire does. That is, it disunites the particles of bodies. This is assuredly the mark of motion, at least in the true Philosophy, in which we conceive the causes of all natural effects in terms of mechanical motions. This, in my opinion, we must necessarily do, or else renounce all hopes of ever comprehending anything in Physics.

Notice how easily Huygens defined "true Philosophy" and incorporated physics within it. The idea of mechanical motions being the foundation of all causes was the metaphysical presupposition shared by him and Galileo, and largely but not entirely, by Newton. Huygens continued:

> Now there can be no doubt that light comes to our eyes from a luminous body by some movement impressed on the matter between the two. This movement is successive and it spreads as sound does, by spherical surfaces and waves. If we examine what the matter may be in which the movement from the luminous body spreads—which I call Ethereal matter—we'll see that it's not the same that serves for sound. For sound travels through air; and if air is removed from a vessel, light still passes through it, but sound *does not.*

Huygens considered light to be like sound in some respects and different in others. They both spread out in waves through some medium—sound in air at a low speed, light at a very much higher speed in something other than air, a material much more subtle than air. Huygens didn't say anything about wavelengths of light.

Contrary to Descartes and Huygens, Newton had a mostly empty universe. So Newton had particles or corpuscles of the various colors of light streaming out from the sun—even more material than Huygens' model for light. And Newton's reputation was so great that his corpuscular theory of light dominated scientists' thinking throughout the 1700s.

## *Thomas Young (1773–1829) and the Interference of Light*

*Thomas Young.*

Around 1800, Thomas Young, in England, began to transform our ideas about light completely. Born into a family of Quaker merchants, Young was a precocious student with a talent for languages, both ancient and modern. He was trained as a physician and elected to the Royal Society at age 21, for an anatomical study of the human eye. For a couple of years, he was professor of natural philosophy at the Royal Institution in London, shortly after its founding. Later, he did valuable work in translating Egyptian hieroglyphics.

Young got to his study of light through his medical interests. In his doctoral dissertation, he studied the similarities between seeing and hearing, which led him to consider the possibility of light being a wave phenomenon like sound. Curiously, he got a clue to this from Newton himself, normally considered to be the great opponent of wave theories.

In his *Opticks,* Newton reported his observations of the colors of thin films. These are the colored patterns you can sometimes see in soap bubbles, or in an oil slick on top of a puddle. Close observation of these colors when produced under controlled conditions shows spectral colors repeated periodically. . Newton found the various colors associated with the varying thickness of the film. (He used an air film between flat and convex glasses, instead of soap or oil, but the principle's the same.) Referring to the thicknesses appropriate to the various colors, Newton wrote:

> And hence I seem to collect that the thicknesses of the Air between the Glasses there…are to one another very nearly as the six lengths of a Chord which sound the Notes in a sixth Major, sol, la, mi, fa, sol, la.

Such a conjunction of measurements in light and sound could not fail to impress Young. He resolved to examine the consequences of explaining the various colors of light by their wavelength. If light is a wave, he reasoned,

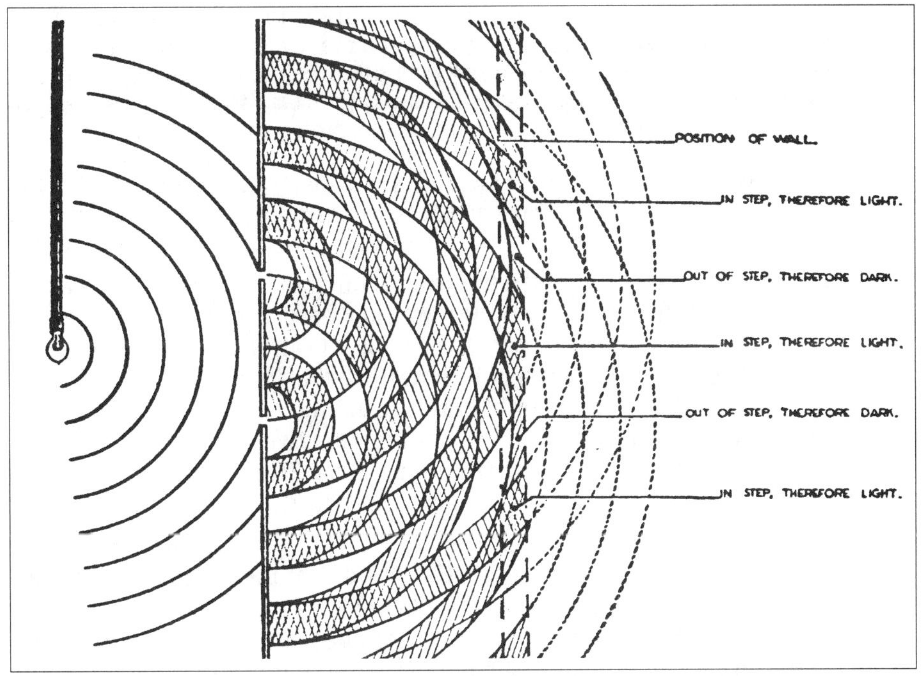


*Thomas Young's two-slit experiment. Light from a single source passes through two very narrow slits placed close to each other. The light diffracts outward from each slit, producing interference patterns that are visible as bands of dark and light.*

then light reflected from the two surfaces of Newton's film of air would experience *interference,* the same phenomenon that determines the pitch of the sound of a stretched string. Think, for example, of the different lengths of the strings of a harp and the periodic repetition of the notes of the octave.

However, Young knew that a mere re-interpretation of a standard phenomenon would not carry conviction. He needed more experimental support. So he sought a means of producing a direct interference of light, rather than using reflected light.

Young reasoned by analogy with waves on the surface of water. Suppose the path of such ripples is split so that they go along two channels of unequal length, and are then reunited. If one channel is half a wavelength longer than the other, crests from one will coincide with troughs from the other, and the water where the channels reunite will be unruffled. This effect of coinciding crests and troughs (whether it's crest and trough adding to make zero wave action, or two crests [or two troughs] adding to make an increased wave action) is called **interference.** All Young had to do was to find an analogous phenomenon in light.

About 1801, he found a way. He started with a very narrow beam of sunlight passing through a pin hole. Then, as he described it:

> I brought into the sunbeam a slip of card, about one thirtieth of an inch in breadth, and observed its shadow, either on the wall, or on other cards held at different distances. Besides the fringes of colours on each side of the shadow, the shadow itself was divided by similar parallel fringes, of smaller dimensions, differing in number, according to the distance at which the shadow was observed, but leaving the middle of the shadow always white. Now these fringes were the joint effects of the portions of light passing on each side of the slip of card, and inflected, or rather diffracted, into the shadow.

**Diffraction** is the term applied to the spreading of light beyond the limits of a shadow as predicted by ray optics. Young clearly recognized diffraction as a consequence of the wave nature of light. In other experiments, he allowed the light beam to fall on a pair of narrow slits very close together. He observed a series of dark and bright bands resulting from the interference of light traveling along different path lengths from the two slits. With a prism, he could illuminate his slits with light of single colors. The interference pattern then was alternate narrow bands of color and black.

By measuring the distances between the light and dark bands, Young calculated the lengths of waves of light. He found that the wavelength of red light was almost double the wavelength of violet, 0.7 μm and 0.4 μm, respectively. With their different wavelengths, colors are distributed along a spectrum, just as in sound, notes are distributed along the scale.

Because of Newton's enormous reputation in England, Young's work was not well received there. So the wave challenge to Newton's authority did not become effective until after Young's work was repeated and made mathematical in France.

*Augustin Fresnel (1788–1827)*

*Augustin Fresnel.*

Fresnel, son of a Normandy architect, was the chief builder of the wave theory in France. He had a solid training in mathematics and civil engineering, the profession he subsequently followed. About 1814, during a period of enforced leisure at the time of Napoleon's return from Elba, Fresnel began his optical researches. He repeated many of Young's experiments on diffraction and interference, without being aware of that English work.

In Paris in 1816, Fresnel reported his research to distinguished members of the *Académie des sciences*, whose Newtonian convictions prevented them from appreciating his work. One of them scoffed that Fresnel's theory implied that there should be a bright spot in the center of the shadow of a tiny ball. Agreeing, Fresnel proceeded to demonstrate it to the skeptic in his laboratory. In 1818, he submitted a paper for a competition of the Academy to determine "all the effects of diffraction of luminous rays."

Fresnel won first prize, the eminent judges having been sufficiently convinced to change their minds.

During the remainder of his short life, Fresnel developed an elegant mathematical wave theory of light, many of the features of which were confirmed experimentally. Fresnel thereby made our knowledge of optical phenomena as fully a part of mathematical physics as Newton had done for motion and gravitation. And like Newton, Fresnel contributed by his talents both in analysis and in experiment. He wrote to Thomas Young that the esteem of the leaders of French science "never gave me as much pleasure as the discovery of a theoretical truth, or the confirmation of a calculation by experiment."

## *Bringing Electricity into Physics*

Newton had brought physics to the stage where the same mechanical laws explained the motions of both cannonballs and planets. With his laws of motion and gravitation, Newton had provided a scheme for analyzing motions and forces. A scheme so effective that other scientists were soon extending it to other forces, like electricity and magnetism. In fact Newton had foreseen this development himself. In the Preface to his *Mathematical Principles of Natural Philosophy*, he wrote:

> the whole burden of philosophy seems to consist in this—from the phenomena of motions on earth to investigate the forces of nature, and then from these forces to deduce the motions of the planets, the comets, and the sea. I wish we could derive the rest of the phenomena of nature by the same kind of reasoning from mechanical principles; for I have many reasons to suspect that they may all depend upon certain unknown forces which either drive the particles of bodies towards one another so they combine in fixed shapes, or cause them to be repelled from one another. Since these forces are unknown, philosophers have been conducting their search of nature in vain; but I hope the principles here laid down will offer some light either to this or some truer method of philosophy.

This program was applied to the study of electricity and magnetism in the late eighteenth century after investigators had been able to clarify the nature of those forces. That clarification began even before Newton's time, mainly in the work of William Gilbert (Chapter 11).

Gilbert published his book *On the Magnet* in 1600. There he explained why one end or pole of a magnet always pointed north. Gilbert also showed that north-seeking poles repelled each other, while a north-seeking pole attracted a south-seeking pole. Likes repel and unlikes attract.

Gilbert also did some experiments in static electricity. He showed that many substances rubbed with fur or cloth could be made to attract bits of chaff. The Greeks had noticed this effect with amber as early as the sixth century BC. Because the Greek word for amber is *electron*, Gilbert used the term **electrification** to describe the state of the materials that attracted after being rubbed. He stressed two main points. First, that electrification was a general phenomenon; it worked on almost everything, not just amber. Second, electrification was different from magnetism. Up till his time, many people had considered electricity and magnetism to be similar magical forces. Gilbert pointed out that magnetism worked only on objects containing iron, while electricity worked on *everything.*

Gilbert described only electrical *attraction.* In 1650, Niccolo Cabeo (1585–1650) in Italy reported that two electrified objects repelled each other. During the following decades, a number of experimenters played with various ways of producing electrical forces. They found that a strongly electrified object would give off sparks. It soon became a parlor game, like electrifying a boy suspended on silken threads and drawing sparks from him. The process of electrification was often called **charging**—in the sense of filling up. So charging means adding electricity to an object, and a "charged object" is one that has been electrified.

In 1735, a French experimenter, Charles Dufay (1698–1739), observed that there were actually two kinds of electric charge, depending on which materials he rubbed. Charges of the same kind repelled each other, and unlike charges attracted.

Through the 135 years from Gilbert to Dufay, the study of electricity progressed slowly. People concentrated on the qualitative effects, without doing much measuring. Then, under the influence of Newton's methods, scientists started to treat electricity quantitatively, by measuring.

But what should they measure? What is a quantity of electricity? First of all, they found that when they rubbed a glass rod with silk, they got one kind of charge on the glass and the other kind on the silk.

And!

The two charges neutralized each other. Whenever they produced one kind of charge, they also produced an equal amount of the other kind of charge. They said: *electric charge is conserved.*

About 1750, Benjamin Franklin (1706–1790) in America took up the study of electricity. He claimed that if one kind of charge neutralized the other, that meant they were merely positive and negative conditions of the same thing. He performed experiments he could interpret like this: every object contains a certain normal but unnoticed amount of electricity. When you bring glass and silk into contact, the glass takes electricity from the silk. That gives the glass a surplus of electricity—it's positive; and the silk is left with a deficit—it's negative. And because the amounts are equal, they add to zero: charge is conserved.

Franklin is also famous for flying his kite in a thunderstorm and drawing sparks from the string. That dangerous trial convinced him that lightning is just a gigantic electric spark. This was something people had suspected for half a century, but Franklin proved it. Luckily for him, he wasn't struck by lightning and killed. Soon after, a European scientist wasn't so lucky. To confirm

Franklin's work, he installed a long metal rod leading from his roof down into his laboratory. When lightning struck the rod, he was killed.

In France in the 1780s, Charles Coulomb (1738–1806) devised an apparatus to measure electricity more precisely than just plus or minus. He figured out a way to measure the tiny forces between electric charges. Now, Coulomb could describe the size of an electric charge by the force it exerted. At the same time, he demonstrated that electric force followed the same rule Newton established for gravitational force. That is, electrical force decreased with distance according to the inverse square law. At double the distance, a quarter of the force; and so on.

Coulomb made similar, careful measurements for magnetic forces. With this sound experimental foundation, analysts were now able to follow Newton's program for these forces. Between 1780 and 1800, they established very detailed mathematics for both electric and magnetic forces.

## *The German Philosophy of Nature*

At the same time, philosophers became so confident that Newton had the key for unlocking the mysteries of nature that they took up and overgeneralized his ideas. Especially in Germany. They created a whole philosophy of nature that's still described by its German name, *Naturphilosophie.* These German philosophers saw the universe as such a coherent, well-ordered system, that they proposed that God had originally invented one single basic force. And that the various forces we experience, like gravity, electricity, light, and so on, are just modifications of that one basic force. This also encouraged them to think that any kind of force could be changed into any other kind.

Then in 1800, Alessandro Volta (1745–1827) in Italy found out how to produce a continuous flow of electricity—the electric battery. Volta piled up copper and zinc plates alternately, separated by cardboard soaked in acid. He found that the pile produced a continuous electrical effect when he connected wires to the top and bottom of his pile. An electric current!

The German philosophers saw the voltaic pile as an example of converting a chemical force into an electrical force. And Humphrey Davy showed the reverse (Chapter 15). He found that the electrical force could produce chemical force when he used electricity to separate salt into its component elements, sodium and chlorine. There was even more. When scientists attached a wire from one end of Volta's pile to the other, the wire got hot. Sometimes hot enough to glow red. So electric force could be converted to heat force and light force. And, of course, chemists already knew of the interconnection of chemical and heat forces. Chemical reactions produce heat, and vice versa. So the philosophers generalized and claimed that all forces were interconvertible, and in some sense identical.

## *Connecting Electricity and Magnetism*

*Hans Christian Oersted.*

This belief in the interconnections among forces strongly influenced the Danish physicist, Hans Christian Oersted (1777–1851). Starting about 1800, Oersted began looking for a way to produce magnetism from electricity. Electricity produced every other force—why not magnetism? For the better part of 20 years, he tried to make a wire carrying electricity deflect a magnetic compass. Since the effect Oersted was searching for turns out to be very readily demonstrated, we must ask why it took him 20 years to find it.

Part of the answer comes from the very same reason that made Oersted search for the effect in the first place. Under the influence of *Naturphilosophie,* Oersted believed that all forces had a fundamental similarity. Now, the best understood force by then was gravitation, which Newton had demonstrated to be a **central force**. That is, the force of gravity between objects acts along the line joining their centers, and is an inverse-square law force. Coulomb's work had shown that electric forces act that way too. Magnetic forces are a little different (because you can't get separate north and south poles—they always appear together), but not enough to alter Oersted's fundamental belief. Thus, as he went about looking for an interaction between electricity and magnetism, he was looking for a central force.

The basic apparatus for Oersted's search was a simple magnetic compass (a light magnet suspended on a pivot) and wires connected to a voltaic battery (a collection of cells in series). Oersted deduced that with the compass aligned in its natural north-south direction, the interaction would go best if the electric current travelled perpendicular to the compass (that is, east-west). Whenever he tried that, he got no result.

Eventually, around 1819, he found that certain orientations of bent wires held beside the compass produced slight but irregular effects. Convinced of the fundamental unity of forces, Oersted arranged a demonstration for some colleagues in the spring of 1820. After explaining his principles in great detail, he performed his unimpressive demonstration. As the discontented whispers got louder, in desperation Oersted tried the one orientation he had always avoided—the one that was entirely contrary to all his ideas about central forces. He held the wire slightly above and parallel to the compass; that is, in the north-south direction. Immediately, the compass needle swung through a large arc (60° or more), tending to align itself perpendicular to the wire. The lecture was saved; the interaction had been discovered—by accident!

It was hard for Oersted to find this transformation because electricity and magnetism do not interact the way other forces do. Electromagnetism is not a polar phenomenon like magnetism or electricity. The magnetic force encircles the wire—there just aren't any poles.

However, as soon as physicists realized the difference, they quickly figured out how to analyze this new kind of electromagnetic force. They performed many experiments to find the various properties of electromagnetism, and they built mathematical theories to account for them. Around 1825, the French physicist André Marie Ampère (1775–1836) produced such a brilliant theory that he was dubbed "the Newton of electricity." Ampère established precise laws for the magnetic force of wires carrying currents, and predicted and demonstrated that two parallel currents exerted forces on each other. This was a kind of magnetic force without a magnet. Thus, Ampère was encouraged to explain the magnetism of iron by the idea that currents circulated naturally inside iron molecules.

*Michael Faraday (1791–1867) and Electromagnetic Induction*

In England, at the same time, Michael Faraday was conducting brilliant experiments. Faraday was the son of a poor blacksmith. By self-education he'd obtained the post of assistant to Humphrey Davy at the Royal Institution in London. He succeeded Davy as its director when Davy died in 1829.

*Michael Faraday.*

When Oersted announced his discovery in 1820, Faraday immediately performed a number of experiments to explore electromagnetism. In one of those experiments, Faraday devised an apparatus to produce a continuous motion from electromagnetic forces. The first electric motor!

During the period from 1825 to 1831, Faraday devised many tests of various theories of electromagnetism, including Ampère's. Among these was an investigation of whether magnetism could induce electricity, one form of a reverse Oersted effect. Although theory and conviction might tell Faraday what to look for, they didn't tell him how. In 1825, he wrapped a coil of wire around a magnet and connected the ends of the wire to an electrical meter. Nothing happened. Faraday could only stare at his apparatus and wonder: "How will it work?" If he'd only known, if he'd only moved the magnet back and forth within the coil, the meter would have registered a current. A fluctuating current. And that was the problem. The effect Faraday was seeking is not continuous. You don't get electricity out of magnetism by a simple reversal of Oersted's experiment. Faraday tucked the problem into the back of his mind and went onto other things.

In the summer of 1831, Faraday returned to that problem by a roundabout route. He began to investigate whether a current in one wire could induce another current in a second nearby wire, not connected to the first one.

He wrapped separate coils of wire around a wooden ring, connected one to a battery and the other to a meter. Nothing happened.

Faraday's next move was nothing short of inspiration. Instead of the wooden ring he used an iron anchor ring, about six inches in diameter. Laboriously, he wound several coils of wire around the ring. Then Faraday connected one coil to his battery and a different coil to his meter. No continuous effect, *but!!* Faraday noticed that the needle on his meter kicked momentarily, every time he connected or disconnected the battery. Connecting, disconnecting—*changing* —that was the clue!

Faraday quickly realized that the current in the first coil produced magnetism in the iron ring, so that by connecting and disconnecting the battery, he was changing the magnetism—which induced a current in the second coil. Then he saw what he'd missed six years earlier. He wrapped a coil around a hollow cylinder and connected it to his meter. When he plunged a magnet into the hollow space of the cylinder, the meter kicked. When he removed the magnet, the meter kicked again—this time in the opposite direction. Faraday had discovered **electromagnetic induction**. Whenever, and however, a magnetic field changes within a coil of wire, that change induces a current of electricity in the coil. It's the change in magnetism that transforms magnetic force into electric force.

If you used a steam engine to push a magnet back and forth in and out of a coil, you could keep the current going, first one way and then the other. An electric generator! Soon inventors found better ways to generate electricity by using rotating machinery.

Throughout these electromagnetic investigations, from Oersted to Faraday, scientists were battling with the legacy of Newton. In particular, with the Newtonian principle that forces acted only on material objects. Ampère tried to cling to that notion by breaking the magnetic forces into infinitesimals that acted centrally, despite the clear evidence that the electromagnetic interaction was not a central force. Faraday, on the other hand, proposed to reify magnetic forces, that is, to treat them as real, but immaterial things. He was bold enough to base his explanations of magnetic interactions on "tubes of force."

And it was this reification of force that led almost inevitably to the idea of energy and its conservation. Although one could deduce the conservation of mechanical energy from Newtonian principles, such a deduction seemed to add nothing to our ability to analyze motions resulting from forces. Energy only became important when it was seen as a means to a grand unification of the forces of Mother Nature.

## *The Principle of Energy Conservation*

Back to the *Naturphilosophen* and their conviction that all forces are merely manifestations of One Basic Force. Although the German philosophers thought they were operating within the Newtonian program, they were actually guilty of the grossest kind of over-generalization of Newton's notions—speculating on a grand scale without resorting to the test of experience.

Nonetheless, the German nature philosophers did receive encouragement from a growing body of experimental evidence, though it was strictly qualitative in nature. As we've seen, at least at a superficial level, a wide variety of phenomena could be interpreted as interactions between forces: mechanical forces, chemical forces, heat forces, electrostatic forces, magnetic forces, electromagnetic forces. Maybe the philosophers were entitled to look at force as some kind of a chameleon, changing its color as circumstances change. Of course, force changes are more fundamental than just a change in color. But is there some one thing that stays the same through all these force changes?

That's a fair question; but to take it out of the realm of philosophic speculation and bring it into the realm of scientific investigation, another question must be asked: Is there some entity we can *measure* that's conserved through all those changes?

This question was asked first by engineers interested in the efficiency of steam engines. A steam engine takes coal to boil the water. The more the engine works, the more steam is needed, and the more coal has to be burned. So, that gave engineers the idea that the engine was changing heat into work. And James Watt claimed that his engine was better than Newcomen's, because it needed less heat to produce the same amount of work—less heat, less coal. In fact, Watt was so confident of the superiority of his engines that he based the price for the licence to operate a Watt engine on the amount of coal it saved. Not until the 1830s did engineers find ways to express the relation of heat to work in unambiguous terms. In the meantime, other developments began pointing in the same direction.

Gradually, the various new conversion processes being brought to light promoted the idea of some connecting link between them. In addition to the ones already mentioned are the thermoelectric effect first reported in 1822, and the action of light in breaking down silver bromide (the basis of photographic films). The possibility of relations among all these "forces" was noted by Mary Sommerville in her book *On the Connexion of the Physical Sciences* (1834), by Faraday in the same year in a lecture on the "Relations of chemical affinity, electricity, heat, magnetism, and other powers of matter," and by William Grove in a lecture series *On the Correlation of Physical Forces* (1843, published in 1846), to give only English examples.

To get to the full generality of the principle of conservation of energy required the recognition of an entity subsisting within all these conversions, *and* the specification of an exact *quantitative* measure of it.

Historians now acknowledge three or four men who arrived at that full generality independently, of whom two will be noted here.

*Julius Robert Mayer (1814–1878)*

In the early 1840s, Mayer, a German physician, had a striking insight. He decided that the oxidation of food in our bodies is a chemical process, supplying both our body heat *and* our muscular work. He was convinced that heat and work were indestructible forces, changing in form, but not in quantity. He knew from the work of Black and Lavoisier that mass is indestructible: why not force too?

*Julius Robert Mayer.*

Mayer's speculative generalizing is clearly shown in passages from the paper he published in 1842, in which he undertook to make the notion of force as clear as that of matter:

> Forces are causes:...In a chain of causes and effects, a term or a part of a term can never...become equal to nothing.

> Two classes of causes occur in nature, which so far as experience goes, never pass one into another. The first class consists of such causes as possess the properties of weight and impenetrability; these are kinds of Matter: the other class is made up of causes which are lacking in those properties, namely Forces, called also Imponderables...Forces are therefore indestructible, convertible imponderable objects.

There's reification for you!

> *...if the cause is matter, the effect is matter; if the cause is a force, the effect is also a force.*

> ...a force once in existence cannot be annihilated, it can only change its form.

Yet along with these grand generalizations, Mayer provided a precise quantitative relation derived from experimental measurement:

> ...the warming of a given weight of water from 0° to 10° C corresponds to the fall of an equal weight from the height of about 365 meters. [This is about 15% below our value for the quantity.]

Earlier, when Mayer had visited his old physics professor to enquire if his idea could possibly be right, the professor scoffed, "Why, if you're right, that'd mean shaking water in a flask would make it warmer—how silly!"

Mayer went away disheartened. Then he decided to try it. It worked. He rushed back to the professor: "It does, it does!" Mayer did not neglect to mention that particular phenomenon in his 1842 paper.

Although Mayer and his contemporaries used the word "force," we now say that heat, light, electricity, and so on, are all forms of **energy**: the "something that stays the same" is energy. As energy is converted from one form to another, its quantity doesn't change: like mass and charge, energy is conserved.

*James Prescott Joule (1818–1889)*

*James Prescott Joule.*

This very problem was also being studied at the same time by James Joule in England. Joule was the son of a wealthy Manchester brewer and had studied with John Dalton—of atomic theory fame. Joule didn't have to work for a living and spent his time in research. He got interested in heat and work by studying electric motors. He found that with electricity flowing, the motor not only spun, it also got hot. If he prevented the motor from spinning, it got even hotter. When Joule realized the electricity was producing both motion and heat, he resolved to measure carefully the quantities involved.

Although electrical quantities were not yet standardized, Joule could measure the mechanical action of the motor and the heat evolved, and arrive at a relation between them. In his first trials, reported in 1843, Joule obtained results ranging from 10% to 50% greater than Mayer's value.

Although Joule did not jump to a grand generalization as quickly as Mayer, he did conclude his report with the words:

> I shall lose no time in repeating and extending these experiments; being satisfied that the grand agents of nature are, by the Creator's fiat, indestructible; and that wherever mechanical force is expended, an exact equivalent of heat is always obtained.

In the following years, while Mayer continued to explore the generalizations of energy conservation (in a useful way), Joule concentrated on narrowing the range of his experimental results to a precise value for the relation of mechanical work to heat. He devised an apparatus to raise the temperature of water by friction; that is, an entirely mechanical device, not involving electricity. He fitted a copper vessel 30 cm in diameter and 30 cm tall with fixed baffles and a set of revolving paddles, designed to produce friction with the contained water with a minimum of turbulence. He arranged for the shaft carrying the paddles to be turned by a cord attached to falling weights.

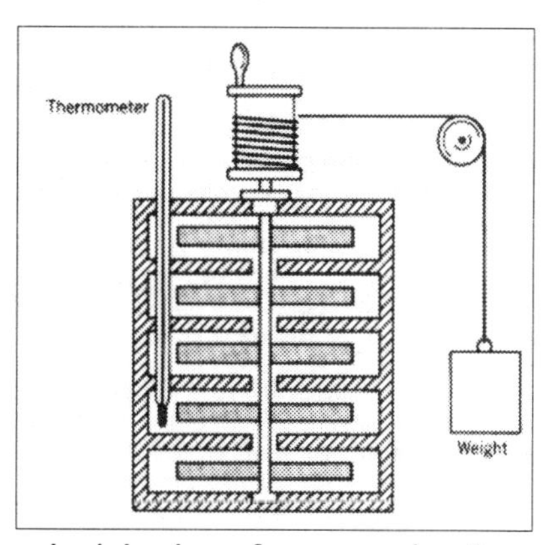


*Joule's churn for measuring the mechanical equivalent of heat by friction.*

Joule measured the temperature of the water with a thermometer he could read to the nearest 0.0050° F. He allowed the 26 kg weights to fall 1.6 m, a total of twenty times. He measured the temperature rise of 6.3 kg of water (including the thermal equivalent of container and paddles) at about 0.31° C. Joule repeated these measurements many times, using several liquids, as well as one series using friction between plates of cast iron. When he reported all his results in a summarizing paper in 1850, he showed how rigorously he had taken account of all imaginable distractions from the accuracy of his results. Joule drew two conclusions:

> 1st. That the quantity of heat produced by the friction of bodies, whether solid or liquid, is always proportional to the quantity of force expended. And,
>
> 2nd. That the quantity of heat capable of increasing the temperature of a pound of water (weighed in vacuo, and taken at between 550° and 600° ) by 10° Fahr. requires for its evolution the expenditure of a mechanical force represented by the fall of 772 lb. through the space of one foot.

By then, it was also clear that whenever heat is converted into work, you always get the same amount of work from a given amount of heat. The problem there, however, is that only part of the heat is changed to work. The heat in a steam engine, for example, makes the engine and its surroundings hotter, as well as driving the piston in the engine. In fact, in steam engines only about 10% to 15% of the heat is converted into work. The rest is just wasted.

Although this all sounds very mechanical and physical, both Mayer and Joule soon applied the principle of energy conservation to living creatures. In fact, they both chose horses. They calculated the amount of work a horse could do and compared it to the heat content of the food the horse ate. They concluded that about 80% of the food energy went to heat, and 20% to work. Mayer even determined that a working horse needs 11 pounds more fodder than one that's out to pasture.

Mayer tied several different forms of energy together when he wrote:

> The vegetable world is the reservoir in which the fugitive solar rays are fixed, suitably deposited and rendered ready for useful application. Plants consume the force of light and produce in its place chemical tension. This tension is a physical force. It is the equivalent of the heat obtained from the combustion of the plant. Animals consume vegetables, and cause them to reunite with the atmospheric oxygen. Animal heat is thus produced—and not only animal heat, but animal motion too.

So the principle of energy conservation was quickly enlarged to cover all of nature. Today, it's every bit as important a principle in biology as it is in physics or chemistry. And today, we say the total amount of energy in the universe is constant. All that ever happens is that energy gets changed from form to form. Energy can't be created—or destroyed.

Although we talk about energy as if it were an entity, a thing, we should rather think of it as a property of physical systems. Energy is more like a medium of exchange, a currency, than a real thing. And the principle of energy conservation is like a bookkeeping system that tells us that our entries of *energy in* and *energy out* must always balance. If they seem not to, we've made a mistake.

The principle of energy conservation was one of the legacies of Faraday's interest in force as an entity distinct from matter. That tied fields of study together—in at least a bookkeeping sense. Another legacy from Faraday's reification of force proved to be equally impressive in unifying physical science; this was the electromagnetic theory of J. C. Maxwell.

## *James Clerk Maxwell (1831–1879) and Electromagnetic Waves*

Maxwell was the son of a Scottish laird and attended universities at Edinburgh and Cambridge. From age 25, he was a professor of physics, first at Aberdeen, then London, and finally Cambridge, where he became director of the newly founded Cavendish Laboratory.

*James Clerk Maxwell.*

Although Maxwell was trained as a mathematician, he was keen to apply his powers to physical problems. He studied Faraday's writings deeply for their descriptions of magnetic and electric actions. Rather than following the French mathematical analysis of Ampère and his colleagues, Maxwell undertook to build Faraday's tubes of force into a mathematical theory. Since Faraday had no training in mathematics, he had used pictorial analogies instead of analysis. Maxwell decided that those pictures could be made mathematical.

Maxwell had not chosen a simple task. From 1856–65, he labored at finding hypotheses for the structure of space appropriate for Faraday's tubes of force. Eventually, Maxwell devised a mathematical model that accounted for all known electric and magnetic interactions. And from the measurements of electric and magnetic quantities, Maxwell deduced that these interactions traversed space at a speed of about 300,000 km/s.

Maxwell was quick to note that this was very close to the value that French experimenters had recently found for the speed of light. He also knew that the mathematical form of his electromagnetic theory had similarities to Fresnel's formulation of the wave theory of light. In 1862, Maxwell wrote:

> The velocity of transverse undulations in our hypothetical medium, calculated from the electromagnetic experiments of MM. Kohlrausch and Weber, agree so exactly with the velocity of light calculated from the optical experiments of M. Fizeau, that we can scarcely avoid the inference that light consists in the transverse undulations of the same medium which is the cause of electric and magnetic phenomena.

Maxwell acknowledged that his medium, often termed the **æther**, was hypothetical for two main reasons. First, no direct experimental measurement of any of its properties had been made; and second, those properties seemed too contradictory to be real. For, Maxwell's medium had an elasticity greater than the finest steel, combined with a density less than the rarest gas. Since Maxwell's equations served well enough, he was inclined to treat his physical hypotheses as a scaffolding that could be discarded once the theory was built. We'll find a fascinating sequel to this story in the work of Albert Einstein in Chapter 19.

At least in theory, Maxwell had provided an account of all electric and magnetic phenomena, with a tie-in to light and optics for good measure. And he desired even more unity in our analysis of phenomena. Despite some embarrassment over the æther, Maxwell did not hesitate to promote a unified mechanical picture of all physical interactions:

> In speaking of the Energy of the field, however, I wish to be understood literally. All energy is the same as mechanical energy, whether it exists in the form of motion or in that of elasticity, or in any other form. The energy in electromagnetic phenomena is mechanical energy. The only question is, Where does it reside? On the old theories, it resides in the electrified bodies, conducting circuits, and magnets, in the form of an unknown quality called potential energy, or the power of producing certain effects at a distance. On our theory it resides in the electromagnetic field, in the space surrounding the electrified and magnetic bodies, as well as in those bodies themselves, and is in two different forms, which may be described without hypothesis as magnetic polarization and electric polarization, or, according to a very probable hypothesis, as the motion and strain of one and the same medium.

Perhaps the most striking feature of this statement is the confidence with which Maxwell asserts that all energy is mechanical energy, so much like Huygens' statement about the "true Philosophy." For Maxwell, energy was more than a mere bookkeeping device: it was a fundamental entity of the physical world, even though it depended on a hypothesis, which he confidently asserted to be "very probable."

## *Conclusion*

Just see how far physics had come in 75 years! Back in 1800, light might have had a spurious mechanical look in the particle theory, but the forces of interaction of light with matter were entirely unknown. Although they were still unknown in 1875, the wave theory of light accounted well for what was known. In 1800, electricity and magnetism were separate fields of investigation. Indeed, current electricity had only just begun with Volta's pile. In the 1830s, Faraday had still thought it important to demonstrate that all kinds of electricity (whatever their source) were of the same basic nature.

The work of Oersted, Faraday, and Maxwell tied electricity and magnetism together into electromagnetism, both experimentally and theoretically.

And the work of Mayer and Joule also linked these phenomena with heat and mechanical action in the principle of the conservation of energy.

We may mention in passing, that heat had for some time been considered to be a special fluid, *caloric*, but, by 1850, was considered to be a mode of motion of the particles of matter. Both the deeper study of thermodynamics and the kinetic theory of gases grew out of the work of Mayer and Joule.

By 1850, scientists in many fields could base their work on at least three conservation principles. Mass is conserved. Electric charge is conserved. And energy is conserved.

With these conservation principles, scientific studies became unified. Not only did the physical world appear unified, the biological world appeared to be united with the physical world, at least at the level of energy exchanges.

*A power generating station in London, England in 1896.*

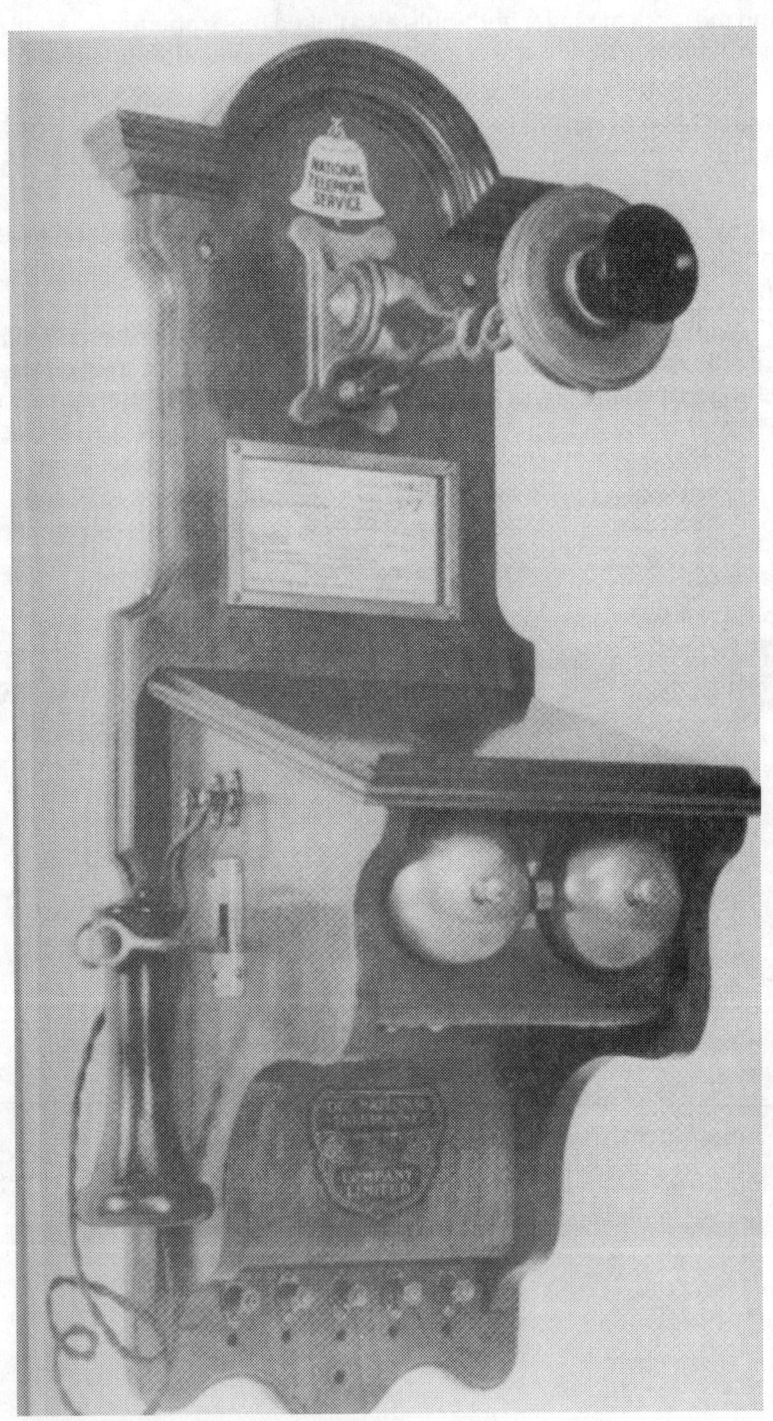


*A telephone, from about 1910.*

## CHAPTER EIGHTEEN

# Electrical Communications

Sending messages over long distances has been a requirement in human commerce for thousands of years. For most of that time, the speed at which a message could travel was the same as the speed of the messenger, whether on horseback or by full-rigged ship at top speeds in the range from 20 to 30 kilometers per hour.

If you can rely on transmission of information by waves instead of having to carry a physical message, you can achieve greater speeds of communication. African drummers and North American smoke signalers used sound and sight to send coded messages. Generally speaking, the variety of messages they could send was rather limited. Often, of course, that's perfectly adequate to the need. In England, Francis Drake signalled the sighting of the Spanish Armada in 1588 by a chain of fire stations, hilltop to hilltop.

Starting in 1792, Claude Chappe (1763–1805) established a network of semaphore stations in France. A pair of movable arms on the top of towers at 25 km intervals could transmit alphabetic messages. They could send messages at about 1000 km/h over a distance of about 250 km. As long as lines of sight are clear between stations, there is no theoretical limit to the range of semaphore; but it is limited by weather conditions.

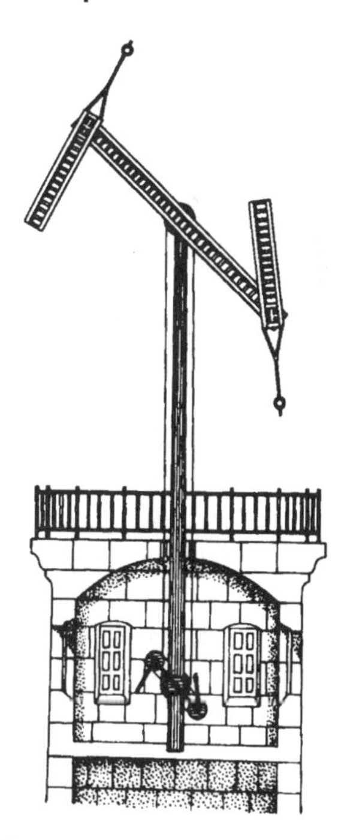

*A semaphore station.*

An indication of the importance people attached to communications is that every time a new electrical phenomenon was discovered after about 1740, someone was sure to suggest trying to use it for sending messages. For a while, no one got past the notion of having a separate line for each letter of the alphabet. First, in the 1750s, someone suggest using static electricity over a long line. Whenever you wanted to send the letter Q, you'd bring a charged rod close to the Q line; and the pith ball at the other end would be attracted, thus signaling "Q." After 1800 and the introduction of Volta's pile, someone else suggested using current electricity and a set of little electrolytic cells. Whenever the Q circuit was closed, the Q vial at the receiving end would show a stream of bubbles. Both methods were slow, inconvenient in needing 26 circuits, and expensive over any significant distances.

### *Electromagnetic Signaling*

Practical electrical signaling had to await Oersted's discovery of the magnetic effect of current electricity. After 1820, with voltaic cells becoming cheaper and more reliable, various attempts were made to use the deflection of compass needles by currents passing near them.

During the 1830s, a variety of electromagnetic instruments were tested with varying results. In Britain, some devices used five circuits which cause pairs of magnetic needles to point to letters arranged in a diamond shape. Others used currents flowing in opposite directions to cause a single needle to move either left or right. Before long, most systems made use of the code devised by Samuel F.B. Morse (1791–1872) in the United States. Although Morse also designed a useful electromagnetic sounder, his greatest achievement was the development of the Morse code—a scheme of dots and dashes to represent letters and other symbols.

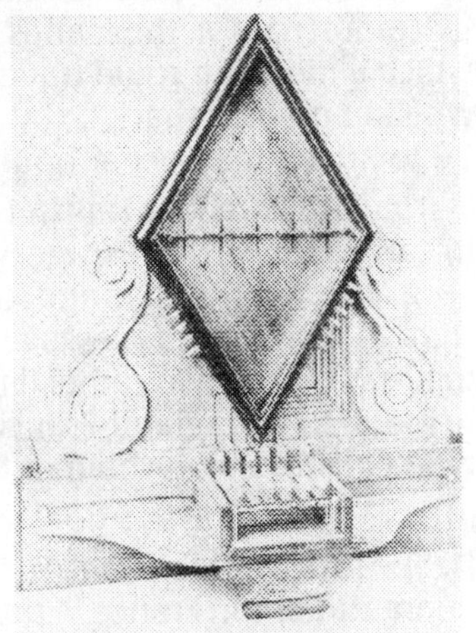

*A five-circuit telegraph.*

Morse graduated from Yale College in 1810 with interests in both science and art. The latter absorbed him first, and he became a prominent portrait painter. Around 1830, Morse's interest in science was revived when he learned some of the fundamentals of electromagnetism. In 1832, he realized that if the presence of electricity could be made visible in a circuit, that would provide a means of "transmitting intelligence." He immediately sketched the design for an apparatus that would pull a pencil electromagnetically against a moving strip of paper whenever a coil was energized.

Morse worked at this instrument for five years and, in 1837, demonstrated an effective device, first over a distance of 500 m, and, within a couple of months, 15 km. Morse was joined now by Alfred L. Vail (1807–1859), who had witnessed the first trial in New York. Vail had just graduated from New York University, but had learned mechanics as a youth in his father's iron works. He offered to assist Morse in the making of the telegraph instruments. While they were developing effective apparatus, Vail and Morse began to seek government backing for establishing a useful telegraph line. They were soon joined in that effort by Hiram Sibley (1807–1888), a banker from Rochester, who'd been an apprentice shoemaker, then had run a successful machine shop, before turning to a financial career.

After some delays, in 1843, Congress voted an appropriation for the stringing of a line from Washington to Baltimore (64 km). This job was supervised by Ezra Cornell (1807–1874), a successful manager of mechanical works in Ithaca, New York. Cornell was assisted by Vail, who, on 24 May 1844, was at the receiving instrument in Baltimore when Morse tapped out the first message, "What hath God wrought!"

Almost immediately, many companies sprang up to connect two distant points, frequently along a railroad right of way. Soon, interconnections were made between the various lines. By 1850, every state east of the Mississippi belonged to the telegraphic network.

In regular practice, the telegraph sounder was the usual receiving instrument. In a simple system, the sending operator switched the electric current (provided by batteries) on and off according to the code, using a telegraph key. The key and batteries at the sending end were connected by a single wire to the sounder at the receiving end. The electric circuit was completed through a ground return, i.e., connected to metal rods driven a couple of meters into the earth at each station. The standard **sounder** consisted of a pivoted soft iron bar that was held against a metal stop by a light spring. Below the bar was the coil, whose current was controlled by the sending key. Whenever the sender depressed the key, the energized coil drew down the bar, making an audible "click" when it struck. When the sender released the key, the bar was released, and the spring pulled it up to hit the stop with a "clack." The receiving operator recognized "click-clack" as a dot, and "click...clack" as a dash.

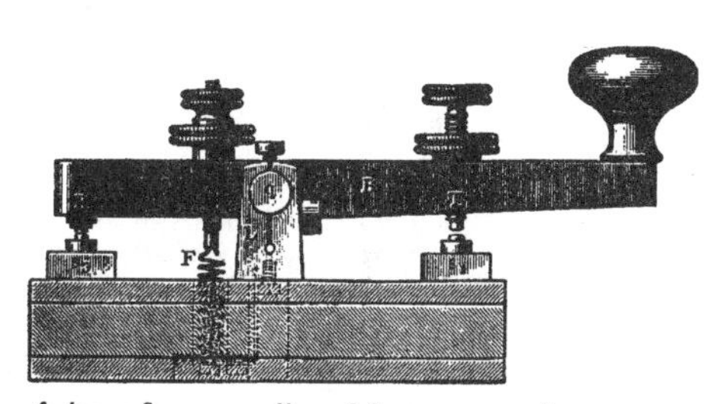

*A key for sending Morese code.*

Of course, there was a limit to the length of line over which one set of batteries could send a current sufficient to operate the sounder. Morse and Joseph Henry (1797–1878) soon devised an electrical relay to extend the distance indefinitely. A **relay** is a device in which the incoming current operates a switch to control the current in a second circuit. Structurally, a relay looked very much like a sounder, except that the bar was made part of the switch in the second circuit. With relays spaced along a line, the sending operator could control a distant sounder through a series of circuits.

The telegraph provided a useful communication system for the railroad, and undoubtedly helped to reduce accidents. However, numerous independent telegraph companies were formed to profit from this rapid method of communicating. In particular, businessmen could send and receive orders much more rapidly than by mail. Soon, newspapers took advantage of the telegraph for the rapid transmission of reports on events of wide interest. In

December 1848, President Polk's message to Congress in Washington was printed in St. Louis, Missouri within 24 hours of its delivery.

In 1851, Hiram Sibley organized the New York and Mississippi Valley Printing Telegraph Company, which built some new lines, as well as incorporating some smaller companies along the way. In 1856, Ezra Cornell cooperated with Sibley in founding the Western Union Telegraph Company by joining a number of companies together. Sibley was the first president of Western Union, which started with 132 offices and had expanded to 4000 offices ten years later. From their accumulated wealth these men contributed to Cornell University in Ithaca, and to its Sibley College of Mechanical Engineering.

In 1860, William F. Cody (1846–1917) became one of the first Pony Express riders for the Central Overland California and Pike's Peak Express Company. Relays of riders, covering 120 km each, could transmit mail 3140 km from St. Joseph, Missouri to Sacramento, California in eight days. A year later, their business was overtaken by the Pacific Telegraph branch of Western Union, which could do the same job in a few minutes. By the way, notice that in this case the telegraph preceded the railway by eight years.

Telegraphic trials were made in Britain at the same time as in the United States. Using apparatus invented in Britain, trials were made in 1837, and a public line opened from Paddington to Slough (35 km) in 1843. A significant value of the telegraph was demonstrated to the public in January 1845: Murder suspect John Tawell was observed boarding the London train at Slough. His description was telegraphed to police in London, who arrested the surprised culprit as he stepped off the train at the Paddington station.

In Britain, the early telegraph lines were used largely for railroad business. Extensions of service occurred in 1850 with the introduction of money orders; and, in 1851, with a submarine cable across the Straits of Dover so that London and Paris stock exchanges were put into direct communication. In the period from 1855 to 1868, the number of telegraph messages in the United Kingdom increased from just over a million to almost six million, an annual increase of 13.3%.

During that same period, there was growing agitation for the telegraph system to be taken over by the British Post Office. Advocates argued that was the only way to give the benefits of the telegraph equally to all citizens. They pointed to the success of government-run telegraphs on the continent, where cheaper rates had increased traffic and profits. They were also in favor of giving equal access to the telegraph to all citizens, whether in metropolitan or rural areas. Eventually, in 1868, the government authorized the nationalization of the telegraph, under the Post Office, to take effect in 1870. Within about ten years, the burdens of increased service required of a government operation led to the telegraph's operating at a deficit, to be made up out of general revenues.

A transatlantic submarine telegraph cable was commenced in 1856 by the American financier Cyrus W. Field (1819–1892) and the British engineer

Charles Bright (1832–1888). The cable, joined together from 1200 sections each about three km long, was ready the following summer. Carried in two ships (one American and one British), the cable was laid for 600 km when it snapped. The following year several false starts were finally followed by a successful connection between Valentia Harbor, Ireland and Trinity Bay, Newfoundland. In August of 1858, Queen Victoria was able to exchange greetings directly by telegraph with President James Buchanan. The queen knighted Bright. However, the cable had been mistreated during testing, and the connection was lost within about a month.

A renewed effort was made in 1865, with improved cable and a single laying ship. The steel steamship *Great Eastern*, launched in 1858, had five times the bulk of any other ship then afloat. Less than 1000 km from Newfoundland, the cable broke. All efforts to retrieve it failed. Next summer, a new cable-laying expedition was fully successful. Then, the engineers were able to find the end of the cable of the previous year, and by early September 1866 there were *two* cables spanning the Atlantic. The mayor of Vancouver celebrated the occasion by sending a telegram 13,000 km to the Lord Mayor of London.

## *The Telephone*

The success of the telegraph did not satiate the public's need for communications. From the 1850s, many inventors proposed schemes for sending more than one message at a time along a single telegraph wire. Some multiplex devices did eventually become practical, but there were great technical problems. Interestingly enough, the two men who eventually made the first telephone instruments did so while searching for ways to multiplex telegraphy. We'll concentrate on the winner, Alexander Graham Bell, with a sidelong glance at the loser, Elisha Gray.

### *Alexander Graham Bell (1847–1922)*

Grandfather Alexander Bell (1790–1865) began his career as a shoemaker in St. Andrews, Scotland. After his marriage in 1814, he went on the stage for about eight years. Later he became a teacher of elocution and eventually added the removal of pupils' speech impediments to his talents. He was followed in this profession by his two sons, the younger, Alexander Melville Bell (1819–1905), being the father of our telephone inventor. Around age 20, Melville spent four years in St. John's, Newfoundland, primarily for his health. Married in 1844, he spent 22 years in Edinburgh, teaching speech and investigating the production of vocal sounds. In 1866, he moved to London to continue the enterprises of his recently deceased father. In 1867, he completed a symbolism for showing how vocables are produced and published his scheme with the title *Visible Speech, The Science of Universal Alphabetics*.

Alexander Graham Bell was the second of A.M. Bell's three sons. Aleck and his brothers were trained in music by their mother, and were destined to follow their father's profession, but the younger one died at age 18 in 1867,

*Alexander Graham Bell inaugurating the first New York-to-Chicago line.*

and the elder at 25 in 1870, both of tuberculosis. By that time, father Bell had established a reputation in both Britain and America. He thought seriously of emigrating with his wife, leaving young Aleck to carry on in England. Then, on the heels of the death of his eldest son he resolved to take his remaining son to the purer air of Brantford, Ontario. Now 23, Aleck had had a regular high school education, as well as considerable scientific encouragement from his father, brothers, and friends.

The Bells settled in Brantford in August 1870. The next year, the elder Bell arranged for appointments for Aleck to teach speech to deaf children in Boston. For the next few years, Alexander Graham Bell spent his winters in Boston teaching, with a couple of months each summer at home in Brantford.

A young man in a strange city, Bell soon found interests at the electrical shop of Charles Williams. Williams' customers included inventors and physicists—Thomas Edison had tinkered there back in 1869. Encouraged by the creative atmosphere in the shop, Bell resolved in the fall of 1872 to try his hand at inventing a multiplex telegraph, still an elusive prize in the industry. His ambition was fired by his belief that his knowledge of music and sound would enable him to contrive a multiplex device on a brand new principle.

Bell's basic idea for multiplexing the telegraph was to send a number of different audio frequencies along a single line. Each frequency could be keyed with a separate Morse message. The trick was to find a device for the receiving end that would sort out the various frequencies, that is, to arrange that there would be a separate sounder responding to each message, and to that one alone. In later radio terminology, the audio frequency was to be the *carrier* signal, which was to be *modulated* by the Morse code message of dots and dashes. For this purpose, Bell required a device to generate the audio frequencies continuously, which meant by some electrical means. Then, he'd need a "demodulator" at the receiving end, which he figured would work by resonance.

To demonstrate **resonance** to his friends, Bell would depress the pedal controlling the felts on the strings of a piano. When he sang a single tone into the piano, the string of that pitch would continue to resound after he stopped singing–*that* string and only that one. This would be Bell's road to wealth: a multiplex telegraph based on harmonic-tone machines.

Bell's first task was to find a way of keeping a reed vibrating at a fixed frequency. During his spare time, from the fall of 1872, he worked intermittently at the problem for a couple of years. Eventually he devised reeds of iron whose vibratory frequency was determined by their length. With electromagnets and make and break contacts, he was able to maintain vibrations of specific frequencies. Then he set up three lines between pairs of vibrators (each pair of a different frequency), with the idea that a receiving vibrator would respond to an electrical signal from a transmitting vibrator by resonance. There were a number of technical problems to overcome, but by the fall of 1874, Bell had a set of apparatus that worked quite consistently.

At this point, the parents of two of Bell's deaf pupils became aware of his telegraphic research. Thomas Sanders (1839–1911) was a prosperous leather merchant, and Bell was living in Grandma Sanders' house. Sanders was willing to support Bell financially for a half share in any revenues that might eventually accrue. At the same time, when Bell described his work to the father of his pupil Mabel Hubbard, the eyes of Gardiner Hubbard (1822–1897), lawyer and promoter, lit up. For he was at that very time trying to establish a competitor to Western Union that would combine post and telegraph to send messages cheaply and quickly. Seeing the advantage of a new scheme for multiplexing, he quickly encouraged Bell's work. Soon, the three men agreed to one-third shares in any profits from Bell's inventions.

Bell reduced his teaching load in order to develop his instruments for a patent application. In January 1875, he began to be assisted by Thomas Watson, aged 20, a mechanic in Charles Williams's electrical shop. With his equipment in good working order by February, Bell travelled to Washington to apply for patents on his harmonic telegraph. There he learned that an application had been lodged for the basic harmonic multiple telegraph system two days earlier, by an electrical inventor named Elisha Gray.

### *Elisha Gray (1835–1901)*

Twelve years Bell's senior, Gray had been a telegraph inventor since 1865. Of less than robust physique, Gray worked for a while on the family farm in Ohio after his father's death. When that proved too heavy for him, he turned to lighter crafts and soon began tinkering with electrical apparatus. Contact with a professor at Oberlin College led him to renew his education, including two years at Oberlin. Gray soon began to make improvements in telegraph relays and other equipment. In 1872, his firm in Chicago became the Western Electric Company for making telegraph instruments, with Gray as its superintendent.

During these years, Gray had discerned the possibility of sending musical tones along wires from electrically produced vibrations. In 1874, he constructed a set of electrical vibrators to produce the notes of an octave–essentially an early electric organ. The principle of electrical vibration (as with Bell) was basically that of an electric buzzer, with the circuit being alternately made and broken as the vibrator was attracted to an electromagnet. Tuning was largely a matter of choosing the right dimensions for the vibrating iron reed.

Each vibrator of Gray's organ turned the current in a circuit on and off at a specified frequency. An electromagnetic device that could respond to electrical changes at such a frequency could be made to give off a sound of the corresponding pitch. For his **receiver** Gray simply mounted a thin iron plate near the poles of a horseshoe electromagnet. When the electromagnet was connected to the organ, the plate (or diaphragm) vibrated with the same frequency as the electrical variation in the circuit. Interestingly enough, this device was never patented, because it was a commonplace among professors of physical acoustics. In those years it was variously known as a tin-box receiver, or wash-basin receiver. The principle of the tin-box receiver found its way into the telephone, where it is still used, except that the core of the electromagnet is a permanent magnet (an improvement made by Bell and Watson about 1876).

News of Gray's organ spread among scientists during the summer of 1874, and in August a newspaper in Hartford reported that "Mr. Gray hopes one day to be able to transmit the sound of the human voice also by telegraph," but added that "towards this curious result nothing seems to have been done."

Bell spent the summer of 1874 at home in Brantford and only heard of Gray's work when he returned to Boston in the fall. He immediately saw that a telegraph inventor like Gray would soon realize the multiplexing implications of his electric organ, if he hadn't already done so. Thus, at the end of February 1875, both Gray and Bell were filing multiple telegraph patents based on the same principle, and Gray won, at least the first round. Bell returned to Boston and resolved to make his system work better than Gray's, with the hope at least of patenting some improved or variant technique.

### *Making the Telephone Hear*

If you compare the telephone to a talking parrot (making it say what it hears), then the tin-box receiver is like the parrot's voice box. The really significant problem for Bell or Gray was to create the telephone microphone or transmitter, which (like the parrot's ear) could convert human speech sounds into electrical vibrations.

In fact, in 1860, a German inventor, Philip Reis, had built an attempt at such a microphone by having a vibrating diaphragm cause an electric circuit to turn on and off at the frequency of its vibrations. When a person spoke into Reis's microphone, the receiver (or speaker) gave out sounds of the same pitch, but didn't produce recognizable speech. Reis's incapacity to find the clue to the problem may be shown in his remark that his apparatus reproduced *consonants* "pretty distinctly," but not the *vowels* "in an equal degree." For 15 years, Reis's apparatus was well known to physicists and students of acoustics; yet none of the experts pointed out the crucial flaw.

Apparently, there was only one man with the combined electrical and acoustical knowledge to find the clue, as well as the imaginative skills and business contacts to make it work—Alexander Graham Bell!

*Item.* In the fall of 1873, Bell was appointed professor at the recently founded School of Oratory at Boston University. He was chosen to give the inaugural lecture, with the title "Speech and the Instrument of Speech." In that lecture Bell described and illustrated the properties of sound and speech, pitch, loudness, and tone quality, including explanations of the ear and vocal chords, and graphs of sound waves.

*Item.* Bell's work on the harmonic telegraph demonstrated his electrical knowledge, and what he didn't know he could check with the experts at MIT (founded in 1865) and Harvard, or at Charles Williams's electrical shop.

*Item.* In the spring of 1874, Bell consulted with Charles Cross at MIT over apparatus for making the wave forms of vocal sounds visible. Bell had a professional interest in such devices as aids to help his deaf pupils make correct speech sounds. Bell also consulted with a Boston ear specialist, Dr. Clarence Blake. They performed some acoustical experiments with the excised ears of a couple of cadavers from the medical school.

*Item.* During the fall of 1873, Bell read in a book *The Wonders of Electricity* about the idea of an electric harp. Each of two identical harps consists of a row of tuned magnetized steel reeds fixed at one end, with their free ends vibrating over the pole of a long, rectangular electromagnet. With the two harps connected together, plucking any reed on one harp would create an induced current of its frequency, which by electromagnetic resonance would cause the corresponding reed on the other harp to vibrate. This note had in fact prompted Bell to use strips of iron instead of tuning forks in his harmonic telegraph apparatus.

In July of 1874, in Brantford, Bell began putting all these items together. He'd brought one of the ears home with him, and used it to construct an apparatus that would draw sound waves onto a moving smoked glass, by a straw connected to the bones of the inner ear. He was struck by the ability of feeble sound waves to move those bones. A cryptic entry in Melville Bell's diary for 26 July is the only clue to where this was all leading: "Electric speech(?)."

After Aleck's return to Boston in the fall, he wrote a letter to his parents, dated 23 November 1874. He talked first of Gardiner Hubbard having discovered that Elisha Gray had already applied for a patent on a form of multiple telegraph:

> It is a neck and neck race between Mr. Gray and myself who shall complete our apparatus first. He has the advantage over me in being a practical electrician—but I have reason to believe that I am better acquainted with phenomena of sound than he is—so that I have an advantage there and I have at hand Prof. Lovering [at Harvard] and Mr. Farmer [an inventor who frequented Williams's shop] as advisers in every electrical difficulty.

Farther on Bell described an idea for which he might be thought insane, namely

> an instrument by which the human voice might be telegraphed without *the use of a battery at all.*

He had recently ventured to discuss it with Moses Farmer and Dr. Blake.

> Although the plan has been in my head for a year or so, I never spoke of it so far as I remember as I was uncertain of the fundamental principle.

Farmer had acknowledged that the scheme to be described would generate fluctuating currents, but had doubts they would be strong enough to be useful over any reasonable distance. Here's Bell's new idea.

> If you take a permanent magnet and place one of its poles near the pole of an electro-magnet a current of electricity is *induced* in the coils of the latter. When the permanent magnet is *removed* from the electro-magnet another current of electricity (of opposite kind to the first) is induced in the coils of the electro-magnet.
>
> These facts have been known for years [since 1831 and Faraday's discovery of electromagnetic induction] but the deduction which I have made from them is new. It is this. If a permanent magnet is made to *vibrate* in front of the poles of an electro-magnet—an induced *oscillating current* will be produced in the coils of the electro-magnet. The oscillations of the electrical current will correspond in *number* and ***amplitude*** with the vibrations of the permanent magnet. Hence if we have a harp of *steel bars* made *permanently magnetized*—arranged over an electro-magnet, and if we have the bars tuned to *minute intervals of pitch*[,] if we talk into one harp certain rods will vibrate with certain amplitudes. Their vibrations will create *electrical* vibrations in the line wire and will force into vibration the corresponding rods of the other harp.

Of course, as we'll see, the harp conception was not entirely feasible and was never implemented. However, now that we know so much about digitizing, it's curious to remark how Bell had seized on the idea of analyzing speech into a spectrum of frequencies. It turned out that wasn't necessary anyway. Bell's letter ended:

> This is the theory. I can never hope to work it out [in practice] myself. [He was wrong there!] Mr. Farmer and Dr. Blake both pronounce it feasible. Please keep this paper as a record of the conception of the idea in case any one else should at a future time discover that the vibrations of a permanent magnet will induce a vibrating current of electricity in the coils of an electro-magnet.
>
> Fond love
> Aleck

Bell delayed trying to "work it out" for some months, because Hubbard was pushing hard to get the multiplex telegraph scheme going. Then, early in 1875, Bell proposed a variant that excited Hubbard. If the Morse receiving instruments drove styluses instead of sounders, Bell imagined that by having

30 of them closely spaced he could reproduce a picture of a document with sets of broken lines—the idea of facsimile transmission born ahead of the technology to realize it. This would be the gimmick to put Hubbard one up on Western Union.

But there were still bugs to be worked out of the basic apparatus. Bell and Watson worked through the spring of 1875, until an incident occurred on 2 June 1875. Between two adjoining attic rooms of Williams's shop in Boston, Bell and Watson connected three electric vibrators by a single line as models of the harmonic telegraph system.

Closing the key of each vibrator in turn would cause the corresponding vibrator in the other room to sound. When one of Watson's vibrators stuck, Bell called to him to pluck it free. When Watson did that, Bell's vibrator began to sound, even though there was no battery connected in the circuit. Bell immediately recognized this as the transmission and reception of induced vibratory currents, such as he'd described in his letter the previous autumn. The steel reed of Watson's electromagnetic vibrator had become slightly magnetized, and its vibrations near the pole of the electromagnet's pole had induced a vibratory current. Bell realized, contrary to Farmer's opinion, that the induced current was not too feeble to produce an effect some distance away.

Then, following a practice that he'd used for tuning the reeds of his instruments, Bell picked up one of his instruments and held it close to his ear, while Watson plucked one of his not attuned to it. Using the reed essentially like a tin-box receiver, Bell heard the pitch of the plucked reed *and* its tone color.

In that magical moment, Bell realized that he could dispense with his harp-like analyzer; that a single diaphragm (like an ear drum) would respond to a whole range of sound frequencies directed to it. And that the same instrument would serve both as microphone and speaker. He sketched an apparatus for Watson to construct, a thin diaphragm to vibrate a steel reed above the pole of an electromagnet.

After a couple of missteps, they'd built

> an instrument modelled after the human ear by means of which I hope...to transmit a vocal sound,

Bell wrote to his parents on 30 June 1875. In early July, with the new instruments on different floors, Bell sang and spoke while Watson listened. He heard the tune, Watson reported, and the voice: "I could almost make out what you said!" Close, but no cigar.

Hubbard now encouraged Bell to abandon the "toy" and get back to serious telegraphy. Then a variety of other matters intervened, and Bell and Watson made no further trials for six months. In the meantime, Bell was sure enough of the principles to begin to write a patent application on "Improvement in Telegraphy." In January of 1876, he added a crucial clause to the effect that a varying electric current could also be created "by alternately increasing and diminishing the resistance of the circuit." Content that his application had covered all possible loopholes, Bell sent it to Hubbard who was in Washington on business. On the morning of 14 February 1876, Hubbard filed the application at the Patent Office. The same afternoon, Elisha Gray filed an intention to apply for a patent for a speaking telephone, based on the principle of variable resistance.

In late fall of 1875, Gray had put together the observation of sound transmitted along a thread between two diaphragms (sometimes called a "lover's telegraph") with his earlier experience of using metal rods dipping into conducting liquids to vary the resistance in a circuit. Since Gray had made no actual apparatus or trials for voice transmission, all he filed was an intention to patent. Bell's priority gave him the patent, number 174,565, issued on 7 March 1876. Three days later, a Bell microphone "heard" speech distinctly for the first time.

Bell had Watson build a microphone with one end of a short brass rod attached to a diaphragm at the small end of a speaking-tube horn. The other end of the rod dipped into a small cup of slightly acidic water. The connections were made through a battery to a receiver two rooms away.

After Bell shouted into the mouthpiece (not loud enough to be heard directly), Watson rushed to the room. He had heard the words distinctly: "Mr. Watson—come here—I want to see you."

Soon, Bell and Watson improved their electromagnetic instruments, particularly by eventually deciding that a thin iron diaphragm clamped around its circumference would serve as the vibrating magnet if the coil of the armature was wrapped around a permanent magnet. Instruments like that were exhibited at the Centennial Exposition in Philadelphia at the end of June. In August, Bell used his instruments to transmit voices 13 km from Brantford to Paris, Ont. In October, Watson took instruments to a town in New Hampshire, 230 km from Boston, and held a two-way conversation with Bell along telegraph lines. A year later the Bell Telephone Company had leased a thousand telephones to infant telephone companies in more than 15 localities. Early in 1878, the first telephone exchange was established in New Haven, Connecticut, with 21 subscribers and eight grounded lines. Hamilton, Ontario had a seven-line switchboard by July of the same year.

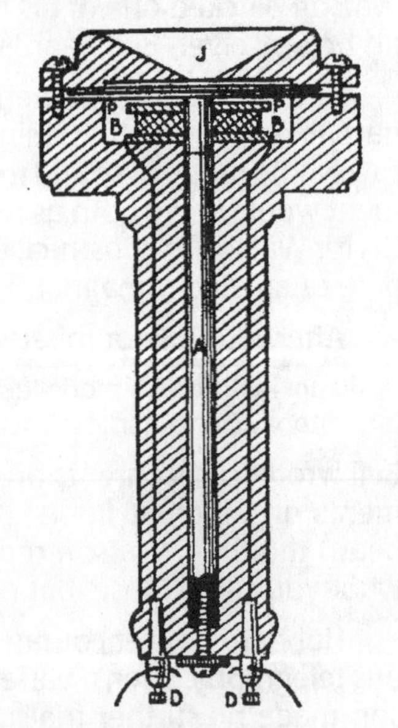

*Bell's telephone earpiece.*

Almost immediately, Western Union hired Thomas Edison to devise a telephone system that did not directly infringe the Bell patents. Edison designed the "carbon button" microphone, in which the diaphragm changed the pressure (and therefore the resistance) of carbon granules loosely packed into a little box. Of course, Bell's patent had included the variable resistance *principle,* and so Edison did infringe it that way. The Bell Company sued Western Union, and in 1879 the latter company withdrew from the telephone business and transferred all its telephone patents to Bell for a share of rental receipts for 17 years. The value of Bell shares doubled within a few days of the settlement.

In 1885, the American Telephone and Telegraph Company was formed to construct and maintain long-distance lines for the interconnection of local offices. Its first president was Theodore Vail (1845–1920), cousin of the Alfred Vail who'd assisted Morse. The younger Vail had been an operator for Western Union before becoming associated with railway mail services, of which he was made federal superintendent in 1876. That brought him to the attention of Gardiner Hubbard, who hired him to be the first general manager of the Bell Telephone Company in 1878. Hubbard made one further major contribution to Bell's welfare—his daughter Mabel. Aleck and Mabel were married in July 1877. Gardiner Hubbard was the founder and first president of the National Geographic Society from 1888 till his death in 1897.

Since telephone relays must be responsive to a fully varying current over a range of frequencies, electro-mechanical relays are difficult to conceive, let alone construct. In fact, adequate relays for voice had to await the invent\\of electron tubes (Chapter 20). In the early 1900s, the maximum distance that could be reached by telephone was the 2482 km from Boston to Omaha, Nebraska. Between 1880 and 1900, the number of telephones in the United States doubled every four years, from 48,000 to 1,350,000.

Telephones invaded Europe early on. On the continent, they were incorporated into government-run telegraph operations, which probably contributed to a slow growth. In Britain, where the Edison Company was active in lighting, Bell and Edison interests combined in the United Telephone Company. Although the Postmaster-General wanted telephones included under his general responsibilities for the telegraphs, Parliament at the time was reluctant to interfere too directly in business enterprises. Nonetheless, both Parliament and local governments refused to grant telephone companies the right to lay cables underground. Various wrangles continued until finally the British government decided that, in 1911, telephones would become a monopoly of the Post Office. While the number of telephones in the United States represented 5% of population, the figures were 1% for Britain and Germany, less for France.

## *Conclusion*

Electrical communications, first the telegraph and then the telephone, greatly increased the speed of signaling. Each symbol travels at close to the speed of light. A trained telegrapher could send fifty words per minute; by telephone you can speak 150 words a minute or more. News, information, propaganda, and gossip can be spread at rates to leave Martin Luther or Marin Mersenne breathless.

Moreover, these new transmission media soon resulted in fantastic increases in the *amounts* of information being transmitted. While daily newspapers had begun in the eighteenth century, they mostly contained stale news until the middle of the nineteenth. And of course, business and personal communications took on whole new dimensions. As Bell's modern biographer, Robert V. Bruce has put it so aptly,

> Without the telephone as its nervous system, the twentieth-century metropolis would have been stunted by congestion and slowed to the primordial pace of messengers and postmen. And the modern industrial age would have been born with cerebral palsy.

Maybe we couldn't even *have* our great metropolises without that telephonic nervous system. Just try to imagine the way city-dwellers would have to operate without telephones. If you need any help, consult your telephone company's advertising department—or just watch their TV ad campaigns.

Another important consequence of our new communications technology is its effect on human skills. First, the telegraph initiated a whole new class of skilled workers—the telegraph operators. As skilled hand-loom weavers were losing employment, a new field was opening up. "Ham-fisted" operators jiggled their Morse keys at amazing rates.

Perhaps more amazing was their skill at reading the sounder. At the height of the telegraph industry, imagine a room of twenty or more operators each at a desk with a clacking sounder. Two of them are carrying on a conversation as their instruments clatter into action. Pausing to roll a cigarette or fill a pipe, the two men eventually turn to their typewriters and copy down the message that had started a couple of minutes earlier. Through the cacophony of sounders and typewriters, they'd casually pick out their own appropriate messages and never miss a word.

Now, telegraph operators are as obsolete as hand-loom weavers. With the telephone, you can be your own operator. Of course, in earlier days, telephone switchboard operators were very numerous. Some of us oldsters can remember the familiar "Number, please" that lasted until the installation of automatic dialing. While there are still telephone operators, they perform rather different functions today, as they sit at computer keyboards looking up numbers or facilitating awkward routings. Of course, without automatic dialing, the number of switchboard operators (at 200 or so lines per operator) would be almost astronomical.

An interesting sideline to the story of these telegraph and telephone operators is that telegraph operators were predominantly male, while telephone switchboard operators very quickly became predominantly female. That occupation was added to the other traditional female roles of nursing, teaching, and clerking to provide another escape route from domestic tasks.

But the march of communications technology has resulted in changing skills. There has also been a certain loss of skills with the decline in the number of telegraph and telephone operators. In their place are new categories, involving, for example, keyboard skills or more technical skills, such as maintenance and repair. At the same time, although de-skilling can have serious adverse effects on the status of workers, there have been some general increases in the skills required of all of us in the general population. Mothers and kindergarten teachers are now concerned that our children develop appropriate telephone manners since four-year-olds are regular users of this new technology. What a thrill the day Mary learns to dial her grandmother!

Developments in the communications industry also have some lessons to teach about how technology proceeds. Consider a few items, in more or less historical order.

The telegraph required the stringing of wires across the countryside. That involved surveying routes, buying rights of way, and organizing a variety of tasks. How to do these things had been partially prepared by the earlier need for rights of way for the canals and railways. We should not be surprised that early telegraph lines often followed rail routes, nor that telegraphy became part of railroading.

When the telephone came along, it could capitalize on the preceding developments of the telegraph. Indeed, as we've seen, Bell, Gray, and others came to the telephone with experience in telegraphy. We can think of the telegraph as being a **pathfinder** for the telephone, both literally and figuratively—as canals had been pathfinders for railways. Another contribution to the telephone, with its need for lines within cities, was the gas-lighting industry. Routing such utilities along city streets had begun in London shortly after 1800. The first city in the United States with gas-lighting was Baltimore, 1816.

Regarding the actual invention of the telephone, you might find it peculiar that an amateur inventor like Bell was able to beat out the more experienced Gray. Historian David Hounshell has described Gray as *handicapped* by his professional credentials, being "an expert and a member of a community of experts." Gray did not pursue his early ideas of communicating speech because he didn't see a market for it.

Hounshell has listed six limiting factors connected with Gray's overcommittment to the telegraph:

> his extensive experience in telegraphy, his sensitivity to problems facing telegraphic development, his association with and respect for the leaders of the telegraph industry, his close personal relationship with his business partner, the pressure from his financial backer, and his trust in the expertise of his patent lawyers.

Bell, on the other hand, had few of these obstacles to overcome. His characteristics are graphically portrayed by Robert Bruce:

> If a computer searching for the most likely inventor of the telephone had run through the punch cards of all the world's people and places at that

> time [1872], one name at least would have been printed out. It would have been that of an enthusiastic, ambitious, intelligent, imaginative young man with unusually keen hearing and sense of pitch; a trained pianist; a man who knew the mechanics of speech and hearing and took a lively interest in telegraphy; one who lived in a society more eager for rapid communication than any before; one who had easy access to leading scientists, skilled technicians, an academic community, and a community of enterprising capitalists.

A tall order for Alexander Graham Bell to fill!

Finally, there's one other curious note. It's often been said that you can't patent a scientific principle. Bell seems to have come about as close to doing that as is humanly possible. Of course, Faraday hadn't had a patent on electromagnetic induction. Yet, when Bell applied that principle to the conversion of sound variations into their electrical analogue, that was almost a scientific principle. It was to this extent—that you can conceive it without necessarily realizing the particular means needed to do it.

That Bell's principle is not obvious is shown by the fact that it had apparently not occurred to anyone else, except perhaps the Frenchman, J. Baile. For it was from his book (translated into English) *The Wonders of Electricity* that Bell got the idea of the "electric harp" he described in his 1874 letter to his parents. Evidently Baile had not patented his harp idea, nor is there any reason to suppose he even ever built it. Reis's "telephone" is clear evidence of how difficult it was for others to conceive of Bell's principle.

The extent of Bell's claims may be found in his Canadian patent No. 7789, issued on 24 August 1877:

> This invention relates to improvements on electric telephony, and consists in the union upon, and by means of, an electric circuit, of two or more instruments so constructed that if motion of any kind or form be produced in any way, in the armature of any one of the said instruments, the armature of all the other instruments upon the same circuit will be moved in like manner and form, and if such motion be produced in the former by sound, like sound will be produced by the motion of the latter.
>
> Also in a system of electric telegraphy or telephony, consisting of transmitting and receiving instruments united upon an electric circuit, the production, in the armature of each receiving instrument, of any given motion, by subjecting said armature to an attraction varying in intensity, however such variation may be produced in the magnet, or the production of any given sound or sounds from the armature of the receiving instrument, by subjecting said armature to an attraction varying in intensity in such manner as to throw the armature into that form of vibration that characterizes the given sound or sounds.
>
> Also, in the combination with an electro-magnet, of a plate of iron, or steel, or other material capable of inductive action, which can be thrown into vibration by the movement of surrounding air, or by the attraction of a magnet.

Of course, in the patent, Bell had to describe a practical application of his principles. And, of course, it's not really a new *principle.* Nonetheless, it was Bell's, and his alone. Just imagine patent attorneys for Western Union trying to advise Edison how to compete with Bell without infringing that patent.

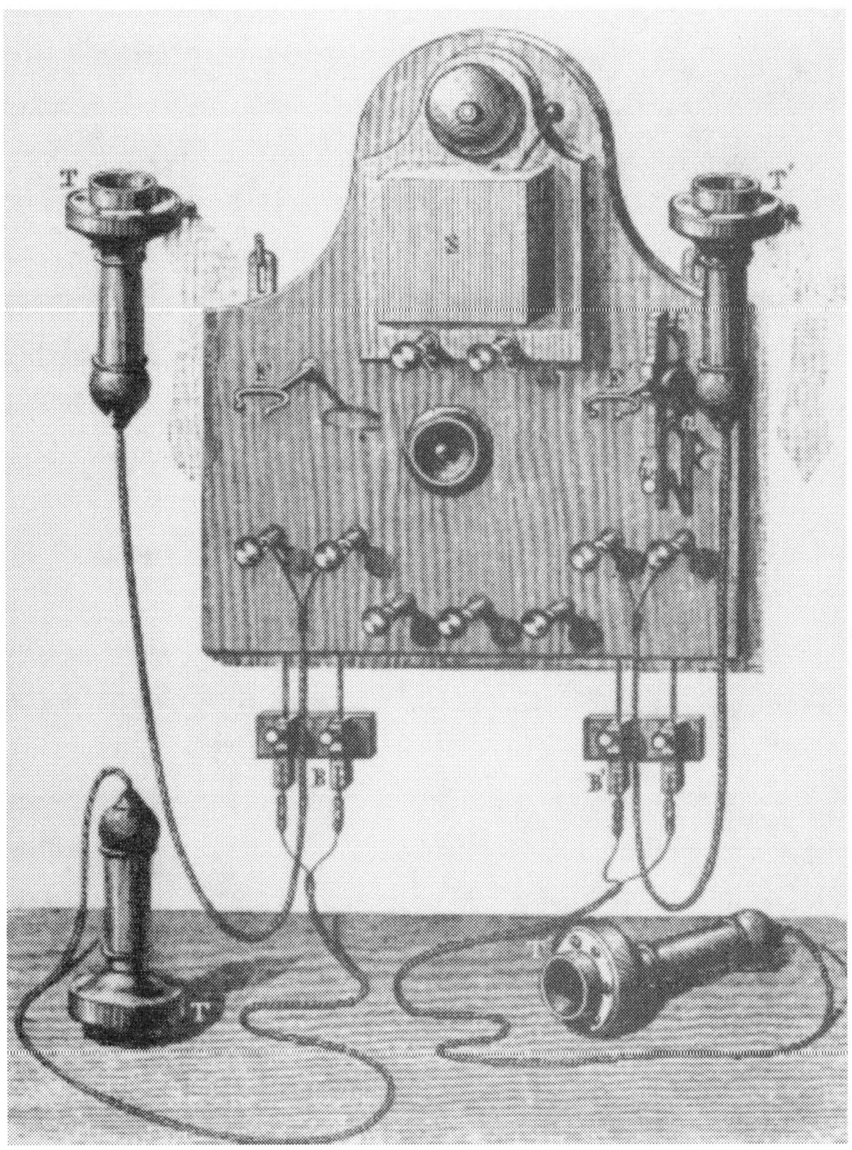


*A 19th century Bell telephone. Note that the microphones and speakers are separate and both are hand-held. This telephone set accommodates two users on the same line at once.*

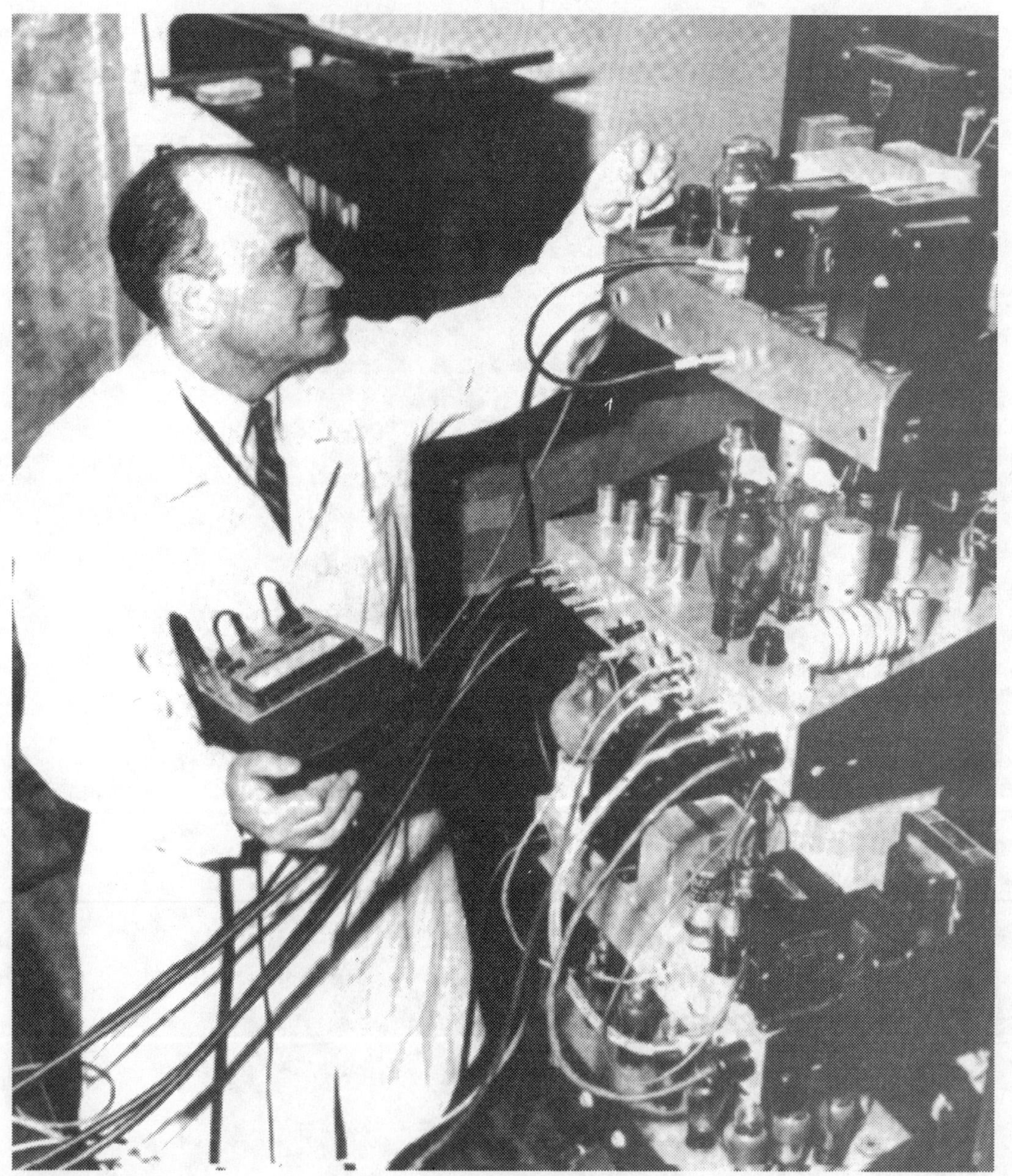

*Enrico Fermi.*

# CHAPTER NINETEEN

## New Revolutions in Physics

The first revolution in physics occurred in the early 1600s when Galileo analyzed measurements he'd made of moving objects. He took the study of motion out of the realm of the philosophy of wordy logic and made it a form of applied mathematics. Seventy years later, Newton solidified the revolution by subjecting (as he said) "the phenomena of nature to the laws of mathematics."

Although some of the developments of physics in the next two centuries could be considered revolutionary (electromagnetism, for example), most physicists from 1700 to 1900 considered that they were merely extending and elaborating the program laid down by Newton (see Chapter 17, electricity). If Faraday and Maxwell were not revolutionary, they were sowing the seeds of revolution.

Those revolutionary seeds began to sprout just at the end of the nineteenth century. Since about 1895, two major revolutionary theories have sent shock waves rolling across the world of physics: the theory of relativity and the quantum theory. Although their originators didn't set out to be revolutionary, they found themselves driven to revolution even as they continued to analyze new observations and measurements in the Newtonian way. The earliest new observations began to challenge cherished theories about the constitution of matter and radiation. By the time physicists had gotten things straightened out, the simple world of atoms of matter and waves of radiation had vanished. Vanished as completely as the center of the city of Hiroshima, Japan, on 6 August 1945 in a stunning performance by those twin tragedians, knowledge and power. For the revolutions in modern physics have created real shock waves too.

In 1900, physicists had a world in which matter came in particles and radiation came in waves. By 1930, with the quantum theory, they learned that waves of radiation also behave like particles, and that atomic particles also behave like waves. They found too that material particles can actually be transformed into bursts of radiant energy—and vice versa. And with the new theory of relativity, they could calculate how much radiant energy they could get from transforming matter into energy.

Until 1900, physicists aimed at depicting the world with absolute certainty. In the twentieth century they've shown absolute certainty to be unattainable. They've shown that we change the things we observe by our very act of observing them. That means we can't know for sure how they were before we observed them. Although our knowledge is now uncertain, we have a good idea of how wide our range of uncertainty is.

## *New Waves and Particles*

In the 1850s, Michael Faraday had studied the conduction of electricity through gases. He sealed metal plates into the ends of a glass tube, created a partial vacuum by pumping some of the air out of the tube, and then connected a high voltage to the plates. As the air pressure went down, he saw fascinating colors in the tube. Later, other experimenters added various gases to the tube, and got different colors depending on which gas they used. Today, we're all familiar with colored advertising signs based on this principle. Neon or some other gas glows brightly in tubular letters when a high voltage is applied across the ends of a tube.

Faraday found that the glow filled the tube when he reduced the pressure to about a hundredth of atmospheric pressure. But when he reduced the pressure further, the glow broke up into alternating bands of dark and color. At still lower pressures the glow diminished even more. That was as far as Faraday could get. About 1870, investigators benefitted from improvements in vacuum-pump technology.

When investigators removed almost all the air from a tube—getting the pressure down to about a millionth of atmospheric pressure—they found that the glow within the tube entirely disappeared. But then, they noticed that the glass at one end of the tube fluoresced green. In England, William Crookes (1832–1919) was a prominent investigator of electricity in gases. Crookes found that magnets deflected his cathode rays in the same direction as wires carrying flows of negative charges. This and other evidence convinced Crookes that the cathode was emitting streams of negatively charged particles: that the cathode rays *were* negatively charged particles.

### *Wilhelm Roentgen (1845–1923)*

While Crookes was deciding that cathode rays were particles, experimenters in Germany were claiming instead that they were a form of electromagnetic radiation. In 1895, Wilhelm Roentgen (or Röntgen) in Würzburg seemed to confirm that idea. He worked with a highly evacuated cathode ray tube, which he covered with cardboard. As Roentgen wrote in his report, whenever he sent electricity through his completely covered tube, he observed

> the bright illumination of a paper screen covered with barium platinocyanide, placed in the vicinity...visible even when the paper screen is at a distance of two meters from the apparatus. It is easy to prove that the cause of the fluorescence proceeds from the discharge tube, and not from any other place.

Roentgen had discovered x-rays—as he named them. And he quickly showed that they were indeed very short electromagnetic waves. By a curi-

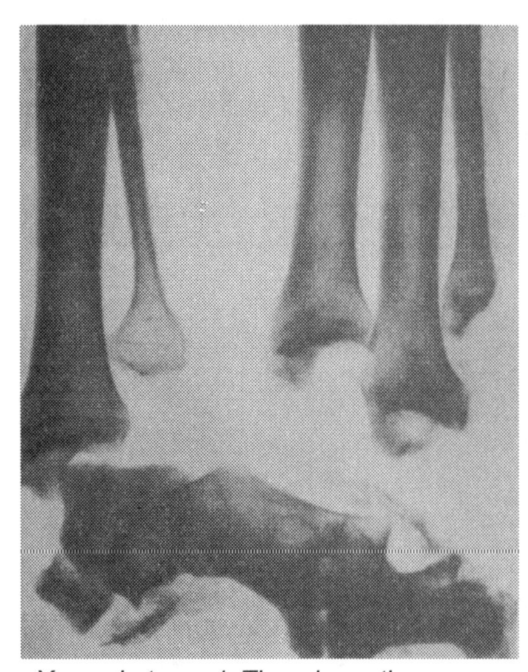

*X-ray photograph. Though a rather abstract discovery of pure physics, x-rays were almost immediately put to practical use in medicine.*

ous irony, William Crookes had observed similar effects of x-rays some years earlier, but hadn't realized it. At least there's a persistent rumor that Crookes had found that photographic film, which was stored in its wrappings near his operating cathode ray tubes, was fogged after he'd used and developed it. Crookes complained to the manufacturers that their film was defective. Only after Roentgen announced his discovery did Crookes realize what must have happened. His tubes had been producing x-rays, which penetrated the wrapping and exposed the film.

At any rate, x-rays were an instant sensation. People immediately realized x-rays could be used to get pictures of the inside of the body, and they quickly became part of medical diagnosis. They also prompted some silly talk that Victorian ladies would have to equip themselves with leaden undergarments, as if some superman with x-ray vision was lurking just round the corner. In the world of physics, x-rays seemed to support the German contention that instead of cathode rays being particles, they were electromagnetic radiations.

Not for long!

*Electrons in Atoms*

Crookes' work was continued at Cambridge University by J.J. Thomson (1856–1940), who was appointed head of the Cavendish Laboratory when only 28. This lab had been founded in 1870 by a member of the family of the scientist Henry Cavendish, famous a century earlier. James Clerk Maxwell had been its first director. In 1897, Thomson made a series of precise measurements of the behavior of cathode rays within a Crookes' tube. By deflecting the rays with electric and magnetic forces, he determined that cathode rays did consist of particles—negatively charged particles, about one two-thousandth of the mass of a hydrogen atom. We now call them **electrons**. When these cathode rays, or electrons, struck the glass at the end of the tube, they generated x-rays in the glass. That's because the decelerating electrons (electric charges) created electromagnetic waves. The extremely rapid deceleration resulted in radiations of very short wavelength (or high frequency).

In his experiments Thomson found that identical electrons came out of every kind of element he could test. They came out of atoms! No longer could atoms be considered to be those tiny, hard, un-cuttable specks imagined by Democritus and Dalton. Atoms have parts!

*Radioactivity*

Not just electrons came out of atoms. In 1896, Henri Becquerel (1852–1908) in France observed radioactivity for the first time. Becquerel was conducting experiments with crystals of uranium salts—crystals which, after being exposed to the sun, gave off light for some time. They seemed to absorb sunlight and then re-radiate it slowly. Becquerel measured this fluorescence by how much the uranium would expose a photographic plate. Then, as he wrote:

> I had prepared several experiments on Wednesday the 26th, and Thursday, the 27th of February, and as on those days the sun shone only intermittently, I kept my experiments all prepared and returned the plate holders to the darkness of the table drawer, leaving the uranium salts in place. The sun not showing itself again for several days, I developed the photographic plates on the first of March expecting to find very faint images. The silhouettes appeared on the contrary with great intensity.

Since the uranium salts had not been fully exposed to the sun, Becquerel had expected the fluorescence to be faint. When it was in fact quite intense, he realized something else must be happening. With further tests, he soon found that the uranium was giving off rays by itself, entirely independent of any previous exposure to the sun. Soon, the phenomenon was called **radioactivity**, that is, the activity of giving off rays, or radiation. Quite naturally, Becquerel surmised that his rays were similar to the x-rays Roentgen had found. As we'll see, he was partly right, but only partly. Becquerel's discovery of radioactivity opened up a whole new field of research for physicists.

In France, Marie Curie (1867–1934), assisted by her husband Pierre (1859–1906), began a systemic search for other elements that might be radioactive. She found that thorium was similar to uranium. Then, by painstaking chemical analysis, she found some new elements in the same ores where uranium and thorium had been found. In 1898, she discovered polonium and radium. She wrote:

> Uranium, thorium, polonium, radium, and their compounds make the air a conductor of electricity and expose photographic plates. From these two points of view, polonium and radium are considerably more active than uranium and thorium. On photographic plates we obtained good images with radium and polonium in a half-minute exposure; it takes several hours to obtain the same results with uranium or thorium.

In further work Madame Curie demonstrated that all elements in the periodic table heavier than lead are naturally radioactive. She also showed that

*Marie Curie.*

the phenomenon of radioactivity was entirely spontaneous, and not influenced by *any* external factors. Not exposure to the sun, not heating–not anything!

> The property of emitting rays is so much the more feeble as the proportion of the active metal in the compound is less. The physical state of the substances seems to be of altogether secondary importance. Various experiments have shown that if the substances are mixed with others, their condition seems to have no effect except as it varies the proportions of the active body and the absorption produced by the inert substance. Certain causes, such as the presence of impurities, which have so great an effect on fluorescence, are here altogether without effect.

Madame Curie was the first woman scientist to achieve worldwide acclaim. She was born Marie Sklodovska in Poland in 1867. After graduating from high school she worked as a governess for six years. Since women were legally excluded from studying at Polish universities, she travelled to Paris to study mathematics and physics. In 1895, she married Pierre Curie, already a prominent physicist. She's the first scientist we know of whose researches were interrupted for a while so she could give birth to a daughter. That was in 1897.

*Ernest Rutherford (1871–1937)*

A different line of research in radioactivity was followed in England by Rutherford, another immigrant far from home. Born and educated in New Zealand, he did graduate work at the Cavendish Laboratory in Cambridge with J.J. Thomson. Then from 1898 to 1907, Rutherford taught physics at McGill University in Montreal. At McGill, he continued the research he'd begun at Cambridge. Rutherford wanted to explore the nature of the rays emitted by uranium. Was radioactivity like x-rays as Becquerel had thought? Or was it something else?

*Ernest Rutherford.*

Rutherford tested the radiations from uranium in various ways. He put absorbing materials of different thicknesses between uranium samples and photographic plates. He deflected the rays by magnetic fields. Eventually, Rutherford was able to show that uranium and other radioactive elements emit three distinct types of rays. He labelled the new rays *alpha*, *beta*, and *gamma*. Further research by Rutherford and others identified the nature of these three rays. Alphas and betas are bits of matter: **Alpha rays** are positively charged atoms of the gas helium—small atoms, next larger than hydrogen in the periodic table. **Beta rays** are electrons, negatively charged particles. But **gamma rays**, like x-rays, are not particles of matter, but electromagnetic waves—like light, but with wavelengths about a hundred thousand times smaller than visible light.

### *Albert Einstein (1879–1955) and Relativity*

Einstein was born in southern Germany in 1879. He wasn't considered an exceptional student in his early years. From 1896 to 1900, he attended a technical university in Switzerland to prepare himself as a teacher of mathematics and physics. However, after graduating, Einstein couldn't get a teaching job and eventually became a patent examiner in Berne, Switzerland. There, in 1905 (at age 26), he published three scientific papers. They covered three separate fields of physics, and all three were brilliant solutions to puzzling problems.

*Albert Einstein.*

Einstein's most famous paper was on relativity, though it was titled "On the Electrodynamics of Moving Bodies." His basic idea was that when we measure the positions and times of events, the values we get depend on how fast we're moving relative to the events we're measuring. Einstein was the first to realize clearly that our measurements are relative, not absolute.

The issue had arisen when scientists tried to measure the speed of the earth through the æther. The æther was the medium that carried light waves at 300,000 km/s. They got results they couldn't understand. If you throw a ball at 20 m/s from a truck travelling at 15 m/s, the ball's speed for a catcher on the ground can be as little as 5 m/s or as much as 35 m/s, depending on the directions of the ball and the truck. But, considering light to be the ball being thrown from the earth, scientists always measured the same speed for the light; it didn't depend on the speed or direction of motion of either the light source or the observer.

Early in his paper, Einstein said he could dispense with the æther. He wasn't going to need the idea of a medium absolutely at rest. Instead, he said, if we want our mathematical equations to describe actual motions, we have to be clear about what we mean by time. The passage sounds like some young smart aleck informing the great physicists of Germany how to tell time:

> If we wish to describe the motion of a material point, we give the values of its co-ordinates as functions of the time. Now we must bear carefully in mind that a mathematical description of this kind has no physical meaning unless we are quite clear about what we understand by "time." We have to take into account that all our judgments where time plays a part are always judgments of simultaneous events. If, for instance, I say, "That train arrives here at seven o'clock," I mean something like this: "The pointing of the small hand of my watch to seven and the arrival of the train are simultaneous events."

Einstein derived mathematical relations from this idea of simultaneity and from two postulates or principles. One of these was simply, if you can't detect the speed of a source of light, then assume the speed of light really is constant and independent of the source's motion.

Einstein called the other the **principle of relativity**: that the formulas we use to describe interactions among objects should depend only on the objects' motions relative to one another. If all we know is that A moves relative to B, then a formula for the force between A and B shouldn't depend on whether A or B is "actually" at rest. And that's why Einstein was considering electrodynamics in this paper, because in Maxwell's formulas for electromagnetism there was just such a dependence.

If you plunge a magnet into a coil of wire, you induce a current in the coil. That was Faraday's discovery. Physically, it doesn't matter whether you move the magnet into the coil, or move the coil onto the magnet. Yet, in Maxwell's formulas for electromagnetism, there was one result if you moved the magnet into the coil, and a different result if you moved the coil onto the magnet.

Einstein said, if it doesn't matter whether you move the magnet or the coil, then the results should be the same for both. Einstein found that his principle of relativity produced a single result. The speed of light was involved in the theory of relativity for two reasons. First, Maxwell had derived from his formulas that light was an electromagnetic wave. Second, as Einstein examined the process of measurement, he recognized that the testing of simultaneity required the transfer of information between two observers (say at the opposite ends of a tape measure, for example). His speed of light principle assumed that the information took time to travel (that is, was not instantaneous).

So Einstein derived formulas from his basic principles that made Maxwell's theory fit experience better. The theory of relativity had other curious consequences, particularly in revising Newton's basic laws of motion. Newton had considered an object's mass to be constant at all speeds. Einstein found in his theory that as an object speeds up, its mass increases. Not much for objects at ordinary speeds like trains or even rockets. But electrons travelling at half the speed of light have a mass that's 15% higher than when they travel at low speeds. And the faster electrons go, the more their mass increases. Their mass doubles at seven-eighths of the speed of light—and their mass would be infinite if they could go as fast as light. But they can't. No material object can attain the speed of light—it's the universal speed limit.

This change of electrons' mass with speed could be measured. Measurements on high speed electrons made up to 1906 in Germany didn't conform to Einstein's formula. Einstein wasn't fazed. While admitting that experiments to date matched another theory more closely than his, he simply announced that his was the only theory derived from general principles.

Within a couple of years, further experimental results confirmed Einstein's relation and identified flaws in the earlier measurements. Since then, this "special" theory of relativity has been confirmed by innumerable measurements, including measurements in places as familiar as your color TV tube. (Electrons accelerated through 25 kV travel at about a third of the speed of light and have a mass increase of about 5%.)

A few months after his first relativity paper, Einstein published another short paper on the subject. It contained a further startling consequence of his theory. This was his famous formula, energy equals mass times the square of the speed of light:

$$E = m c^2$$

Einstein found that his formulas for the relativity of motion contained the following idea. If an object—let's say an atom—gives off energy in the form of light, then the mass of the atom will decrease slightly. And this decrease in mass will be equivalent to the amount of energy emitted. Because the speed of light is so great, a tiny change in mass represents a tremendous amount of energy. That's why 35 years later, physicists would realize that a tiny mass change, like that in nuclear fission, would release super-explosive amounts of energy.

On the theoretical side, though, it looked as if we could now think of matter as congealed energy; or energy as liberated matter.

## *Quanta and Photons*

In another of his 1905 papers, Einstein developed an idea first proposed by Max Planck (1858–1947)—a new way to describe the radiation and absorption of light. Five years earlier, Planck had just been appointed professor of theoretical physics at Berlin University. There he studied a problem that had physicists baffled. It had to do with the way objects emit light as they're heated up. Think of putting a poker in a hot fire. At about 500°C it gives off a dull red glow. As its temperature increases, the poker glows brighter, and eventually gets to be white-hot.

*Max Planck.*

Physicists had good experimental measurements relating the temperature of the poker to the color and energy of the light it emitted. But they couldn't explain the "why" of the relation. They couldn't find any theory to match those experimental results. Then Planck made a bold hypothesis. He assumed that light was not being given off continuously, as you'd expect from the wave theory, but in little packages, which he called **quanta**. He said the energy content of each quantum of light depended on its color—actually its wavelength. Quanta of blue light, with a short wavelength, have more energy than quanta of red light, which has a longer wavelength. Planck's assumption allowed him to construct a formula that perfectly fitted the experimental relation of temperature to light's wavelength and energy. A stunning result!

If you could accept this quantum idea, you could explain the relationship. But not everyone could accept it. The German physicist Max Born wrote about this work in his obituary notice for Planck when he died in 1947.

> It has been generally acknowledged that the year 1900 of Planck's discovery marks indeed the beginning of a new epoch in physics. Yet during the first years of the new century very little happened. It was the time of my own student days, and I remember that Planck's idea was hardly mentioned in our lectures, and if so as a kind of preliminary "working hypothesis" which ought of course to be eliminated. Planck himself turned to other fields of work.

Now Einstein was only four years older than Born. But Einstein took Planck's idea much more seriously. Not just as a "working hypothesis," but as a completely new way of thinking about light. In 1905—that crucial year—he analyzed Planck's quantum idea and wrote a paper on it. Right at the start, Einstein wrote that it appeared to him that the emission and absorption of light could be explained by assuming

> that the energy in light is distributed discontinuously in space... [and that] light consists of a finite number of energy quanta...which are absorbed and emitted only as units.

Physicists soon began to use the term quantum generally for any little packet of energy. So they coined the term **photon**, specifically for a quantum of radiant energy. Photons are little packets of radiant energy—that is, light, x-rays, and gamma rays.

Einstein used Planck's quantum idea to explain various light phenomena in a new way, a way that could now be checked experimentally. For example, when light strikes a metallic surface it can eject electrons, as in the photoelectric cells that are often used in elevator door openers. Einstein worked out a formula relating the energy of the ejected electrons to the energy of the incident light. In 1916, Robert Millikan (1868–1953), in the United States, made experiments that confirmed Einstein's formula. In fact, Einstein won his Nobel Prize five years later, in 1921, more for his photoelectric formula than for his theory of relativity. Even in 1921 relativity was considered too radical by some scientists. Millikan won the Nobel Prize in 1923 for his work. It took Einstein

16 years from the time of his original 1905 papers to get the Nobel prize—more than twice as long as it took for Millikan. In the twenties, scientists were slow to absorb revolutionary theories.

### *The Nuclear Atom*

Yet physicists did accept the next revolutionary theory—the nuclear atom—very quickly. They accepted it because it came from experiments—and, I think, because they were already beginning to suspect that atoms must have some kind of structure. In 1910, Ernest Rutherford in England set up an experiment to shoot high-speed alpha particles at thin foils of solid gold. The alpha particles came from a radioactive source. After the alphas struck the foil, they hit a fluorescent screen, where Rutherford observed their impact. That way he could tell how the alphas scattered as they passed through the gold foil. Most of the alphas passed straight through the foil—undeflected—showing there was a lot of empty space within those solid gold atoms. But occasionally, Rutherford observed that an alpha particle would rebound from the foil as if it had hit something very hard. Think of shooting bullets at the side of a barn; it's as if the barn had a paper-thin wall with occasional small lumps of steel embedded in it. Those small lumps represent the nuclei of the atoms.

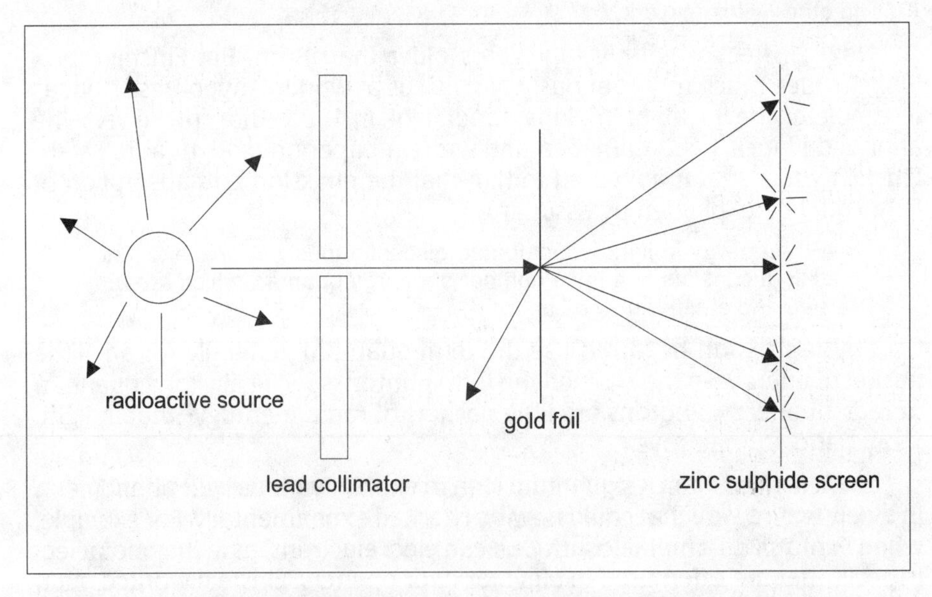


*Rutherford's experiment. Relatively heavy alpha particles were fired at a sheet of gold foil. Most went straight through it and were detected as they hit a zinc sulphide screen on the other side of the foil. A small number were deflected and bounced back toward the source of the alpha rays, indicating that they had struck something solid and heavy.*

Rutherford then proposed that atoms had more than 99 per cent of their mass concentrated in extremely tiny nuclei. Compared to the size of the whole atom the nucleus is like a pea in a football field. With a positively charged nucleus, Rutherford could imagine electrons—negatively charged—circling about on the track at the edge of the field. But he didn't say how those electrons were arranged way out there beyond the nucleus.

## *Bohr's Atomic Model*

The Danish physicist Niels Bohr (1885–1962) was keen to find out how those electrons were arranged. In 1913, Bohr was working in Rutherford's laboratory at Manchester University. Bohr had the same kind of deep theoretical insight that Einstein had. Indeed, a recent writer has called Bohr and Einstein "the two great masters whose work formed the basis of most phases of modern physics." Bohr puzzled over how electrons could be arranged around the nucleus of an atom.

*Niels Bohr.*

He decided to build a model of the atom that would look like the solar system. He put the nucleus in place of the sun and had electrons in "planetary" orbits around it. Instead of gravity acting between them, he had the electrical force of attraction between the positive nucleus and the negative electrons. But, according to Maxwell's electromagnetic theory, the electron in such a system would emit radiation continuously and spiral into the nucleus in less than a microsecond. Such an atom would be unstable. That didn't deter Bohr. He knew atoms were stable and didn't radiate continuously. So he just assumed his orbits were stable. He'd worry about why later—in true Galilean fashion.

But atoms do radiate energy sometimes. Bohr had to explain that. He considered the light given off by hydrogen gas in electrified tubes—like the light from neon gas tubes in electric signs. Back in the 1880s, scientists had used prisms to analyze the light from electrified hydrogen tubes. A prism spreads light out into a spectrum. But, instead of a continuous spectrum—red, orange, yellow, green, blue, violet—like you get from the sun, the hydrogen spectrum contains just a few narrow lines, one red, one blue-green, and a few violet lines. Bohr calculated the energy of the photons of each of these colors, using Planck's quantum idea as Einstein had done.

Then Bohr assigned a set of orbits to the single electron of a hydrogen atom. He assumed that the electron could whirl around in any one orbit without any change in energy. Changes in energy would only occur if the electron jumped from one orbit to another. Bohr calculated the energies for this set of orbits, assuming only the force of attraction between the nucleus and the electron, and that the orbits were "quantized." That is, the orbits were like spaced-out grooves on a track, with distances from the center being whole

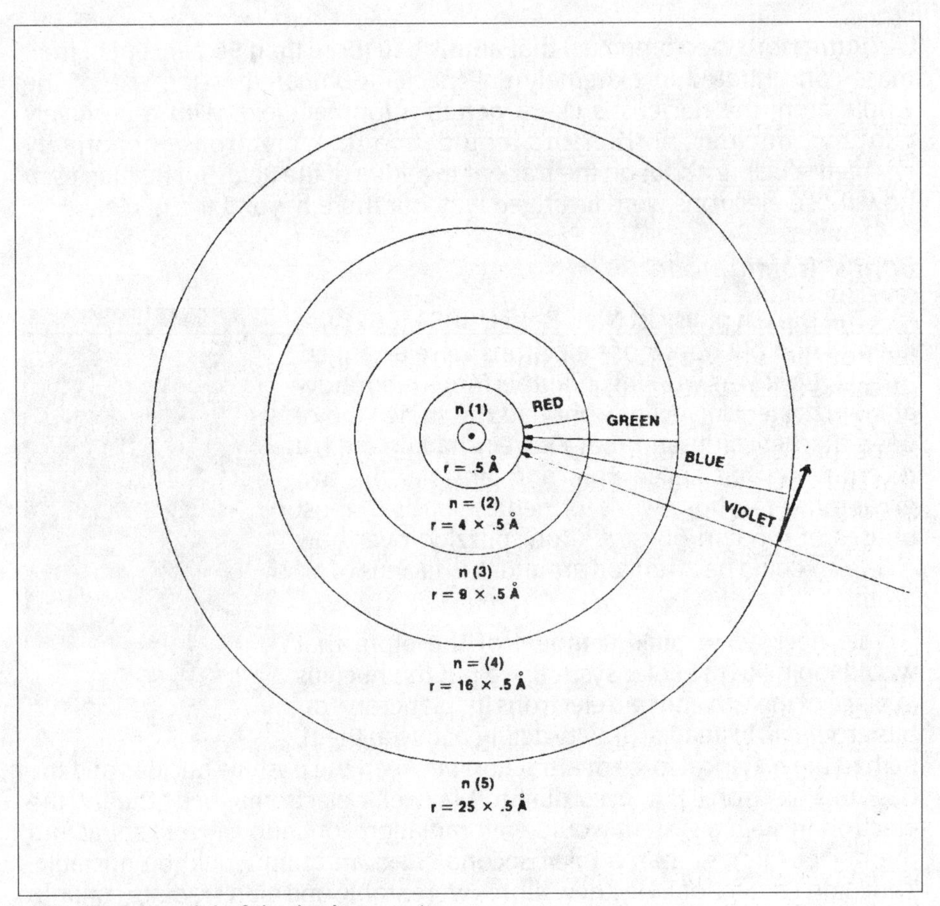


*The Bohr model of the hydrogenatom.*

number multiples of a value that contained the constant from Planck's formula. Bohr then calculated the energy of the electron in each orbit. If he was at all on the right track, the differences between those energy values should correspond to the energies of the photons that a hydrogen atom can emit or absorb.

Bohr got values that exactly matched the energies of the photons of the various colors that are emitted by electrified hydrogen gas. A truly brilliant result!

The next year, 1914, two German physicists measured the energy levels in gases experimentally and got precisely what Bohr's model predicted.

Soon, chemists were using Bohr's model to work out the arrangements of electrons in many different atoms. All they had to do was assume that two electrons would fit in the first orbit, eight in the second, eighteen in the third,...and presto!—they had the arrangement of elements in the periodic table. Thus the white magic of physics explained the black magic of chemistry!

## *Quantum Mechanics*

But the physics itself still needed some explaining. What's going on down there at that deep sub-atomic level? What determines the precise energy levels of the electrons? The startling answer came in 1924. Electrons are waves—not electromagnetic waves like light, but matter waves.

Louis de Broglie (1892–1987) in France suggested the idea of electron matter waves. He simply reversed Planck's formula for quanta of light, i.e., photons. Planck had calculated the energy of a photon from its wavelength. So, de Broglie said, why not calculate the wavelength of an electron from its energy? He did that using the energy levels of electrons in Bohr's atomic model. De Broglie got wavelengths that exactly fitted the radii of the orbits in Bohr's atomic model.

*Louis de Broglie.*

To picture Bohr's model with electrons as matter waves, think of the strings of a harp. Each string produces its own note; the wavelengths of the notes correspond to the lengths of the strings. Longer strings produce the deeper notes, which have longer wavelengths. Now imagine forming the strings into circles and putting them concentrically round a nucleus. Each string is just the right length to carry its resounding wave round and round the orbit: constructive interference. A different wavelength for each orbit.

Electron waves between the resonant orbits cannot be sustained; their interference is destructive. So electrons can only exist in those wave orbits, each with its particular energy. When an electron jumps from one orbit to another, its energy changes by emitting or absorbing a photon of light. And that photon's energy exactly equals the electron's change in energy .

Now this is a very crude picture. The precise picture can only be given in mathematical equations. But, while we're still talking pictures, it's fair to ask—does this picture match reality in any way? How can you tell if electrons are waves, or rather "behave like waves"? The answer takes us back to 1800 and the experiments Thomas Young did to show the wave nature of light. You'll remember Young shone light through narrow slits and got a pattern of alternating bright and dark bands. Anything showing that phenomenon of interference is a wave. When experimenters sent beams of electrons through appropriately narrow slits—lo and behold, the electrons displayed that same interference pattern of alternations. Material particles do behave like waves!

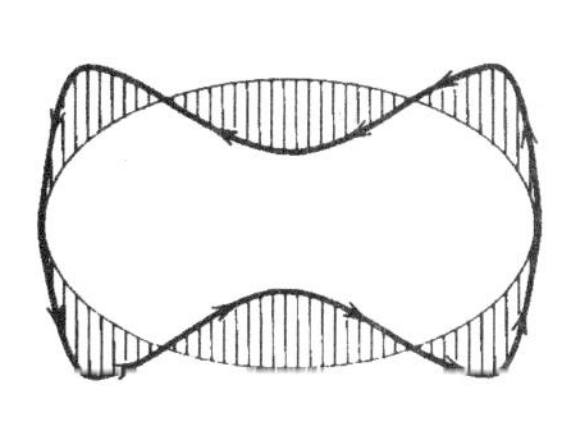

*De Broglie's model of matter waves.*

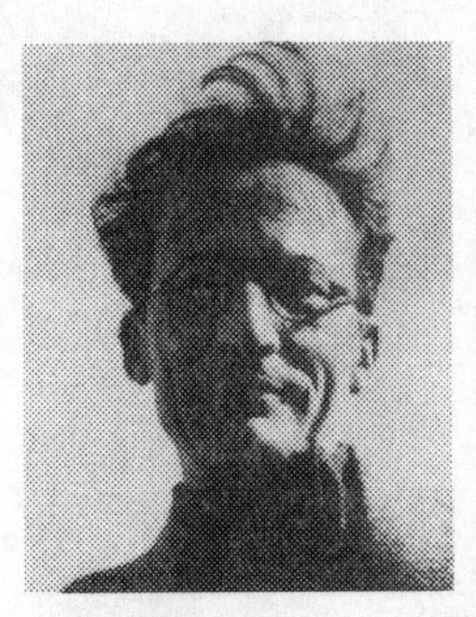
*Erwin Schrodinger.*

The mathematical equations of the precise picture of atoms in the new theory were developed about 1927 by two German physicists, Erwin Schrödinger (1887–1961) and Werner Heisenberg (1901–1976). With their equations, our crude pictures fade away. We're no longer entitled to imagine electrons as tiny spheres whirling around the even tinier sphere of the nucleus. Nowadays physicists want to avoid pictures they can't check. All they do is give us an equation that says, in effect: "Tell me what you're shooting in, and I'll tell you what'll come out. But don't ask me what's going on inside, 'cause we can't see that."

*Werner Heisenberg*

*The Principle of Indeterminacy*

Why can't we see inside atoms? Partly, because they're too tiny. So you'll say—get a more powerful microscope. At this point Heisenberg steps in with what he called the indeterminacy principle. The principle of indeterminacy or uncertainty is an exact statement of what we can't know—what we can't determine.

Technically, as Heisenberg expressed the relation, when the location of a particle is pinned down to within an accuracy of $\Delta q$, then the accuracy of its momentum, $\Delta p$, cannot be better than the value of $h/2\pi\Delta q$ ($h$ is Planck's constant). That is, the more accurately you know the position, the smaller $\Delta q$ is; and the smaller $\Delta q$ is, the larger $\Delta p$ has to be. (In modern metric units, the value of $h/2\pi$ is about $10^{-34}$ kg.m$^2$/s.) In a typical situation, say of trying to find an electron in a hydrogen atom, if you want to pin it down to within $10^{-10}$ m, then you can know its momentum to within $10^{-24}$ kg.m/s. But according to Bohr, the momentum of an electron in a hydrogen atom is about $10^{-24}$ kg.m/s. The result is that the uncertainty is as big as the value you're trying to specify. It's as if you knew the speed of a train is 100 km/h with an uncertainty of 100 km/h.

So, the indeterminacy principle says if you want the precise location of an electron, you'll have to be satisfied with a fuzzy speed. And if you want its precise speed, you'll get a fuzzy location.

Before you suspect that Heisenberg just made up the uncertainty relation in his imagination, perhaps to annoy his colleagues, be assured there is a physical way to understand it. To see anything, you have to shine light on it. Light comes in photons. The tinier the thing is, the shorter the photons' wavelength has to be. But photons of very short wavelength have very high energy. So to see a very tiny object you have to use very energetic photons. The more energetic the photons, the harder they'll hit the object. And the tinier the object is, the more it'll be buffeted by the photons. With those photons knocking the object about, you'll get a fuzzy picture. If you shine less energetic photons of longer wavelength on the object, they'll just sweep past the object almost undisturbed, and you'll get a fuzzy picture.

For something to be seen, it must interact with photons of light. The tinier the thing, the more energetic the interaction will be, and the fuzzier your picture becomes. There's just no way out. We can't see electrons in atoms. Physicists have destroyed certainty. The very act of observing an electron changes it.

Take an example from a very different field of investigation—far from physics—to illustrate the meaning. Suppose an anthropologist wants to know how the natives on Bora Bora interact with each other in the course of their normal everyday life. She goes to Bora Bora to observe them. Can she observe them without them observing her? Try to imagine it. How much could she learn without ever interacting with them?

But if our anthropologist does interact with those natives, how can she tell that hasn't changed the natives' regular behavior? How can she ever find out how unobserved natives behave? The act of observing changes what's observed.

Of course, the indeterminacy principle doesn't affect ordinary observations. You can watching geese flying north or the stars in their courses, and be confident that your watching doesn't change them. If everyone on earth watched Halley's comet at the same time, that wouldn't affect the comet one iota. But there are enough situations, like trying to observe the precise location of an electron, where the indeterminacy principle does apply and has to be taken seriously.

And taking indeterminacy seriously has made scientists more humble. In the twenty-first century there are clear limitations to what we can know. And those limitations are firmly built into the physics we now use.

## *Nuclear Energy*

Between 1900 and 1910, Ernest Rutherford and other scientists had worked out a scheme for the radioactivity of elements that emit alpha particles. Elements heavier than lead in the periodic table. Lead occupies the eighty-second place in the periodic table—we say its atomic number is 82. All elements with atomic numbers greater than 82 are unstable; that is, they're naturally radioactive. Elements like radium, thorium, and uranium. When these radioactive elements emit an alpha particle, they're said to decay.

Suppose you start with uranium of atomic number 92. When a uranium nucleus emits an alpha particle, it decays: it loses a little piece of itself. So, what's left behind is no longer uranium. It drops two places in the periodic table: uranium of atomic number 92 becomes thorium of atomic number 90. Later, the thorium nucleus will decay. When thorium emits an alpha particle, it drops from atomic number 90 to 88: thorium becomes radium. In a similar way, radium decays to radon, then to polonium, and finally to lead. That final lead is not radioactive. The series of decays ends at lead.

In decaying, each radioactive element follows its own time schedule. It takes a lump of uranium more than a billion years for half its atoms to decay. That's pretty slow! Radium on the other hand takes only 1600 years for half its atoms to decay. That makes radium much more radioactive than uranium; it shoots out alpha particles much more frequently.

Scientists had the radioactive decay scheme worked out by 1910, the same year Rutherford had discovered the atom's nucleus by shooting alpha particles at a gold foil. Later, starting about 1920, scientists shot alpha particles at targets of every element they could get hold of. Three laboratories dominated in this work: the Cavendish lab at Cambridge where Rutherford was now the head, Madame Curie's Radium Institute in Paris, and the University of Rome in Italy, headed by Enrico Fermi (1901–1954).

*Neutrons*

*James Chadwick.*

At Cambridge, Rutherford was assisted by James Chadwick (1891–1974). They began to bombard *light* elements with alpha particles. And they found that the bombardment induced ordinary elements to become radioactive. Instead of the natural activity of radium or uranium, they'd produced artificial radioactivity. And more than that! When they shot alpha particles into nitrogen, they got hydrogen and oxygen coming out. That's nuclear chemistry—one element can be changed into another. Not like ordinary chemistry where all you can do is make atoms trade partners. In this new alchemy, scientists were actually transforming elements. The alchemists of old had dreamed of transforming lead into gold. That could be done now, by nuclear chemistry, but it would be far more expensive than mining gold, even if the miners were very well paid.

Soon, Rutherford dubbed the hydrogen nucleus the **proton**, a positively charged particle with two thousand times the mass of an electron. Now we had two elementary particles: protons and electrons.

In Paris, Madame Curie's daughter Irene (1897–1956) conducted similar research along with her husband Frederic Joliot (1900–1958). They were married in 1926 and adopted the combined surname Joliot-Curie (they were sometimes nicknamed the Jolly Curios). When the Joliot-Curies bombarded another light element—beryllium—with alpha particles, they found a new phenomenon. The beryllium emitted an intense radiation which they assumed was high energy electromagnetic waves, that is, gamma ray photons.

Back at Cambridge, in 1932, James Chadwick was bothered by the results the Joliot-Curies reported for the bombardment of beryllium. The effects they described seemed contradictory to him. Chadwick showed that if the emitted rays were photons, this would violate the principle of energy conservation. He found he could overcome the difficulty by assuming instead that the emitted energy was carried by a particle with mass, not a pulse of electromagnetic wave energy like a photon. So Chadwick said the beryllium was emitting a new kind of particle, not electromagnetic pulses. He labelled the new particle the **neutron**, a particle with about the same mass as a proton, but with no charge. Now we have three kinds of elementary particles: electrons, protons, and neutrons.

Neutrons were quickly accepted because physicists found they could also use neutrons to explain the structure of nuclei. They found that, within nuclei, neutrons exerted a strong force of attraction on protons: they called it the nuclear force. Neutrons provide the nuclear glue that holds the nucleus together by preventing protons with their positive charges from repelling each another. But this nuclear force has a very short range. It acts *only within* nuclei.

To see where neutrons fit, let's look at the atomic structure of several elements. Start with the third element in the periodic table, lithium. The chemical properties of lithium are determined by its having three electrons. To balance that charge of −3, the nucleus of a lithium atom has a charge of +3, supplied by three protons. To make up lithium's atomic weight of seven—the three protons are glued together by four neutrons. Now look back one to helium. The helium atom has a nucleus composed of two protons and two neutrons. Remember that when a helium atom is stripped of its two electrons, the bare nucleus is an alpha particle. So an alpha particle is a unit of two protons and two neutrons.

For one more nucleus, look at argon at the end of the third row. Argon has an atomic number of 18. That means it has 18 electrons filling the space around the nucleus. So the nucleus has 18 protons to make the charges balance. Since the atomic weight of argon is 40, the mass of the 18 protons is supplemented by 22 neutrons, which glue the protons together. That's the bare bones of nuclear structure.

Scientists immediately began using neutrons, as well as alpha particles, to bombard elements. Being uncharged, neutrons weren't repelled from positive nuclei the way positively charged alpha particles were. Neutrons just dove right in! And at very close range, the nuclear force could actually capture the neutron into the nucleus.

In 1934, Enrico Fermi and his team in Rome tested every element in the periodic table with neutrons; that is, they bombarded or irradiated the elements with neutrons. To produce the neutrons they used small glass vials containing powdered beryllium mixed with the radioactive gas, radon. When alpha particles from the radon struck the beryllium, it emitted neutrons, as Chadwick had shown. The activities of Fermi's team have been vividly described by his wife Laura in her (auto)biography, *Atoms in the Family* (1954). The team included Emilio Segré (1905– ), Edoardo Amaldi (1908– ), and Gian Carlo Wick (1909– ). Laura Fermi wrote (pp. 89–90):

> The activity of radon decayed in a few days, and fresh radon was extracted once a week. The small glass tubes that could be sealed without breaking were brought to the second floor and used as neutron sources to irradiate all the elements that Emilio had procured. Irradiated substances were tested for radioactivity with Geiger counters. The radiation emitted by the neutron source would have disturbed the measurements had it reached the counters. Therefore the room where the substances were irradiated and the room with the counters were at the two ends of a long corridor.
>
> Sometimes the radioactivity produced in an element was of short duration, and after less than a minute it could no longer be detected. Then haste was essential, and the time to cover the length of the corridor had to be reduced by swift running. Amaldi and Fermi prided themselves on being the fastest runners, and theirs was the task of speeding short-lived substances from one end of the corridor to the other. They always raced, and Enrico claims that he could run faster than Edoardo. But he is not a good loser.
>
> While these experiments were going on, a respectable-looking Spanish scientist in black suit and white shirt came one day to the physics building and asked to see "Sua Eccelenza Fermi." Emilio Segré, who happened to be in the hall on the first floor, told him absent-mindedly: "The Pope is upstairs," and upon noticing the other's puzzled expression he added: "I mean Fermi, of course."
>
> As the visitor reached the second floor, a rosy-cheeked youth and a short-legged man, both in dirty gray coats, tore madly by him, holding strange objects in their hands. Bewildered, the visitor wandered around a while, found nobody in sight, and came back to the hall. Again the two madmen tore by him. At last he found Gian Carlo Wick, a soft-spoken, refined young man, who was making a name for himself in theoretical physics. Wick was in little sympathy with bustle and fuss.
>
> "I'm looking for Sua Eccelenza Fermi," the Spanish man said; "could you show me his office?"
>
> When the visitor and Wick stepped out into the hall, the two strange men were having their third race.
>
> "Enrico," Wick called as loudly as his good manners permitted him. "This gentleman is here to talk to you."
>
> "Come along," Enrico shouted; and disappeared.
>
> The interview took place in front of a counter—as all Enrico's interviews did, both with students and with visitors—between readings, while he jotted down figures on bits of paper. But the Spanish visitor could not conceal the depth of his thwarted expectations.

With all their mad rushing and hard research, Fermi's team made many elements artificially radioactive by adding neutrons to them. Their work showed clearly the nature of isotopes. The term **isotope** (Greek for "same place") had been coined in 1913 to refer to varieties of atoms of a single element. Back then it had been known that atoms of the same element could have different weights. This was now explained with neutrons.

For example, all carbon atoms have six protons in their nuclei. While most of them have six neutrons as well, occasionally a carbon atom may have one or two extra neutrons. Carbon's atomic weight is normally 12, but with one or two extra neutrons it becomes carbon 13 or carbon 14. They're all carbon because it's the six protons that determine carbon's chemical nature. Even with extra neutrons, all the isotopes of an element fit into the same slot in the periodic table. Their atomic weights are different, but their atomic numbers are the same. Many of these artificially produced isotopes are unstable—that is, they're radioactive. The carbon 14 isotope is radioactive—just as the isotope cobalt 60 is radioactive. Many of these artificial radioactive isotopes, like cobalt 60 or iodine 131, are used in medical therapy.

When Fermi bombarded uranium with neutrons, he got strange results. Because uranium is so unstable to begin with, the resulting atoms didn't exhibit the properties he'd expected. Fermi didn't know what he had. Actually his bombardment probably caused some of the uranium nuclei to split almost in half—nuclear fission. But nuclear fission wouldn't be recognized for four more years, near the end of 1938.

*Nuclear Fission*

That "fissioning" of uranium nuclei was entirely unexpected. Till then, all radioactive transformations had involved only minor shifts of one or two places in the periodic table—with the release of an alpha particle here, or the capture of a neutron or a proton there. In 1938, nuclear chemists Otto Hahn (1879–1968) and Fritz Strassmann (1902– ), in Berlin, analyzed more carefully the products that came from bombarding uranium with neutrons. Among those products they found one they thought might be radium, with atomic number 88. Now, radium belongs to the same chemical family as a lighter element, barium, with atomic number 56. That is, radium and barium have very similar chemical properties. This makes radium and barium difficult to tell apart, but there are ways. Yet, try as they would, Hahn and Strassmann couldn't actually prove they'd produced any radium. So when they published their results early in 1939, they said, in effect, "if we're to be true to chemistry we really ought to call the product barium. But as nuclear chemists we just can't bring ourselves to take such a drastic step." To do that, they'd have had to admit that bombarding a nucleus of uranium of atomic number 92 had created a nucleus of atomic number 56—that is, barium. Impossible!

Well, not impossible. Scientists soon found that uranium nuclei bombarded by neutrons really do split into two chunks; that is, two complete nuclei, each smaller than uranium. In the process, some of the mass is converted to energy—a tremendous amount of energy—a million times more than in an ordinary chemical reaction. The process fits Einstein's famous formula: $E = m c^2$, energy equals mass times the speed of light squared.

Of course, Hahn and Strassmann were working with such minute quantities of uranium—micrograms in fact—that they weren't in any danger of blowing up.

Curiously enough, other scientists besides Fermi had also experienced nuclear fission without recognizing it. Some physicists reported later how they'd bombarded uranium with neutrons and tried to measure the energies of the products. They used instruments that measured on a scale from, let's say, zero to ten—expecting results in the middle of the range. Occasionally, the needle banged way off scale past ten. Instead of recognizing this as a great discovery, they just attributed it to a faulty instrument. Actually, they were getting pulses of energy from fission fragments twenty times more energetic than they'd expected. It's easier to blame the equipment than to make a new discovery.

Once scientists had absorbed Hahn and Strassmann's discovery of fission, they quickly knew what they had: a new source of tremendous energy—nuclear energy! More than a hundred papers on uranium fission were published in 1939. And in 1939, with war clouds on the horizon, many scientists pondered its military potential. What if nuclear energy could be used to make bombs?

*New Bombs for Old*

The eruption of World War II, in September 1939, changed everything. Soon physicists imposed a moratorium on the publication of their nuclear research. For the duration of the war they operated in the world of military secrecy—not a comfortable world for people whose careers had depended on open publication. Physics was no longer confined to the ivory tower.

In their early work, physicists were just motivated by plain curiosity, exploring the properties of uranium and the conditions needed for nuclear fission. They found out a lot, particularly the features of what's called a **chain reaction**. Suppose you have a chunk of uranium. When a spontaneous neutron comes along and strikes a uranium nucleus, it'll split that nucleus. In the collision, if a couple more neutrons are released, then those two will split two other uranium nuclei; two will beget four, then eight, sixteen, and off you go. That's a chain reaction. And the splitting of each nucleus releases its own intense blast of energy.

So physicists experimented and calculated: How many neutrons are actually released in each fission? Two or three—enough to keep a chain reaction going. Do all uranium nuclei behave the same? No—the uranium isotope U 235 is more unstable and fissions more readily than the other isotope, U 238. But natural uranium is mostly U 238. It contains only 0.75% U 235. How big a lump of U 235 would you need to sustain the chain reaction long enough so most of the uranium nuclei would split before the neutrons escaped through the surface of the lump? To make a bomb? About 20 kilograms. Because it's a dense metal, 20 kilograms of uranium will easily fit into a breadbox. A small bomb and a gigantic explosion: roughly a million times the explosive power of dynamite. With much smaller quantities of uranium, of course, there'd be no explosion, because there wouldn't be enough to sustain a chain reaction.

*Enrico Fermi (left) and J. Robert Oppenheimer.*

When they'd figured this all out, allied scientists became afraid the Germans might be ahead of them. After all the Germans had discovered fission, and Germany had many competent physicists. Hitler would be sure to recruit them to make uranium bombs. Physicists in Britain and America were determined not to lag behind. Particularly those who were refugees from Nazi tyranny.

When Hitler came to power in 1933, he made anti-Semitism national policy, dismissing Jewish officials, scholars, and scientists. Many escaped to Britain and America. Albert Einstein went to Princeton. In 1938, the anti-Semitic purge hit Italy too. Since Laura Fermi was Jewish, Enrico Fermi took his family to New York, to Columbia University.

Curiously, Hitler never realized the full potential of nuclear energy, but the Allied forces did. So physicists were recruited in Britain, Canada, and the United States to work on nuclear energy and to continue their basic research on uranium, nuclear fission, and chain reactions. And to develop a weapon of mass destruction. Both science and technology.

Scientists informed President Roosevelt of the destructive possibilities of nuclear energy and the state of scientific activity in Germany. He made funds available for research early in 1940. Progress was slow for a couple of years. Then, with the American entry into the war at the end of 1941, an all-out effort was mounted. In August 1942, the Army Corps of Engineers was authorized to supply work on nuclear weapons in a division named the Manhattan Engineering District. In charge was Brigadier General Leslie Groves (promoted from colonel on his appointment), who had been a prominent engineering officer.

A number of the scientists had already been recruited to work on nuclear energy in university and industrial laboratories. Now, General Groves (1896–1970) was charged with building new facilities in remote parts of the country. For the gigantic task of designing the bomb itself, Groves recruited J. Robert Oppenheimer (1904–1967). Oppenheimer was a native New Yorker, trained in Germany. In 1929, he returned to the United States and as professor of physics in California soon established a coterie of bright young theoretical physicists in Lotus Land. He accepted Groves's assignment, and recommended that the bomb laboratory be built in an area of New Mexico where he'd vacationed—a few miles from Los Alamos.

The tasks of the Manhattan Project were enormous. The engineers and scientists were starting almost from scratch. They hadn't even produced a chain reaction yet. And they knew that separating the minute quantities of fissionable U 235 from the more abundant U 238 would be very difficult. General Groves had to prepare industrial sites without being entirely clear how they'd be used.

By 1942, scientists had found that when uranium 238 is bombarded with neutrons some of it turns into plutonium. That's an element two places *above* uranium in the periodic table. The scientists calculated that plutonium should fission even more readily than U 235. And they figured they could make plutonium using a controlled chain reaction—in a nuclear reactor. So they set about building their first nuclear reactor. They did this for two reasons—first to see if they *could* produce and control a chain reaction, *and* (if they could) to produce plutonium. Enrico Fermi was in charge of that project—at the University of Chicago. He and his team built their reactor as a structure consisting of lumps of uranium surrounded by blocks of graphite—to slow down the neutrons. They called the structure a pile because they had to put the graphite blocks on top of one another, in a pile about 7 m high.

Then, if everything worked, scientists would have two sources of explosive power, U 235 and plutonium. But they still had to figure out how to separate U 235 from U 238. Chemical separation of isotopes of a single element was impossible, so they had to use extremely difficult physical methods—that is, methods based on the minute difference in mass between the two isotopes. Getting plutonium was easier. When the irradiated lumps of uranium were removed from the pile, plutonium could be separated by chemical methods—since plutonium is a chemically distinct element, not just an isotope of uranium.

When General Groves started looking for industrial sites, the first experimental pile was still under construction. Fermi and his team achieved the first sustained chain reaction on December 2nd, 1942. By then Groves had already found a tract of land near Oak Ridge in Tennessee for the next phase.

Construction began in January 1943, and a production plant was in operation by the end of the same year. The nuclear pile at Oak Ridge generated five thousand times the power of the one at Chicago. Even that wasn't enough. Another pile at Hanford, Washington, went into operation in September 1944, with twenty times the power of Oak Ridge. In less than two years, scientists and engineers had gone from the first man-made nuclear pile to one that was a hundred thousand times more powerful! And this was made possible by the pressures of war and the abundant wealth of North America—Canadian uranium and American industrial strength—two billion dollars' worth!

These plants produced U 235 and plutonium to be used in bombs. Bomb design and construction was entrusted to Oppenheimer's team at the Los Alamos facility. By the middle of 1945 they had enough U 235 and plutonium for three bombs. July 16—the first bomb was tested in the New Mexico desert. One nuclear explosion equivalent to twenty thousand tons of TNT. By comparison, the blockbuster bombs used in the thousand plane raids over Germany were only about 5 t of TNT each. August 6—the second bomb was dropped on Hiroshima, Japan. August 9—the third bomb was dropped on Nagasaki. August 14—Japan surrendered. World War II was over.

Arguments over whether we should have used the bomb on Japan are still going on. More importantly, we now live in a world where the major powers have enormous arsenals of nuclear weapons. A nuclear war would be very different from anything we've ever experienced before! That's the legacy of nuclear science. Curiosity about the innermost workings of matter has led to weapons of mass destruction. We have seen the power of science, and scientists have had responsibilities thrust on them that they were little prepared for. Oppenheimer in a meeting with Harry Truman said, "Mr. President, I have blood on my hands!" Truman was annoyed. He said to an aide, "All *he* did was make the bomb. *I'm* the guy who fired it off!"

## *Conclusion*

The two revolutions of relativity and quantum theory grew in ground prepared by Faraday and Maxwell. On the experimental side, work deriving from Faraday's electrified vacuum tubes led Roentgen to x-rays and Thomson to electrons. x-rays soon became important as tools of analysis within physics itself, as well as in medicine. Thomson's electrons provided clear evidence that atoms were not the "uncuttable" entities everyone had imagined. From about 1900 to 1935, the main lines of atomic structure were established as experiment and theory continued their traditional *pas de deux*.

Still on the experimental side, the phenomenon of radioactivity was wholly unexpected, coming as it did out of studies of mineral phosphorescence. However, Becquerel was quick to relate radioactive rays to x-rays; and Marie Curie was quick to extend the work into chemical analysis and the discovery of new elements. The discovery of particle emissions (*alpha* and *beta*) in radioactivity provided a new tool for the analysis of matter, and was adeptly employed by Rutherford for the new alchemy.

Maxwell's theory of electromagnetic waves was fertile ground for both the quantum and relativity. In reworking the theory to account for the energy-temperature relations in luminous radiations, Planck found himself forced to attribute a corpuscular character to a wave phenomenon. Aged 42, Planck was never comfortable with the quantum—truly a revolutionary in spite of himself. It was different with Einstein five years later, at age 26, who (for a while at least) was willing to follow the quantum wherever it beckoned.

Yet even Einstein did not approach physics with revolutionary fervor. In his work on relativity, he considered that he was tidying up a defect in Maxwell's theory. That he did it by starting back at first principles simply seemed to him the most reasonable way to proceed. By the time Einstein was finished he'd redefined "mass" every bit as drastically as Copernicus had redefined "planet" 400 year earlier.

Bohr started a new revolutionary drama in 1913 when he found a role for the quantum on the stage of Rutherford's nuclear atom. This play was concluded about 1929 with Schrödinger's quantum mechanics and Heisenberg's principle of indeterminacy.

In relativity, Einstein taught that what we measure depends on how we're moving. In the indeterminacy principle, Heisenberg taught us that we change what we measure by the very act of measuring it. Relativity, quantum, and indeterminacy are intertwined. They give us a view of the world very different from the neat, predictable particles of matter and waves of light that we inherited from Newton and Maxwell. Now we also have particles of light and waves of matter. More than that, in the 1930s, scientists observed collisions between pairs of material particles where the particles disappeared and were replaced by high energy photons. Matter and radiation not only are more similar than we suspected, they also seem to be interchangeable. And yet...

And yet we and our material world continue; the sun still sends its warmth to the earth; foxes still eat rabbits; and your toe still hurts when you stub it. The new, uncertain world our physicists describe is, so to speak, deep beneath our daily world. The Canadian physicist Harry Duckworth tried to put things in perspective in a book he wrote about the new physics, by calling it *Little Men in an Unseen World*. Physical theory may have provided us with a satisfying intellectual structure to explain how everything's related, but it didn't change the brute facts of daily life—until the physicist actors left their theoretical stage and taught us how to split the unsplittable.

A sequel to the Bohr-Heisenberg drama began about 1930 with names like Dirac, Feynman, Hawking, Gell-Mann, Glasher, Oppenheimer, Pauli, Salam, Schwinger, Weinberg, and Yang figuring prominently. These physicists produced a new *quantum electrodynamics* penetrating deeper into matter with strange particles like quarks and bosons, and also with attempts to produce a grand unification of physical theory. Their aim is to find a single theory that will account for all the various forces, such as the nuclear force, electromagnetic forces, and gravitation. They are writing the history of physics in the twentieth century in a language most of us will never understand.

What we can understand, or at least appreciate, are the outcomes of the work in nuclear physics. In 1932, Chadwick, following Rutherford's research program, found the neutron. Then, at the end of 1938, Hahn and Strassmann opened the nuclear age with their discovery of fission. The political climate made it inevitable that bombs would follow. That was bad enough. As a new source of energy, nuclear fission unlocks vast quantities of heat at something like a million times the intensity of chemical reactions. So, the wartime re-

*Hiroshima, Japan, after the atomic bomb was dropped.*

search in nuclear piles led almost as inevitably to nuclear thermal power stations for generating electricity. They in their turn now lead to concerns about thermal pollution and the disposal of radioactive wastes (see Chapter 22).

World War II called the physicists out of their ivory tower. By putting unlimited quantities of energy at the disposal of presidents and generals, they have irreversibly altered the balance of power between leaders and those they lead, whether in democracies or in dictatorships. Political and military decisions have to be made in the shadow of the nucleus by experts whose loyalties may be unfathomable. As the story of King Midas demonstrates, having an unlimited amount of anything can be dangerous. Energy sources can be creative forces, but they're destructive too. The tragedy of knowledge and power is that destruction can be the consequence of well-meaning acts.

Unfortunately, while the physicists brought us knowledge of virtually unlimited power, they brought us no wisdom about how to use it. We're still stuck with horse-and-buggy politics in an age where one crack of the whip can unleash a maelstrom. We don't know how to put the nuclear genie back in his bottle, and the political controls needed to keep us safe from him may be as frightening as the genie himself.

*An early television set.*

## CHAPTER TWENTY

# Electronic Communications

As the twentieth century dawned, the world was spanned with innumerable telegraphic connections—more than two million kilometers of lines, of which a fifth were submarine cables, including a link across the Pacific Ocean from Canada to Australia. Although communication among fixed land stations was now quick and convenient, ships at sea had no access to wired telegraphy. Yet, by 1910, many ocean-going vessels did have access to a wireless telegraphy network through the equipment and operators of the Marconi Company.

Communication by radio was born in the late 1890s, initially as a supplement or competitor to the telegraph. The early equipment was wasteful of energy and "air space," but it provided an environment in which many inventors contributed the multitude of innovations that created the electronics industry—an industry to rival the leading industries of earlier centuries: cotton in the eighteenth, steam and steel in the nineteenth, oil and autos in the earlier twentieth century.

The pace of change in radio and electronics has been astounding. Ship-to-shore telegraphy dominated the first couple of decades. The introduction of the vacuum tube gave distance to the telephone and voice to the radio. From the twenties to the fifties, radio broadcasting filled the airways with soap operas and vaudeville performers. After that, the new technology of television tubes provided instant salvation for stay-at-homes, while transistors provided melodic accompaniment for the restless. For, the broadcasting industry spawned a lusty offspring in the recording industry: from shellac discs to tapes, CDs and videos. The electronics of radio also provided the foundation for the computer industry. In contrast, the makers of the earlier mechanical calculators had hardly even considered themselves an industry separate from business machines.

Today, for less than a day's wages you can hold in your hand more computing power than could be squeezed into a space a million times larger only 40 years ago. Digital communications now have their own wired and wireless networks for the rapid transmission of data worldwide. And cellular telephones bring radio back full circle to the point-to-point communications of the Marconi Company.

The history of electronics is as fascinating as the technology itself. That's primarily because you might expect this branch of electricity to be derived more from science than most other technologies. After all, radio inventors were continually at the frontiers of the unknown; if they couldn't find some ready-made science they could research their own new knowledge. Sometimes they did, and sometimes they just went ahead and tinkered until they solved the problem of the moment. So, contrary to expectations, there were a number of innovations where the technology led the science, or just went blithely on without it. Of course there were also some features of radio and electronic invention that depended directly on science. None more so than the first one.

### *Heinrich Hertz (1857–1894) and Radio Waves*

*Heinrich Hertz.*

James Clerk Maxwell developed his electromagnetic theory in the 1860s (Chapter 17). By that theory Maxwell coordinated a large amount of the current knowledge of electricity and magnetism. In addition, he made two startling predictions: that light was an electromagnetic wave in space, and that oscillating electric charges produce such waves traveling at the speed of light, but differing from light in frequency and wavelength. (A red light has a wavelength of about 0.7 μm and a frequency of about 0.4 PHz [μ (micro) = $10^{-6}$, P (peta) = $10^{15}$], and their product is the speed of light, about 0.3 Gm/s, or 3 x $10^{8}$ m/s. Hz is the symbol for the **hertz**, or cycle per second.)

Hertz was born in Hamburg, Germany, and was educated in a technical "high school" (more like our colleges than our high schools) and the University of Berlin. He received his doctorate in 1880 for electrical studies under Hermann von Helmholtz and Gustav Kirchhoff. In the late 1880s, while professor of experimental physics at a technical high school, Hertz set up apparatus to test Maxwell's prediction of electromagnetic waves.

What Hertz needed was a device for generating radio waves, and another for detecting them. He eventually arranged to have the waves reflect off a metal sheet at one end of his laboratory in order to create a stationary wave pattern by constructive interference. For his transmitter he used a battery-operated induction (or spark) coil similar to the device still used in gasoline engines. The coil produced a series of sparks, each one of which consisted of a burst of oscillating electric charges. The frequency of the oscillations depended on the electrical geometry of the wires in the antenna circuit (technically its capacitance and inductance). From physical measurements, Hertz used Maxwell's theory to calculate the frequency; he couldn't measure it directly. He measured wavelength by finding particular locations in his laboratory where tiny sparks crossed a gap in a circular loop of wire a few centimeters in diameter.

Over a series of trials, Hertz's apparatus gave frequencies ranging from about 35 MHz to 500 MHz (roughly the band used today by television), with corresponding wavelengths from 9 m to 0.6m. With a variety of uncertainties in calculations and measurements, Hertz did well to get values for the speed of waves within about 15% of the speed of light.

Hertz went on to perform a series of trials to demonstrate that his radio waves behaved as expected from Maxwell's predictions (that they could be reflected and refracted, for instance). His untimely death by blood poisoning prevented Hertz from further exploitation of radio either scientifically or technologically.

At the same time as Hertz, Oliver Lodge (1851–1940) had performed similar radio experiments, but did not publish his early results. In 1894, Lodge, professor of experimental physics in Liverpool, demonstrated his apparatus to audiences in connection with lectures on theories of light and vision. Incidentally, he probably made one of the earliest wireless transmissions of Morse code over a distance of about 50 m between two buildings. But Lodge seems not to have thought of commercial exploitation of radio until 1897. In that year he took out patents on his design of a complete system of radio telegraphy. By then, however, the commercial prospects were being vigorously pursued by another.

## *Guglielmo Marconi (1874–1937)*

Marconi's father was a well-to-do merchant and landowner in Bologna, Italy. His mother, Annie Jameson, belonged to a family of Scots-Irish distillers, and also was related to the families of Ballantyne and Haig. Guglielmo received a sporadic education, sometimes with tutors, but mostly following his own interests. In his later teens he became interested in physics and electricity, and was assisted in his studies by Augusto Righi (1850–1920), a physics professor at the University of Bologna. Righi had become interested in Hertz's work and communicated it to the young Marconi.

*Guglielmo Marconi.*

At age 20, Marconi became enthusiastic about the possibilities of signaling with Hertzian waves. He began in his attic with apparatus copied from the equipment Righi was using in his university lab. But while Righi, like Lodge, concentrated on the scientific study of these new radio waves, Marconi soon indulged an urge to extend the distance of working. Starting with ringing a bell by radio across his attic, Marconi was able to extend the range to about 3 km by early in 1896.

Here we should pause to consider some of the technical aspects of early radio. Radio waves are transmitted into space by the rapid oscillations of elec-

*A Marconi spark transmitter from 1897.*

trons in an antenna. In Hertz's apparatus, those oscillations originated in high-voltage sparks jumping across a narrow gap. This was basic to most radio transmitters until about 1920.

The frequency of sparking is usually quite low, not more than a few hundred hertz. But the frequencies of radio transmissions are in the range of hundreds of kilohertz to hundreds of megahertz. Those higher frequencies are contained within each individual spark (which, like a single lightning flash, is actually composed of many alternations of electric charge). In fact, a sparking device generates radio waves over a very large range of frequencies (all the way from a few kilohertz to many megahertz). You can notice this on radio or TV if someone runs a sparking motor (say, an old electric drill) nearby.

Eventually, Lodge and Marconi gained some control of the frequency transmitted by adjusting the circuits containing the spark gap and the antenna; that is, they tuned the circuits. **Tuning** is like adjusting a piano string to the pitch of a tuning fork. It's accomplished by matching the values of the electrical properties of two circuits, in particular their capacitance (storing energy in a charge field) and inductance (storing energy in a magnetic field). But an understanding of how to do that adequately, especially with spark generators, came rather slowly.

In the early work, starting with Hertz, the principal transmitted frequency depended on the physical dimensions of the antenna (its "natural" capacitance and inductance); the smaller the antenna, the shorter the wavelength and the higher the frequency. The frequency to which the antenna was tuned was not the *only* one transmitted, but it predominated to some degree. Now, as physicists investigated the properties of radio waves in their laboratories, they naturally used apparatus of small dimensions, and therefore of high frequency (500 MHz and beyond). When Marconi started reaching for distance, he used larger antennas and, consequently, lower frequencies, soon below 1 MHz.

At the receiving end, the apparatus was even more primitive by modern standards. The **detector**, or device for converting the radio wave signal into intelligence, was called a coherer. Various types were invented by several different men. The basic coherer was a small glass tube containing a heap of metal filings. When a faint electrical signal passed through the filings, they stuck together. Some kind of electrically operated tapper had to be operated to shake them loose again. When the coherer detected a signal, it allowed the small current to control another circuit, which made a mechanical or audible response. Eventually, a telephone receiver was commonly used to hear the dots and dashes that were transmitted by keying the spark gap on and off.

In 1896, Marconi and his mother travelled to England. He gave demonstrations of his apparatus over a distance of about 3 km at a frequency of about 150 MHz. He patented his apparatus (applying before Lodge thought of doing so) and contacted the British Post Office for assistance. In the meantime, his mother's family helped Marconi found the Wireless Telegraph and Signal Company in July 1897. Because Parliament had given the Post Office control of communications, the Post Office had the right to licence or operate radio stations. In the early years, Marconi's company was the only commercial firm licensed to operate radio stations. The company also supplied equipment to the navy.

The Marconi Company established a network of shore and ship stations, and hired and trained the men to operate them. By 1903 they were operating 25 shore stations, communicating with 70 ocean-going vessels. Within five years every large transatlantic liner was carrying a Marconi operator for its radio equipment. To transmit over ever greater distances, Marconi developed his apparatus along two lines. One was to supply ever greater power to the transmitter. He soon used a regular alternating current generator (at about 50 Hz) to drive the spark gap, and he made his antennas larger and larger. Each enlargement of the antenna resulted in lower frequencies being transmitted. By 1905, he'd installed an antenna at Glace Bay, Nova Scotia, with wires strung out over a circle of 0.8 km, with a transmitted frequency in the neighborhood of 50 kHz, the band we now call "low frequency."

In December 1901, Marconi conducted a trial transmission across the Atlantic, from Poldhu in Cornwall to Signal Hill in St. John's, Newfoundland. Listening at the latter station at pre-arranged times for a pre-arranged signal, Marconi claimed to have picked the "dot-dot-dot" of the letter "s" out of the static. There is now considerable doubt whether that trial over a distance of 3400 km was in fact a success. Indeed, in systematic trials at sea the following year, they found that regular transmissions from Poldhu reached only about 1100 km in the daytime and 2700 km at night. Nevertheless, with shore stations on both sides of the Atlantic, ships could keep in touch with the shore over most of the crossing.

*Signal Hill in St. John's, Newfoundland.*

The dominance of the Marconi Company in ocean radio continued until after 1918. By then, whole new systems of transmission and reception had been developed. Although Marconi eventually moved with the times, most of the new technology was devised by others, many of them in the United States.

Before turning to those innovations, we should survey the radio scene in the Marconi context up to about 1910. There are two main issues of technical significance. The first has to do with tuning. In 1897, Oliver Lodge had obtained a patent for apparatus to ensure that the transmitting and receiving antennas were tuned to the same frequency. In 1900, Marconi obtained a patent for apparatus to ensure that the circuits connected to the antennas were also tuned to the same frequency. Since Lodge's patent was earlier and more fundamental, he eventually began litigation against the Marconi infringement. Finally, in 1911, the Marconi Company bought the Lodge patent.

Although we should not suppose Lodge and Marconi to be ignorant of physical principles, their tuning ideas were at least as much empirical (what worked) as theoretical (what ought to work). Indeed, the subsequent development of sophisticated theoretical circuit analysis depended a lot on the empirical experience of the early radio inventors. The other technical development also has to do with the interaction between technology and science. It is the invention and discovery of the ionosphere.

Since, according to Maxwell and Hertz, radio waves are simply longer wavelength (lower frequency) electromagnetic waves like light, they were expected to travel in straight lines. Today we talk of "line of sight" communications for television and FM. Thus, people expected transatlantic radio signaling to be impossible. And of course, Marconi used tall antennas to extend the line of sight. Between antennas 1 km tall, the line of sight grazes the earth when the antennas are 225 km apart. Theoretically, there should be no reception beyond that distance. As we've seen, that distance was greatly exceeded—and with antennas much less than 1 km high.

In 1893, the Irish mathematician G.F. Fitzgerald (1851–1901) had surmised on theoretical grounds that ultraviolet light from the sun could ionize atoms in the upper atmosphere. When radio signals were received beyond line of sight, Oliver Heaviside (1850–1925) in England and Arthur Kennelly (1861–1939) in the United States suggested that such an ionized region could refract radio waves downward to account for long distance transmission. From about 1925 onward, systematic vertical transmissions of radio waves by Edward Appleton (1892–1965) of Britain, and others, demonstrated several distinct layers of free electrons and positive ions starting at about 60 km above the earth's surface. Since these ionizations are caused both by ultraviolet light and charged particles emitted from the sun, the degree of ionization in the layers depends on the condition of the sun (e.g., sunspots), the time of day, and geographic location (since the charged particles follow paths dictated by the earth's magnetic field). The layers account for long distance radio reception, not only at low frequencies (say 100 kHz), but also at high frequencies (3 to 30 MHz). Beyond 30 MHz, radio waves mostly go right through the ionosphere without being returned to earth.

Here again, radio experience led the science of radio wave propagation for almost 50 years.

## *Electron Tubes*

Of course, those scientific investigations of the ionosphere used the new radio equipment based on electron tubes and precise tuning of circuits. For, the transmissions from spark apparatus were inherently "dirty"—that is, they were spread over a very broad band of the radio frequency spectrum. An intriguing example of the effect of spark's broad band comes from attempts to report the progress of the America's Cup race off New York in 1901. Marconi had previously made successful reports when his was the only transmitter. But in 1901, rivalry among news gatherers resulted in there being three radio transmitters reporting the race. They interfered with one another so effectively that no messages were received on shore.

*An early electron vacuum tube.*

At that event, one of the transmitters was Marconi's; another had been designed by Lee de Forest (1873–1961), one of the prominent early American radio inventors. De Forest received a PhD from Yale University for a thesis on "Reflection of Electric Waves of Very High Frequencies at the End of Parallel Wires." Born in Iowa, de Forest was the son of a Congregational minister. He spent most of his youth in Alabama where his father was president of a teachers' college for blacks. In his youth, Lee was intrigued by the mechanical gadgets around him. He took to Yale an ambition to be an inventor and a millionaire. His engineering studies were intended, he said, "to direct and temper my genius."

After graduation, Lee de Forest worked for a couple of years testing dynamos for Western Electric in Chicago. In his spare time he scanned the scientific

*Lee de Forest in Hollywood.*

literature for any clues to an improved detector of radio signals. He also followed an empirical "cut and try" procedure to produce a detector better than the balky coherer. Eventually he settled on an electrolytic type of detector which, for several years, was an improvement. However, in 1906, he lost a patent litigation in connection with the device he manufactured.

From 1901 to 1906, de Forest was associated with a variety of radio telegraph companies in New York, which he founded with several businessmen. Unfortunately, the ethics of his business partners left much to be desired. They were essentially stock promoters and expected de Forest to make radio demonstrations that would induce people to invest. The companies lived off new investment money and had little interest in making a profit. Thus when they did sell equipment, they could easily underbid their competitors.

In 1906, the United Wireless Telegraph Company was formed without de Forest, and took over the network of radio stations he had established. After the patent loss on the electrolytic detector, the company bought the rights to one of the early solid state detectors—like the ones familiarly known as "cat's whiskers" (see later in chapter). By 1911, the company's business practices resulted in it having a dominant position on the Atlantic seaboard and Great Lakes, with 70 shore stations communicating with 400 ships. They also resulted in the two principals in the company being sentenced to terms in the penitentiary. And because of an infringement of Marconi patents, the company's assets were transferred to the Marconi Wireless Telegraph Company of America.

Meantime, de Forest worked at finding a further improvement in detectors. Because of a chance observation, he developed the idea of a detector based on the conducting properties of ionized gases, in a burning gas flame. In theory, the flame acted as a relay; when the incoming signal caused a current in the antenna, the flame was made more conducting, which altered the current flow in a second circuit containing a battery and a telephone receiver.

In a couple of steps of development, de Forest moved to a hot filament in a partially evacuated glass tube, with two other electrodes—a plate and a grid. A small signal on the grid affected the flow of charge from grid to plate. De Forest patented his "audion" in 1907 as "a device for amplifying feeble electric currents." Though you may recognize this as a device which emits electrons from the filament to the plate, de Forest claimed to have visualized the filament as heating the residual gas in the tube to make it conducting. Probably the two modes were jointly effective.

De Forest's conception had patent implications because John A. Fleming in Britain had patented a two element "valve" in 1905, with the idea that the filament in a highly evacuated tube was emitting electrons. This valve (or rectifier) would also serve as a detector in radio receivers, though it was not obviously superior to the cat's whisker detectors.

Indeed, although the future of radio lay with those vacuum tubes, a few years elapsed before radio people began to take seriously the problems created by a technical system that depended on many patents. Out of suit and counter-suit, in 1915 came the result that de Forest couldn't sell triodes without a licence from Marconi, who held the Fleming patent; and Marconi couldn't use triodes without permission from de Forest.

By 1915, de Forest, among others, had recognized the amplification usefulness of his audion, and licensed it to American Telephone & Telegraph for use in trans-continental telephone circuits. In the process, the telephone engineers turned the gassy, low-power audion into a high-vacuum, high-power triode. By then, several investigators had also recognized the ability of the triode to act as an oscillator to generate narrow band radio frequencies (clean signals compared to the spark transmitters). Near the end of 1915, AT&T engineers demonstrated both features of triodes by using a bank of 500 triodes to feed a radio signal into the US Navy's antenna system at Arlington, Virginia. They modulated the signal with voice and music that was heard as far away as Paris to the east and Honolulu to the west. Spark transmitters were on their way out—as also were alternatives that had seemed promising, namely arc oscillators, and high frequency rotary alternators. By 1920, everyone knew that the future of radio belonged to electronic vacuum tubes.

However, that future was cloudy as long as a lawsuit resulted every time one company made use of principles already patented by another company. By 1920, every sensible radio system derived from a bewildering array of patents. The problem was solved by a series of cross-licensing agreements in the years 1919–1921. First, General Electric purchased the American Marconi Company and formed it into the Radio Corporation of America. That gave GE control of the Fleming patents. Then RCA and AT&T (which held the de Forest patents) signed a cross-licensing agreement. For ten years the two companies had access to all of each other's radio patents. During the same period, the Westinghouse Company had acquired some clever circuit patents of Edwin H. Armstrong (1890–1954). They joined the patent group with an agreement that RCA would purchase 40% of their radio apparatus from Westinghouse, the other 60% being supplied by GE.

With the patent picture clarified, these companies were free to engage in the research, development, and market expansion that marked the succeeding decades.

## *Radio Broadcasting*

From 1900 to 1920, the whole emphasis in radio had been on point-to-point communication: sending messages from one station to another. Of course, with telegraph and telephone lines spanning continents and oceans, the radio pioneers concentrated on communications with ships at sea. The navies of the great powers were major customers of Marconi and his rivals.

### *Reginald A. Fessenden (1866–1932)*

*Reginald A. Fessenden*

Also, through this first 20 years, with spark transmitters dominant, most messages were sent by Morse code. Voice modulation of the "dirty" spark signal was infeasible. However, early on, some innovators began to search for an alternative method of generating high frequency radio waves—particularly a method for producing a *continuous* wave. This was the motivation of Reginald Fessenden. Born and educated in Canada, Fessenden was teaching mathematics at Bishop's College School in Quebec at age 15. He developed an interest in science and technology by reading the journals of the day. In the late 1880s, Fessenden spent three years in the employ of the Edison Company in New York where he gained experience in electric power.

For the next 20 years Fessenden held a variety of positions with electrical manufacturing companies, a couple of universities, and the United States weather bureau. During that period he combined his knowledge of electrical power with a new-found interest in radio waves. From what he learned about alternating current theory, Fessenden decided to generate radio waves with a rotary alternator. Of course, to get to sufficiently high frequencies, he knew he'd need the high-speed rotation of many coils. A series of machines were built for him by General Electric in the period 1903–06.

The alternator Fessenden put into service in the fall of 1906 reached a maximum frequency of 76 kHz, with an output of less than 500 W. For a detector, Fessenden had designed a liquid device that operated either thermally or electrolytically. It did provide continuous detection, of which the coherer was incapable. He also developed the heterodyne principle, which eventually became central in the operation of radio receivers. It was needed because of an inherent difference between spark transmissions and continuous waves.

The spark signal carried its own modulation, that is, at the frequency of the sparking. So, an operator listening to the signal heard that frequency in the earphone attached to his detector. The pure continuous waves Fessenden produced carried no such modulation. In the **heterodyne** principle, Fessenden arranged a spark oscillator in his receiver that could generate a signal at a frequency a few hundred hertz different from the one being received. The difference or beat frequency was audible in the earphones, even though the two frequencies being mixed were far above the audible range. Later, vacuum tubes were used to produce the second frequency, leading to the kind of squealing you can hear in one receiver if someone tunes through the band on a second receiver nearby.

Now with "clean" continuous wave transmission, Fessenden could use a microphone to modulate his signal to transmit speech. He first experimented between two stations on the Massachusetts coast. He established regular voice communication over a distance of 18 km at the end of 1906. And, in a technical *tour de force* he announced (in Morse code) to radio operators on ships off the east coast to stay tuned for an important holiday message. Christmas and New Year's greetings, with voice and music, were transmitted and received by the startled operators.

For the next few years, radio engineers concentrated on improving point-to-point communications to achieve greater distances, more reliable working, narrower band signals, etc.

### *Early Commercial Broadcasting*

Count the number of receivers of broadcast information in your home today: three TV sets, the stereo tuner, the clock radio, the kitchen radio, and five personal portables, not all working. Now, think back to 1920 and try to imagine how to get this industry started. It's a kind of a chicken-and-egg situation. If you're the boss of a radio company, and an engineer suggests spending the money for a studio and transmitter to broadcast news and entertainment, how will you respond? "Be serious, Mr. Jones. Pray tell me, to whom shall we broadcast?"

He might tell you how cheap it would be to produce a little crystal set with an earphone. And you'd wonder what on earth would induce people to spend $5 for a receiver when there's nothing yet to receive. Since we know the industry did get started, we may wonder how they broke out of that vicious cycle.

There's a dual answer. The first is that there was already a hardy band of radio amateurs who had built their own crystal sets to listen in to an increasingly noisy radio spectrum. The second is that the radio companies were spending money to experiment on new forms of radio transmission. In 1920 there were a number of broadcast trials to audiences larger than Fessenden's back in 1906.

On 23 February 1920, the Marconi Company in Britain broadcast music for an hour a day from an experimental transmitter at its research plant near Chelmsford. Listeners included ships' telegraph operators and radio amateurs. Broadcasts later in the year were received in northern Europe and sent along telephone lines to homes in several cities. In Canada, the Marconi Company began broadcasting on a regular schedule from Montreal in May 1920.

In the United States, the Westinghouse Company set up radio station KDKA in Pittsburgh in November 1920, in time to announce the results of the presidential election of that year. Soon the public was clamoring for receivers,

*A radio drama being broadcast live.*

and the radio broadcasting industry burgeoned. In the United States, early programming was sponsored by department stores—commercial radio was off and running. In Europe, radio broadcasting tended to become the domain of government, more as a medium for information than for entertainment. Within a decade, radio was playing a crucial role in German politics.

## *Later Developments*

### *Television*

Television broadcasting did not become a common commercial enterprise until after 1945. However, researchers had been actively developing TV systems from 1920 onwards. Indeed, curiously enough, the earliest research dates from before the experiments of Heinrich Hertz on electromagnetic waves. Given the complexity of the system, we should not be surprised that 60 years intervened between the first glimmerings and the final realization.

The technical task consists of dissecting an image into a large number of tiny segments or dots (now called *pixels*), unraveling them in a regular sequence, transmitting that sequence, and then reconstructing the image at the other end. To produce a rough replica you need about $200^2 = 40$ k pixels; to get reasonable fidelity, you need 250 k pixels. For a black and white replica you need to attach a number for the intensity of the pixel; say, 0 for white and

*John Baird's experimental "television camera" based upon the Nipkiv photo-mechanical system of rotating discs.*

7 for black. Just imagine composing that information by hand; even if you could send the information at the rate of 100 pixels per second, it would take your 40 minutes to describe a single picture in detail. To show moving pictures you actually need to send 30 complete pictures per second—seven million pixels per second! A task not to be imagined by hand, nor even mechanically. It takes electronics.

Actually, the first scanning devices were mechanical. In the 1880s, a student in Berlin, Paul Nipkov (1860–1940), used a spiral of holes in a rotating disc to dissect a picture crudely. Light from the scene passing through the holes fell on a primitive photoelectric cell. The intensity of light that struck the cell through each hole determined the strength of the electric current in the cell's circuit. The brightness of a neon glow tube connected to the cell varied with the current. Nipkov could look at that tube through a similar spinning disc synchronized with the first one. Because of the persistence of vision, his eye reconstructed a very rough replica of the original scene. In the 1930s, when European radio companies were experimentally transmitting Nipkov pictures, radio amateurs with similar primitive apparatus could experience the thrill of seeing pictures sent through space.

However, the receiving apparatus would have to be greatly improved to make TV into a commercial enterprise. The cathode ray tube was invented in 1897 by C.F. Braun (1850–1918). He derived it from the tubes used by William Crookes, J.J. Thomson, and others (Chapter 19) in their investigations leading to finding electrons. Braun's tube is the forerunner of all subsequent cathode-ray tubes, right down to the picture tube in your TV or computer monitor. For many years its principle application was in the cathode-ray oscilloscopes used by radio engineers and technicians. Braun, a professor at the University of Strassburg and consultant to the electrical manufacturers, Siemens and Halske, was active in a number of areas of radio research. Braun was awarded the Nobel Prize for physics in 1909, jointly with Marconi.

In 1907, Boris Rosing in St. Petersburg used a Nipkov scanner to transmit pictures along wires and displayed them on the face of a Braun tube. However, his work, and that of later experimenters, did not bear fruit until after 1930. By then, the large radio firms in the United States were putting resources into the development of a commercial television system. While they had the Braun tube as the effective display device in the receiver, they still lacked a satisfactory electronic image scanner.

That lack was supplied by a former pupil of Rosing's, who emigrated to the United States in 1919. Vladimir K. Zworykin (1889-1982) worked for the Westinghouse Company, where, in 1928, he patented his *iconoscope*. Two years later, Westinghouse and General Electric transferred their TV researchers to RCA. There, Zworykin completed his system, and RCA soon began experimental broadcasting in New York. Similar trials were occurring at about the same time in Britain and Germany, using essentially the same equipment.

The heart of Zworykin's **iconoscope** is a screen composed of a mosaic of many tiny pixel-like photocells. The charge on each cell is proportional to the light falling on it, focussed there from a camera lens. Then, a beam of electrons controlled electrically, much as in the Braun tube, is directed across the screen in a regular pattern. As each cell is scanned, a variation is produced in the output current, proportional to the charge on the cell. From a single scan, completed in 1/30 second, the output current is an electrical replica of the light intensities across the scene. For transmission purposes, the TV signal must also contain information that will control the location of the beginning of each picture on the tube in your receiving set.

All the information needed for a reliable picture is many times greater than that for voice, or even music. For this reason the frequency range assigned to television channels is much larger than for radio. The total radio frequency spectrum is a scarce resource, which must be apportioned fairly among all competing users. Assignments of radio frequencies are made by international agreement through the International Telecommunications Union.

Between about 0.5 MHz and 1.5 Mhz you'll find the standard AM broadcast stations. Since they come at 0.01 MHz (or 10 kHz) intervals, there's room for a hundred of them. The ITU also assigns the strength and direction of transmissions so the same frequency can be used several times (e.g., in Quebec, Illinois, and California). This is less important for FM and TV, which are limited to little more than line-of-sight distances. Since TV stations come at 6 MHz intervals, they have to be assigned to a much higher band in the radio spectrum. There are five TV channels between about 55 MHz and 85 MHz, seven more up near 200 MHz (channels 2–13 are called VHF), and 70 UHF (ultra high frequency) channels at even higher frequencies. You should realize how all this systematic development depends on the precise tuning of radio signals, which was almost impossible back in 1905 with Marconi's spark transmitters.

*Transistors and Solid State Devices*

Solid state devices depend on the motions and distributions of electric charges within or between pieces of solid matter. They can be contrasted with vacuum tubes in which we imagine streams of electrons traveling through an evacuated space and being pushed around by electric and magnetic fields. Nowadays, about the only familiar vacuum tube left is the cathode-ray tube. All the other tubes of the earlier days of radio and electronics have been replaced by transistors, chips, and integrated circuits.

The development of solid state electronics is a fascinating story of a complex interaction between the observing of new phenomena and the framing of theories of matter, as well as between technology and science.

One of the early observations of a solid state behavior was made by Michael Faraday. In 1833, he noticed that the electrical resistance of silver sulfide decreases at higher temperatures; whereas the normal behavior of metals is to have higher resistance at high temperatures. Another observation was made by C.F. Braun in 1874. He noticed that a circuit in which a pointed conductor touched the surface of a crystal of lead sulfide (galena) had a higher resistance in one direction than the other. This behavior is known as rectification, and is one of the methods of detecting radio signals. Indeed, Braun's little gadget was the forerunner of the cat's whisker detector used in the early days of radio, working on currents of less than 1 mA.

A mark of these phenomena is that they are non-linear; that is, they do not behave in the "regular" way expected from Ohm's law—in which the current through a device is directly proportional to the voltage across it. Later researchers found non-linear behavior at the junctions between different materials. The junction between copper and copper oxide began to be used in the 1920s as a rectifier of currents in the range of a few amperes.

Heinrich Hertz noticed in his experiments that some of his apparatus behaved differently when a surface was illuminated. He investigated this photoelectric effect experimentally. Albert Einstein used Planck's quantum hypothesis to provide a theoretical explanation of photoelectricity, which was fully confirmed in careful measurements by Robert Millikan. Although early photocells were vacuum tubes, the phenomenon is essentially a surface (i.e., solid state) activation of electrons by photons of light, for example, at the surface between a thin layer of selenium on an iron base.

Until the 1920s, most of the development of solid state devices was quite innocent of any theoretical understanding of what was going on at the atomic level. Researchers just kept trying various combinations and arrangements until they got the results they sought. Solid state theory came along in the wake of the theory of atomic structure by Niels Bohr and those who followed him (Chapter 19). Major contributors to solid state theory included Arnold Sommerfeld (1868–1951) in Germany, Felix Bloch (1905–1983) in Switzerland, and Alan Wilson (1906–1976 ) in Great Britain.

During the 1930s and 1940s, research into the solid state continued on both an empirical and theoretical basis. It was some time before anyone understood how to use the theory to achieve practical results. Because of the inherent reliability of solid state devices (with no filaments to burn out), researchers at the Bell Telephone Laboratories were particularly interested in developing solid state amplifiers to be used in telephone lines. And they proceeded both empirically and theoretically.

Two major lines of research at the Bell Labs eventually bore fruit. Walter Brattain (1902-1987) investigated the distribution of charge on the surface of highly purified germanium (a metallic element called a semi-conductor). He made measurements of currents with two fine wires touching the germanium surface at different distances apart. Brattain's measurements were interpreted according to solid state theory by his co-worker John Bardeen (1908–1991). In 1947, they realized how to use that combination as a triode. They called it a point-contact transistor (from *trans*fer res*istor*).

Meantime, William Shockley (1910–1989 ) was engaged in a purely theoretical design of a device consisting of a sandwich of two different forms of semi-conducting material. He worked out what the charge motion and distribution would be at the junction between the different slices. His work resulted in a practical *junction transistor* in 1951. The point-contact transistor was in commercial production by then, and was used for a while in such devices as hearing aids. Before long, the junction transistor, in a variety of forms, was shown to be clearly superior. Its principles formed the basis for the subsequent development of integrated circuit chips, in which innumerable transistors are connected together within a tiny space about the size of a gnat's wing.

While those three Bell researchers were awarded the Nobel Prize for physics in 1956, subsequent developments have involved the combined efforts of workers in many fields—not only experimental and theoretical physics, but also in chemistry, metallurgy, and electrical engineering. Techniques have been developed for growing, refining, and doping pure crystals of germanium and later silicon. Since silicon has a much higher melting point than germanium, it is more difficult to handle. Much of the early work on silicon was done by Gordon Teal (1907– ) at the Bell Labs. In 1953, he moved to Texas Instruments where the first silicon transistor was soon produced commercially. Texas Instruments was the first new firm in the semiconductor industry. (The earlier firms were already manufacturers of vacuum tubes.) Texas Instruments produced the first transistor radio in 1954. In 1959, Jack Kilby (1923– ) of Texas Instruments filed a patent for the first integrated circuit.

## *Electronic Computation*

To get to the beginnings of calculation—simple arithmetic—you have to go back at least to neolithic shepherds keeping track of their flocks by matching pebbles in a leather bag with members of their flocks. Pebbles and fingers may rival language as the first stages of arithmetic. However, with language and memory, you can do quite a bit of calculating without the aid of any technology. Eventually, of course, you'd expect to run into the first mathematician who said the equivalent of "Have you got pencil and paper?"

The point is that the technology of computation goes back a long way. This is not the place to rehearse the whole history from pebbles to your handy pocket calculator. Yet, you could find an intriguing story in all the stages of calculating mechanisms, including the abacus, the Antikythera (a geared, Greek device), astrolabes, Galileo's geometrical and military compass, Napier's bones, the slide rule, Pascal's geared adding machine, Leibniz's stepped-cylinder multiplier, Babbage's difference and analytical engines, Herman Hollerith's punched card accumulator, comptometers, and cash registers, down to 1944 and the IBM automatic sequence-controlled calculator.

All of these machines were mechanisms of one sort or another, with the more recent versions incorporating at least an electric motor, if not electric switching. And we can acknowledge progress in speed and accuracy of operation, and in the capacity of the machines. Yet, only with the truly electronic calculators do we begin a new history, which in fifty years compressed a million-dollar computer of two-kilobytes computing capacity from a room of 100 cubic meters volume into a $50 pocket computer of 100 cubic centimeters volume (a millionth the size), while also increasing the capacity a thousand times, not to mention increasing the computation speed phenomenally!

The capacity and speed of computing machinery has been combined with great advances in programming so that our digital computers don't even feel like mathematical machines any more. Not only do we indulge our accounting fantasies with spreadsheets, we also play complicated space-war games, design large buildings and machines (including the next generation of computers), manipulate gigantic data bases, process words, and publish books from our desktops.

The first digital calculating machine using electronics was built at the University of Pennsylvania for the United States Army. It was called the Electronic Numerical Integrator And Computer (ENIAC for short). Completed in 1946, ENIAC's eighteen thousand vacuum tubes and associated circuitry occupied a room 12 m by 6 m. ENIAC was digital, meaning that it operated in the binary number system, which uses only the digits zero and one. These are appropriate for electrical circuits, with zero represented by "off" and one by "on." Each vacuum tube switch represented one digital place. In order to represent numbers and manipulate them, the tubes had to be wired into circuits in particular ways.

*This photo of the ENIAC computer was used again and again in news stories and advertisements, including in a U.S. Army recruitment ad, featuring the computer as the latest in whiz-bang technology. The army was offering technical training on the latest electronic devices to attract recruits.*

The wiring in ENIAC organized the vacuum tube switches in a particular pattern appropriate for some sequence of calculations. To solve a different problem, which required a different sequence, the very numerous connections among the tubes had to be changed. All done by hand—very laborious, and prone to error. Soon, the engineer on the ENIAC project, J.P. Eckert (1919–1995), and the mathematical designer, J.W. Mauchly (1907–1980), realized that there were ways of coding instructions to the machine in the same kind of digital numbers they were using for their data. With the assistance of John von Neumann (1903–1957), a brilliant mathematician, they devised the first numerical programming code. (Von Neumann typically dubbed one of his models the Mathematical And Numerical Integrator And Computer, i.e. the MANIAC.) Programming provided the means to vary the sequence of operations without having to change the wiring every time.

In the early 1950s, Eckert and Mauchly designed a machine for commercial production by Remington Rand, the UNIVAC, or universal variable automatic computer. IBM also entered the field with its Selective Sequence Electronic Computer (1948), and the Naval Ordnance Research Calculator (1954), designed by astronomer W.J. Eckert (1902–1971). These early ma-

*J. W. Mauchly (left) and J. P. Eckert.*

chines occupied large rooms, needed substantial amounts of electric power both for operating and for cooling, and required constant maintenance in replacing tubes as they burned out. A typical early system had data and programs contained on punched cards. These were fed into a card reader where electrical contacts were made through the holes in the cards. Each card provided an instruction or a number that was stored on magnetic tape and fed to the vacuum tube processor in appropriate chunks. The results could be output onto punched cards or printed on ingenious high speed line printers. Such machine systems were the "mainframes" that dominated the early years of the computer industry. Large computers are still called mainframes, but they've incorporated all the innovations (and more) that you can find in your hand-held calculator or your desk-top personal computer. And you can hold in the palm of your hand more computing power today than a mainframe of the 1950s.

Miniaturization and increased speed and memory came about through numerous technical improvements. One of the more significant improvements was the invention of the integrated circuit chip, which contains within a minute space a million semiconductor diodes (switches), and a suitable set of interconnections. A number of these arranged and inter-connected on a "mother board" with a power supply and sockets for connecting to peripheral devices constitute the heart of your personal or home microcomputer. Usually the main box also contains a mass storage device, typically a magnetic disc. A single "floppy" disc can hold a million characters of information; a "hard" magnetic disc can hold a hundred million to a billion or more characters depending on its size. And if that is not enough, you can back up your entire system to a central server elsewhere via the internet. Typical peripheral devices include a keyboard for entering data and instructions, a cathode-ray-tube screen for reading what's being input and output, and a printer to produce "hard" copy.

It's difficult to think of another industry in which such great advances have been made within such a relatively short time-span. And the computer industry has not only transformed its own way of doing business, it has invaded many other industries and businesses. Not just in making computation easier, but also in controlling electrical and mechanical operations. Even if you have no calculator or computer in your home, many electrical devices bought within the past couple of years have computer chips in them. The VCR and its remote control box, the microwave oven, the fuel injection system of your automobile are only a few of the more obvious examples. Even clinical thermometers are available based on semiconductor technology.

### *Data Communication*

A curious result of our computerization of the western industrialized world can be seen in communications. More and more we find that our nations' communications networks are being used to transmit machine-readable information. The wires are humming with digital data, the modern equivalent of Morse code. Now, instead of dots and dashes, we send zeros and ones. And where an old telegraphist was doing well at 250 characters per minute (that's about 50 words per minute), our machines now send data whistling down the wires at up to 500,000 characters per *second*. These data communications channels are everywhere. The Toronto *Globe and Mail* is printed in several plants thousands of kilometers apart, providing the same pages across the country by sending the information via satellite from the editorial staff to the printers.

Banks have been large users of data communications since the late sixties. The clearing of checks, for instance, does not depend any more on the physical arrival of those pieces of paper back at your home bank. The required information can be transmitted between branches and central offices overnight. Your account is debited long before the cheque arrives. It was only natural for this process of electronic funds transfer to expand to interface with you and me. Instant tellers provide the convenience of 24-hour banking without full-time human attendance.

The traffic signals of large metropolitan cities are now tied together by a traffic control computer. When it's properly programmed, it can aid a steadier flow of traffic. When it crashes (as it may do from time to time), it can generate monumental traffic snarls.

All in all, the pervasive intrusion of microchip technology into many earlier technologies can give you ample food for thought. Are you better off because of it? If you're not, is anybody? How can this technology affect the quality of your life, for better and for worse? What will you do about it?

## *Beginnings*

We're at the end of the first century of radio. As historians usually measure time, they might say that it is too soon to determine the significance of the multitude of social effects of instantaneous worldwide communications that impinge on us citizens of the twentieth century. We are living through revolutionary times, and the final outcome cannot yet be determined. It *can* be imagined, and many self-styled prophets have given us their views of the future. Only time can judge how valid any of them were.

So in this section I avoid drawing hard conclusions, choosing rather to identify some tentative issues that might bear watching in the years ahead. How much are ordinary citizens held in thrall to succeeding generations of whiz kids? Who is responsible for determining the social utility of each new technical innovation? Who's in charge around here?

Before the advent of radio, the quickest way to communicate information to large numbers of people was through the newspapers. Until the introduction of steam-powered rotary presses in the nineteenth century, newspapers were produced on hand presses quite similar to those of Gutenberg (Chapter 7). By the end of the nineteenth century, industrialized countries had numerous metropolitan daily newspapers; with widespread compulsory elementary education, a majority of citizens could read them.

On 15 April 1865, John Smith in Chicago read of the assassination of President Lincoln the night before in Washington. The news was telegraphed from Washington to Chicago in time for the morning editions.

On 22 November 1963, John Smith IV in Chicago saw the newsreel footage on television of the assassination of President Kennedy in Dallas within an hour of its occurrence. And next morning he watched his TV with fascinated horror as he saw Jack Ruby shoot Lee Harvey Oswald, live to the nation.

When will doctors of the mind be able to tell us with assurance how this kind of instantaneous information affects our psyches and our behavior? We have Marshall McLuhan's image of the "global village," but maybe we shouldn't take too seriously the idea of simulating face-to-face relations with five billion people. A global megalopolis, maybe. For surely "village" conjures a more idyllic picture of morning coffee klatches, afternoon bridge parties, all-night poker games, and a Sunday morning soccer scrim on the meadow. That is, "village" is un-technological almost by definition; no one should expect technology even to try to simulate it.

Rather, then, we may more likely expect advancing communications technology to destroy the kind of village life we've been used to. Yet there are those who see in modern communications a way to escape from the ills of Megalopolis without being deprived of its benefits. With high quality equipment, you can get concert hall entertainment in your own home. So, you can move out of the urban jungle with its high crime rates and equip your log cabin in the woods with the latest electronic gear.

It should be evident that attitudes, emotions, and values are involved in assessing the costs and benefits of the cornucopia of electronic gear available today. Since those human feelings are not technological, it should also be evident that decisions about technology must be controlled by non-technological agencies. Whether it's done in the marketplace or the ballot-box, the people who take charge should be aware of the various subtle ways technology can twist human nature.

Recently, some people have expressed glee at the prospect of every person on earth being equipped with her and his own personal cellphone. Whether that represents the apex of the global village or the nadir of the global asylum depends on your point of view. Whose point of view will prevail? What are the features of the technological humans who'll be comfortable in this world of gadgets? Who will take the blame or the credit?

*James D. Watson (left) and Francis Crick demonstrating their original wire model of the DNA molecule just after their discovery of its structure in 1953.*

# CHAPTER TWENTY-ONE

# Genetics and Molecular Biology

The history of genetics is a story of fascinating complexity. Consider these four images. First, in a quiet monastery garden in Austria in 1860, Gregor Mendel, an Augustinian monk, is tending a patch of garden peas, and cross-breeding them. Second, about 1910, in a laboratory at Columbia University in New York, T.H. Morgan and his graduate students are breeding fruit flies in milk bottles and examining them under a microscope. Third, in a physics laboratory in London, England, in 1952, a young physical chemist, Rosalind Franklin, is taking x-ray photographs of a crystallized cellular material. And finally, also in 1952, in an office at Cambridge University, English physicist Francis Crick and American biologist James Watson are trying to fit together models of some complicated organic molecules.

Stretching halfway round the world and spanning most of a century, these four different images provide the settings for the most exciting biological discovery of recent times: the structure of the molecule of heredity—DNA—deoxyribose nucleic acid. To find that structure, scientists had to bring together observations and theories from many different fields: breeding experiments, the microscopic examination of cells, the organic chemistry of cells, the statistics of inheritance, the x-ray examination of organic crystals, the theory of the structure of organic molecules.

Many of these observations and theories began separately, isolated from one another. It took the genius and insight of many scientists to tie them all together.

## *Gregor Mendel (1822–1884)*

Johann Mendel was born in Silesia (now central Czechoslovakia) of peasant stock. He attended the local elementary school and showed enough promise to continue his education despite financial hardship. In 1843 he entered the Augustinian Order of priests at Brno in Moravia, changing his Christian name to Gregor.

After completing his theological studies, Mendel began to teach natural science in the local high school. He continued teaching from 1849 to 1868 when he was elected abbot of his monastery. However, Mendel was never able to pass all the examinations needed to obtain a state teaching certificate. For two years in the early 1850s he even attended the University of Vienna to prepare himself better for the teaching exams. He seems to have been a better student of mathematics and physics than of biology. Nonetheless, his biology professors were leaders in research in plant breeding at the time.

*Gregor Mendel.*

After failing the exams again, Mendel settled into his teaching duties. Then, starting about 1856, he undertook a prolonged series of plant breeding experiments. His approach seems to have been uncharacteristic of biological experimenting at the time. At least, as we read his final report, we can get an impression of a very carefully planned research strategy, and a statistical analysis of results more usual in the physics of the time than in biology. Early in the report, Mendel wrote that no previous experiments had been conducted

> to such an extent and in such a way as to make it possible to determine the number of different forms under which the offspring of the hybrids appear, or to arrange these forms with certainty according to their separate generations, or definitely to ascertain their statistical relations.

Mendel selected seeds of a number of varieties of the garden pea. He found seven characters (which he called "forms") in his varieties that displayed alternatives, such as the color of the seeds' interior (albumen yellow or green), the shape of the seeds (round or wrinkled), or the height of the mature plant (tall, about 2 m, or short, about half a meter). Mendel planted each variety separately. By allowing the plants to fertilize themselves, he tested succeeding generations to be sure that each variety was breeding true. That is, he made sure that all tall plants had only tall offspring, short only short, and so on. Mendel continued this testing for about three years.

Once Mendel was sure that each variety bred true, he performed cross-fertilizations among varieties that differed in a particular character. He created hybrid forms by interbreeding the varieties that had been breeding true. Let us, with Mendel, call the original seeds the parental generation. They were planted and cross-fertilized by hand. That is, with all the varieties carefully labelled, before the flowers were ripe enough to self-fertilize, Mendel carefully opened each flower and removed all its stamens. Then, he used the stamens from one variety to fertilize the stigma of a different variety. He performed reciprocal crosses by dusting pollen from the stamens of tall plants on the stigma of short plants, *and* the reverse.

When the flowers of these plants produced seeds, Mendel collected them carefully and noted their forms. In the cross fertilization between round and wrinkled seeds, the first generation all had smooth seeds. And in other seed characters, Mendel found only one kind of each—the other had "disappeared."

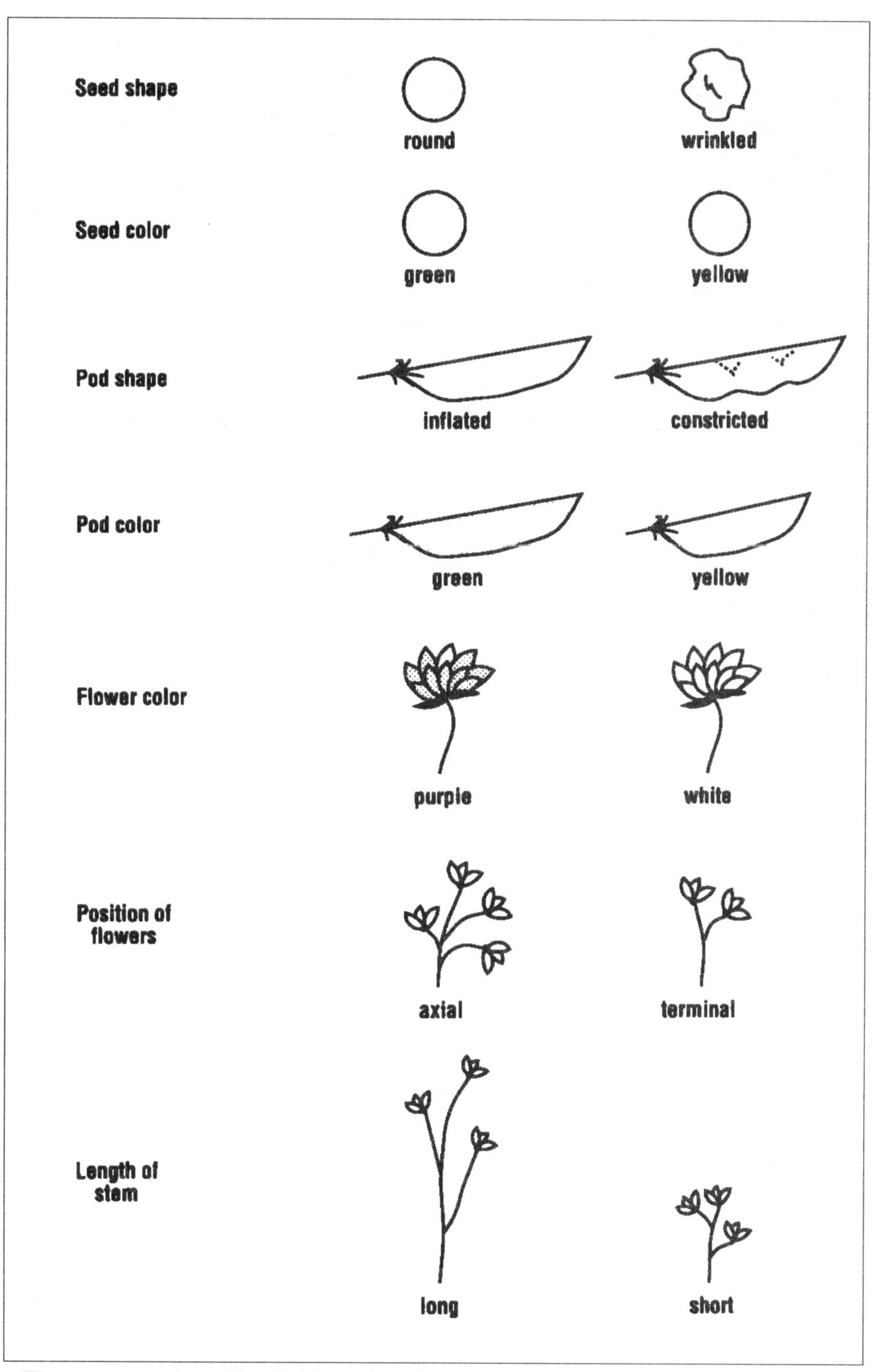


*The seven pairs of heritable traits of the pea plant studied by Mendel.*

Mendel then planted the seeds he'd collected. In the first generation, offspring of the cross between tall and short were all tall. Indeed, for all of the seven characters he'd identified, only one form appeared. Each of these forms, whether tall in stem, or round or yellow in seed, Mendel called **dominant.** The alternative forms (short, wrinkled, or green) not seen in the first generation of offspring, he called **recessive**. Then, Mendel proceeded to a second generation of offspring by collecting the seeds and planting them, and then allowing the plants to self-fertilize.

That next generation turned out differently. Although three of every four offspring of those tall hybrids were tall, one was short. The recessive form had reappeared—it hadn't been lost, only hidden. Also, one-fourth of the seeds from the smooth-seeded hybrids had wrinkled seeds. Mendel got the same statistics for the two forms of each of his seven characters. There were no intermediates; no half-wrinkled, half-smooth seeds; no half-tall, half-short plants. No blending—just either smooth or fully wrinkled, either tall or short, in ratios close to 3:1.

Mendel showed that every offspring gets two factors for each characteristic, one factor from each parent. That is, each parent contributes either a tall or short factor, and a smooth or wrinkled factor, and so on, for all the other features that make a pea plant. But, the factors tall and smooth are dominant over the factors short and wrinkled, which are recessive. Even if an offspring gets only one dominant factor, from either parent, that's the one that'll be expressed. But even though that hybrid offspring is expressing the dominant factor, it's also carrying the recessive factor. For the recessive factor to be expressed, the offspring must get that same recessive factor from both parents.

Here's an example of dominant and recessive that most of us can relate to. In human eye colors, brown is dominant and blue is recessive. We carry two factors for eye color, getting one from each parent. Even if you carry one recessive blue factor, along with the dominant brown factor, your eyes will be brown. So, if you're brown-eyed, you may either have both factors for brown eyes, or only one. Two brown-eyed parents can have a blue-eyed child, if both parents are carrying one recessive blue factor. On average, you'd expect only one child in four of such parents to have blue eyes. But since blue is recessive, both color factors have to be blue in order to be expressed. If you're blue-eyed that means both your color factors are the recessive blue. And so, two blue-eyed parents can have only blue-eyed children. (Human eye color is actually somewhat more complicated than this; but the general principle is sound.)

Now, back to Mendel and his peas. Once he'd checked his counts, Mendel repeated his crossing experiments to be sure of his statistics. He continued some for as many as six generations. He also tested the ratios to be expected of combinations of characters, and found them well enough confirmed. In fact, some statisticians in this century suggested that Mendel's expectations were too well confirmed. That is, they didn't show the kind of deviation from perfection that you'd expect from random sampling. Although there's no reason to suppose that Mendel cheated, it is entirely likely that with some of his larger quantities, he neither used modern sampling techniques nor counted every seed. We may imagine him counting a number of random handfuls and watching the totals fluctuate above and below 3:1. Toward the end, without counting absolutely every seed, he could have just chosen to stop counting when he reached 6022 yellow seeds and 2001 green ones. Of course sampling statistics don't *demand* the result to be more like 3.11 to 1; it's just that 3.01 is less likely (like being dealt a straight flush in poker).

Mendel completed his report in 1865 and presented it to the local natural history society in Brno. As far as we know, the report excited no particular interest in the audience. If you read the report today you might remark on the heavy mathematical symbolism in a biology report, most unusual for the time. For example, he represented the round seed form by *A* and wrinkled by *a*; albumen yellow by *B* and green by *b*. Notice he began the practice still used of representing the dominant factor by an upper case letter. Then, if round-yellow parents are mated to wrinkled-green ones, what distribution of factors should you expect in the second generation (the first being all dominant)? Mendel wrote

> the offspring of the hybrids, if two kinds of differentiating characters are combined in them, are represented by the expression
>
> AB + Ab + aB + ab + 2ABb + 2aBb + 2 AaB + 2Aab + 4AaBb.
>
> This expression is indisputably a combination series in which the two expressions for the characters A and a, B and b are combined. We arrive at the full number of the classes of the series by the combination of the expressions:
>
> *A* + 2 *Aa* + *a*
> *B* + 2 *Bb* + *b*.

This is pure combinatorial probability, where you have an equal chance of choosing *A* or *a* and *B* or *b*. The only difference we'd make today is to double the letters when they appear singly (e.g. *AA* + 2 *Aa* + *aa*). Mendel expressed the results, again confirmed by his trials, in terms of nine distinct genetic classes:

> the forms AB, ab, resemble the parental forms, the next two present combinations between the conjoined characters...Four classes appear always twice, being constant in one character and hybrid in the other. One class appears four times, and is hybrid in both characters.

Because of the effects of dominance, there are only four different visible forms. According to the theory, they will appear in the following ratios: one green wrinkled, three green round, three yellow wrinkled, and nine yellow round. Mendel's counts in these four categories were 32, 108, 101, 315 respectively; which theory would predict to be 35, 104, 104, and 313. They represent deviations of 8, 4, 3, 0.6% respectively. Not too shabby!

Mendel's report was published in his local society's journal in 1866. Some hundred copies were distributed to libraries in Europe and the United States. No reaction! I suspect that the few biologists not repelled by the algebraic expressions would still doubt that the evident complexities of organic inheritance could possibly be treated by the simple exactness of Mendel's mathematical approach. At any rate, there was no significant response to Mendel's report for 34 years.

Yet with his theory of dominant and recessive factors, Mendel had provided the clue to inheritance that Darwin and Wallace had missed in their theory of evolution by natural selection. Mendel showed that characteristics are transmitted from parent to offspring in *units*. They don't merely blend and average. The variations that are naturally selected must come from variations in the characters transmitted at the moment of conception. While the environment can work on the differences that are expressed in the growing offspring, the environment can't create the differences. To make the theory of evolution really work, scientists needed a proper theory of inheritance.

About 1870, the term genetics—from the Greek root *gen* meaning birth—began to be applied to the study of the principles of inheritance.

## *Chromosomes and Nucleic Acids*

That was all academic as long as Mendel's report lay gathering dust in the libraries of Europe. In the meantime, other lines of investigation were providing other clues to inheritance. Increased activity in the study of light and optics after 1800 led to significant improvements in microscopes. Around 1850 biologists made a number of discoveries by examining plant and animal tissues under high-powered microscopes. They saw that tissues are composed of cells, and that cells have structure. They saw that tissues grow because the tissue cells divide—a mother cell splits into two daughter cells.

By about 1875, some biologists were concentrating on a structure within the cell called the nucleus. To improve their observations they stained the cells with aniline dyes. And within the nuclei of dividing cells, they found tiny threadlike structures, which had selectively absorbed the dye. They called these colored threads, chromosomes. Then biologists noted how these chromosomes changed during cell division. The chromosomes themselves doubled to form identical pairs. These pairs of chromosomes in the nucleus of the mother cell then separated into two identical sets to become the nuclear material of the two new daughter cells. Making each daughter cell an identical replica of the mother cell.

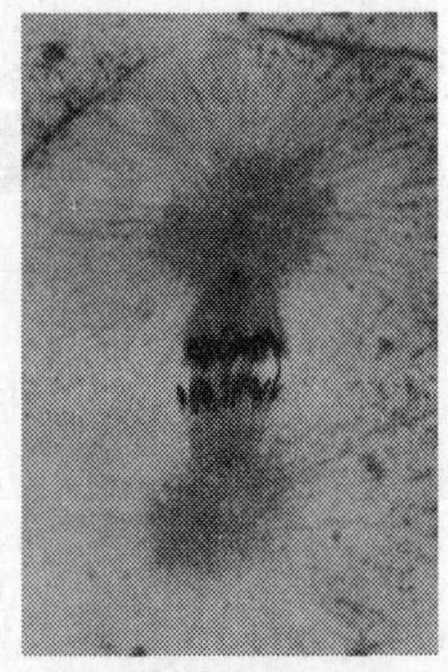

*Chromosomes separating during cell division.*

Biologists also noted that different species have different numbers of chromosomes in their cell nuclei. But there is a constant number in any one species. We humans have 46 chromosomes in our cells. Pea plants have 14; dogs have 56.

Biologists were becoming pretty certain that the chromosomes were where the action was in inheritance. They saw that egg and sperm cells have only half the regular complement of chromosomes. So, at fertilization the first cell of the new offspring gets half its chromosomes—half its genetic factors, whether dominant or recessive—half from its mother and half from its father. From that moment on, its genetic characters are fixed. And this keeps the number of chromosomes constant in the species. Of the 46 chromosomes in our cells, 23 came from our fathers and 23 from our mothers.

But how exactly could chromosomes carry genetic information? Well, chromosomes are bits of organic material; they're made of molecules. So scientists needed to know what kind of molecules, whether molecules could carry information, and how they could pass it on.

Organic chemists were already probing into cell nuclei. In the 1870s they'd collected the nuclei of many cells and analyzed their chemical composition. They found that the nuclei, like all other tissues in the body, contained protein, but that they also contained a different class of substances, which they dubbed nucleic acids. For a brief period, some biochemists wondered if the nucleic acids could be the molecules by which the chromosomes determined the structure of growing organisms. However, on further analysis they found nucleic acids to be too simple for that role. That is, they decided that if the nucleus determined the structure of the many protein tissues of an organism, the molecule to do that would have to be very complicated—much more complicated than nucleic acids appeared to be.

By 1900, scientists had laid a number of separate strands into the warp of the fabric for a molecular theory of inheritance. But it would take these scientific weavers 50 more years to find all the required weft strands—all the interconnections between their various isolated studies—in genetics, cell biology, and organic chemistry. It would take many ingenious experiments and insights in other fields as well.

## *Discovering Mendel*

The warp strand Mendel supplied for the fabric of a theory of inheritance lay dormant for almost 35 years. During the 1890s, Hugo de Vries (1848–1935) performed breeding experiments in Holland with a variety of plants. In some trials of single characters he obtained ratios of dominant to recessive similar to Mendel's 3:1 ratio. By 1899, he had tested a considerable number of plant species and seemed ready to announce his conclusion that genetic characters behaved that way in general. Early in 1900, when de Vries was preparing reports of his work, a friend sent him a copy of Mendel's paper, suggesting he might find it of interest. Some historians think de Vries already knew of Mendel's paper, but chose not to reveal it. Now, his hand was forced, and he did mention Mendel's paper.

Within a month the German botanist Carl Correns (1864–1933) published a paper supporting de Vries's experiments and noting that he too had discovered Mendel only a few months before. By 1900, knowledge of cell division and chromosomes made the Mendelian ideas much easier to accept than had been possible in 1866. Immediately, genetics researchers acknowledged Mendel's priority, and worked actively to discover the range of organisms where inheritance followed the Mendelian ratios. There are numerous multigenic characters (human skin color is one) where several genetic factors are combined, so that the simple ratios are harder to distinguish. There are other reasons for deviation from Mendelian ratios, but his findings provided the basis to start from—as Newton's first law did for mechanics.

The delay of almost 35 years suggests that Mendel was the only one in the nineteenth century who had a firm grasp of the hypotheses about the combining of parental factors in offspring. It did not become evident to de Vries and others until they had substantial experimental evidence for the 3:1 ratio.

An early weft strand for molecular genetics came from the study of the beginnings of new life, the fertilization of eggs and the growth of embryos, called **embryology**. In the 1890s, embryologists were arguing about whether early growth depended just on the chromosomes in the nucleus, or on the rest of the cell as well, the material called the cytoplasm. The American geneticist, T.H. Morgan (1866–1945), began his scientific career in embryology.

*Thomas Hunt Morgan.*

Morgan was convinced that characteristics of embryos were determined at least partly by material in the cell outside the nucleus. He performed experiments with jelly fish, which seemed to confirm his hunch. In the early 1900s, Morgan learned of work that suggested that there was a specific chromosome in insects that determined sex. Soon, a biologist at Bryn Mawr, Nettie Stevens (1861–1912), investigated a large number of organisms, and showed not only that there was a chromosomal difference between males and females, but that often the sex determinant was in the male sperm.

In humans, for instance, the female eggs all contain only an X chromosome, while the male sperm may contain either an X or a Y. The combination XX makes a female, while the combination XY makes a male. This work convinced Morgan to switch from embryology to genetics.

## *Genetics of Fruit Flies*

After Mendel was rediscovered, geneticists tested and confirmed his laws with many organisms. Soon they coined the term **gene** to mean "a particle of heredity." This helped them to clarify the meaning of words like character and factor—to distinguish between the expressed characteristics of a plant and the actual hereditary material that caused those characteristics. So a tall hybrid pea plant could be described as having a dominant gene for tallness, and a recessive gene for shortness. Genes were clearly unitary factors (particles, perhaps molecules) which specified the details of an organism's genetic makeup. A set of genes determined the whole organism.

Then came the questions: What are genes? Are chromosomes genes? Or are genes parts of chromosomes? Or are genes something else that just act the same way that chromosomes do? Into the laboratory!

Find a creature that breeds fast enough that we don't have to wait forever for results. One whose chromosomes can be examined. Perform Mendelian breeding tests and look at the chromosomes. Can you find any connection between the appearance of the chromosomes and the characters of the creatures? That is, if the genes are part of chromosomes, do you see anything within the chromosomes you can relate to characteristics you see in the creatures themselves?

Enter *Drosophila melanogaster*—the common fruit fly! Morgan set out to follow the above program with his students in the "fly room" at Columbia University in New York. He was a fine person to have chosen this field, because he began with considerable skepticism—he was not setting out to prove that inheritance was determined by molecules.

Starting in 1909, Morgan and his team bred fruit flies and examined their physical characteristics under a microscope. They found giant chromosomes in the salivary glands of the fly larvae, whose features could be matched with the physical forms of the adult flies. For a while nothing untoward happened in succeeding generations. Then, one day, Morgan noticed a white-eyed male fly among all his red-eyed brothers and sisters. The exception was immediately placed in a bottle with red-eyed females, and the researchers awaited the results with anticipation.

Sure enough, it seemed Mendel had been right, Morgan now admitted. For, while he found the offspring in the first generation were all red-eyed (that being the dominant character), their intercrossing produced three red to one white in the next generation.

But, there was a surprise. All the white-eyed flies were males. Not only was white the recessive eye color, it was **sex-linked**. That is, the gene for eye color must reside on the sex chromosomes, the X and Y (similar in fruit flies to humans). While eye color is not sex-linked in humans, color blindness is. Continued revelations like this made Morgan a strong advocate of Mendelian genetics.

*Morgan's "Fly Room" at Columbia University.*

Soon, Morgan and his team were inducing and observing other deviations from normal in fruit flies—such factors as various wing shapes and various body colorings. By 1912, the Morgan team was sufficiently convinced of the regularity of Mendelian ratios in normal breedings that they could remark on deviations from them. They soon realized that the deviations were more than the result of statistical variation. They were able to show that during cell division, chromosomes sometimes broke and recombined in new ways. That resulted in unusual combinations of characters in the offspring. And it enabled the Morgan team to plot maps of characteristics along chromosomes. For, by seeing which characters stayed together, they could estimate how much of a chromosome had crossed over to another one. Repeated tests allowed them to confirm various map points—that is, a quite precise location for a particular gene on its chromosome.

By the 1920s, fruit fly genetics had become pretty routine. Bands on the chromosomes could be mapped against the various characters of the flies. A chromosome of a red-eyed fly could be distinguished from that of a white-eyed fly. It seemed clear that the genes—those particles of inheritance—were on the chromosomes.

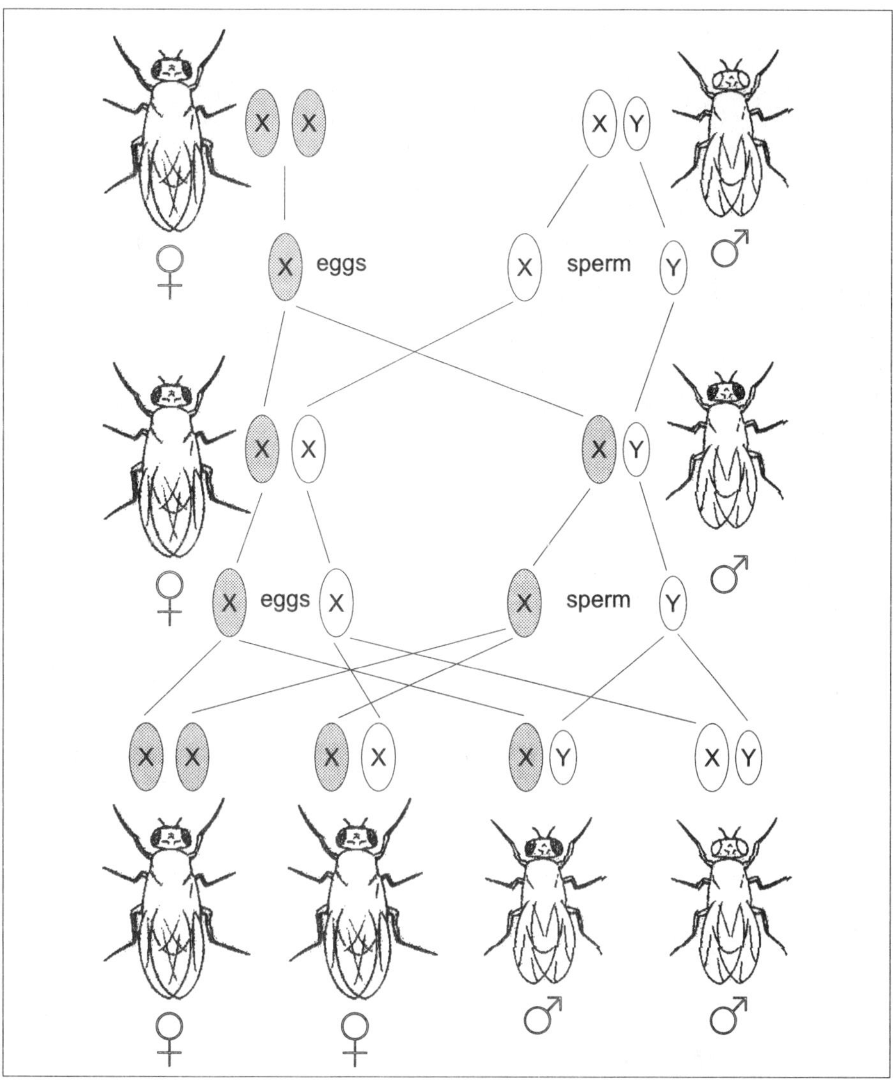


*A demonstration of sex-linkage. The genes for eye color in fruit flies reside on the chromosomes that determine sex, the X and Y chromosomes. White eyes in fruit flies result from a combination of an affected X chromosome and a Y chromosome, which only occurs in a male. Here, the possible combinations of the mating of a normal female fly with a white-eyed male fly (the top row) are traced. In the first generation, both sexes carry an affected gene from the male parent, but neither male nor female exhibit white eyes because each also carries a normal gene from the female parent. For the next generation, there are four possible combinations of genes, two male and two female. Only one of the males will exhibit white eyes, though two of the remaining three will carry an affected gene.*

## *DNA and Genetics*

The next question was: Are the chromosomes made of nucleic acids? A tentative yes was given to that question in 1944. In that year a team of American medical researchers performed an intriguing experiment. They used two strains of a particular bacterium: a lethal form that causes pneumonia, and a benign variant that doesn't. Previous breeding experiments had shown that the gene for lethal was dominant and the one for benign was recessive. When the bacteria were crossbred, the offspring of the first generation were all lethal. Now doctors Oswald Avery (1877–1955), Colin Macleod (1909–1972), and Maclyn McCarty (1911– ) killed the *lethal* strain, extracted its nucleic acid, and fed only the nucleic acid to the benign strain. They found that the nucleic acid, all by itself, could make the benign strain lethal. That non-living chemical, the nucleic acid, had produced a genetic change in the living strain of previously benign bacteria. Nucleic acids really are where the genetic action is. In particular, deoxyribose nucleic acid—DNA for short. At least, we now acknowledge that result from the work of Avery and his team. At the time, some biochemists expressed doubt that the doctors' techniques had in fact eliminated all the protein from the nucleic acid "soup" they'd fed the bacteria.

Chemists had known the components of DNA for some time. DNA is composed of a particular kind of sugar—deoxyribose—along with phosphate, and four particular small molecules built on rings of carbon and nitrogen atoms. These molecules, called bases, are adenine, thymine, guanine, and cytosine. Let's just call them by their initials. A and G are of one type, and C and T are a different type.

*Erwin Chargaff.*

By 1950, another intriguing fact was known. Scientists found that DNA from all organisms contained the same ratios of sugar to phosphate to bases, no matter how many chromosomes they had in their cell nuclei. Yet the DNAS were not entirely identical. From species to species, those four bases A, G, C, & T differed in their proportions to each other. For example, mouse DNA might have more A & T than G & C, while cow DNA had more G & C than A & T. But Erwin Chargaff (1905– ) found by experimental measurement that in all the species he tested, whatever the amount of A was, the amount of T was about the same. And the amount of G roughly equalled the amount of C. Could these DNA similarities and differences be related to a genetic action?

### *The Structure of* DNA

To answer that question, scientists began to examine the actual molecular structure of DNA—how the sugar, phosphate, and bases are situated in the molecule. To do that they had to use another branch of science: X-ray crystallography. About 1910, physicists had found they could get regular patterns of x-rays reflecting off crystals. Patterns they could use to describe the internal physical shape of the crystal. In the 1930s and 1940s, biochemists used this x-ray technique on organic molecules they could crystallize, such as certain proteins like the hemoglobin in our blood. This way they got some idea of the actual structure of the hemoglobin molecule. They used structural chemistry to determine the compounds in the protein and x-rays to find their shape.

*Maurice Wilkins.*

DNA could also be crystallized. So, about 1950, crystallographers began to analyze the molecular structure of crystallized DNA. Rosalind Franklin (1920–1958) and Maurice Wilkins (1916– ) worked on the problem at King's College in London. Franklin had previously done x-ray analyses of coal, and was brought to King's in 1951 to apply her experimental talents to DNA. Wilkins had worked on radar and nuclear energy during the war, and was just beginning to develop skills in biophysics. He and Franklin were soon working in parallel on the structure of DNA, rather more competitively than cooperatively. This was the result partly of a misunderstanding and partly of personality differences.

The x-ray analysis of DNA could only show its basic physical structure, that is, its size and shape. By the end of 1952, Rosalind Franklin had evidence that the DNA molecule was twisted into the shape of a long spiral—a helix—of two or more strands intertwined. Other work had shown that the helix had a backbone composed of the sugar and phosphate. Two main questions remained. One, was the sugar-phosphate backbone on the inside or the outside of the helix? And two, how were those bases A, G, C, & T arranged? Franklin showed that the backbone was on the outside. But she hadn't tackled the location of the bases yet.

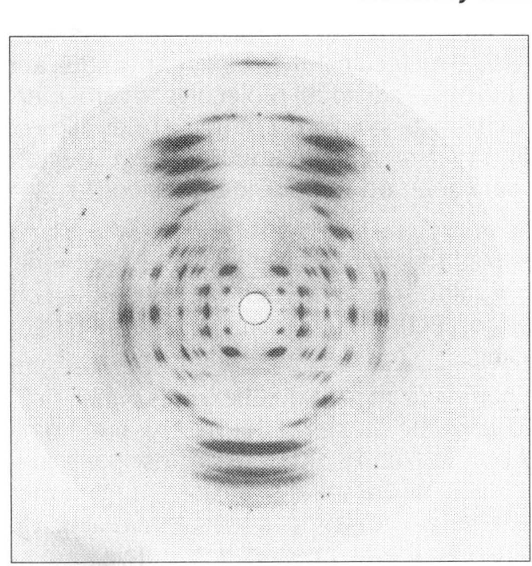

*A X-ray of crystallized DNA in the A form by Rosalind Franklin, which was strong evidence that the molecule was helical in shape.*

Meanwhile, James Watson (1928– ) and Francis Crick

*Francis Crick (left) and James D. Watson at Cambridge University.*

(1916–) in Cambridge were playing with models of the DNA molecule. Watson, with a fresh PhD from the University of Indiana, was looking for the secret of life. Crick, a physicist turned biophysicist like Wilkins, was still working on his PhD on the structure of proteins. Both Watson and Crick were enthusiastic about a small book written in 1944 by Erwin Schrödinger, *What is Life*? Schrödinger discussed living processes from a physicist's point of view—such things as the bonds between atoms in molecules and how much energy (say in the form of ultraviolet light) was needed to break them, and how information could be coded in the molecules of chromosomes by some scheme like the Morse code. This helped to spur Watson and Crick to discover the structure of DNA. For now, they wondered if the apparent simplicity of DNA might somehow contain a code. Maybe it wasn't necessary for nuclei to contain actual models of the proteins that cells construct.

They used small balls to represent atoms and joined them together with wires into the shapes of the individual molecules—the sugar, phosphate, and bases. Then they kept trying to fit these individual molecules together into forms that fit the rules of organic chemical structure. But those rules allowed many possibilities. When Crick and Watson learned about the DNA structure from Franklin's work—with the backbone on the outside—that quickly narrowed the range of possibilities.

Watson and Crick finally realized that the sugar-phosphate backbones were like the outside railings on a spiral staircase. Then they began to consider how those bases were arranged. Perhaps they came in pairs stretching like steps between two spiral railings.

Watson got the idea of base pairing from reading Chargaff's paper. Crick checked with a biochemist friend what the shapes of the bases were. Finally the two realized that A would link only to T, and G only to C. In those pairs, they formed steps between the two railings of the spiral staircase. The two men also realized the force joining each half of the base-pair to its own railing was much stronger than the force between the two halves of the pair forming each step. So, if the two strands of the double helix split apart, each strand would keep its own set of bases attached to it. But there would be an exact correspondence in the sequence of bases along the two strands. Because each base could only match its own particular corresponding base: T with A, C with G. One strand was a kind of mirror image of the other.

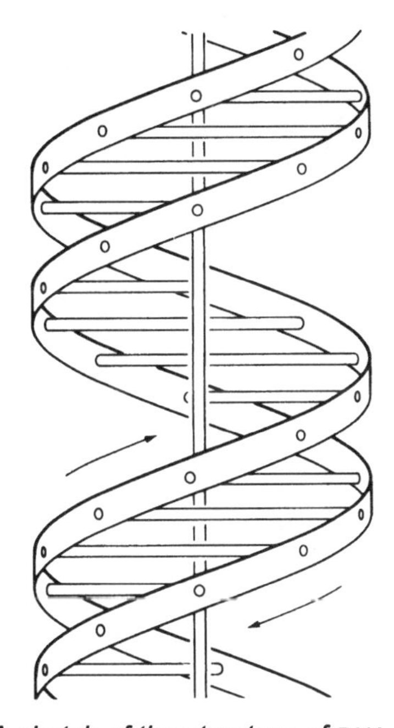

*A sketch of the structure of DNA showing the two sugar-phosphate backbones of the helix running in opposite directions with the bases attaching to the backbones and to each other inbetween. From Watson and Crick's 1953 paper.*

So, for example, if the sequence of bases along one helix was C A G T, the corresponding sequence on the other would have to be G T C A. That made it possible for Crick and Watson to suggest two features of this DNA structure that have startling biological consequences. First, that the DNA can behave the way chromosomes do. That is, when the two strands split apart, each will contain a full set of genetic information. Second, that two additional strands of DNA can be constructed from smaller molecules by attaching As to Ts, Cs to Gs, and so on, by forming alongside the two original strands, which act sort of like templates. Then each new strand could form a spiral staircase with its complement original strand, so that two identical double helix DNA molecules could be formed from one original.

So the chromosomes are made of DNA, and the double helix model explains how chromosomes split and reduplicate themselves. At the molecular level! But how can the bases along the DNA spiral actually code information for directing life processes?

That was discovered in the burst of activity of molecular biologists in the twenty years following Watson's and Crick's announcement of their model in 1953. What happens in cells is that the DNA controls the chemical composition of cell parts and of the enzymes that make the parts work.

These constituents are mostly proteins, long molecular strings consisting of hundreds of small building blocks called amino acids, of which there are only about twenty different ones. How can you use just those four nucleic bases, A, G, C, & T, to encode long strings of the twenty different amino acids?

Think of the problem as one of constructing a language. You have a four-letter alphabet, A, G, C, & T. You need about twenty words—one to encode each amino acid. The code for a protein is like a sentence—with the words for the amino acids strung out along a strand of DNA. How many letters in each word? Well, with our four-letter alphabet, one-letter words would only code four amino acids; two-letter words only 16; and three-letter words, 64. That's enough! More than enough. So we can have words to represent the various amino acids, such as AAA, TGG, CTG, GTC, and so on—up to 64 words.

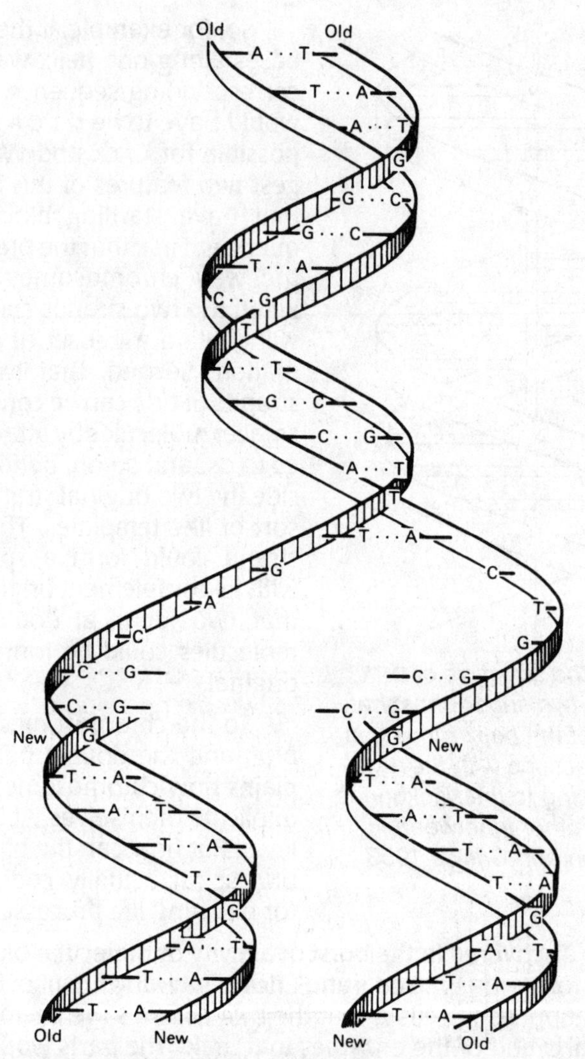


*A sketch of* DNA *replication. During cell division, the two backbones separate, carrying the attached bases. Another backbone forms alongside the exposed piece of backbone, carrying the complementary bases which then bond to each other, forming two identical molecules of DNA.*

Now that's all just simple human logic. Is it the way nature actually arranges things? Yes indeed! Molecular biologists soon found they could make these strings of three-letter words with A, G, C, & T on DNA in test tubes; and actually make particular kinds of proteins using the strings of words along the DNA that coded the various amino acids in the proteins. Since 64 words are more than enough for 20 amino acids, it turns out that most amino acids are coded in several different words in the language of that four-letter base alphabet. The amino acids corresponding to the words above are: for AAA, lysine; TGG, tryptophan; CTG, leucine; GTC, valine.

So, genes are messages that are chemically coded into the sequence of bases along the spiral of the DNA molecule. The DNA is the source of the genetic information in the chromosomes.

*"Honest Jim" and "Rosy"*

The story of the discovery of DNA has gained notoriety because Watson described it in an autobiographical novel, *The Double Helix*, published in 1968. Parts of the book read like Watson's diary at the time, and parts of it read like a nineteenth-century melodrama. It needs to be mentioned primarily because of its treatment of the character of Rosalind Franklin, who died in 1958. Of all the persons in the book, she was the only one unable to defend herself. Watson's account needs to be balanced by Anne Sayre's biography *Rosalind Franklin and DNA* (1975). One of Sayre's motives, besides defending her friend, was the distortion she saw Watson conveying regarding women in science. She reported having been at a school board meeting where a father demanded that girls be excused from studying science because he didn't want his daughter "to grow up like that woman Rosy-what's-her-name in that book."

Rosalind Franklin.

"Honest Jim" was the working title of Watson's book while it was being written. Its ambiguity implies both that Watson was telling the story of scientific research "like it really is," and also that some observers were skeptical that it was the unvarnished truth. "Rosy" is the nickname Wilkins and Watson used to refer to Rosalind Franklin, though not to her face. The episode has been thoroughly treated in various places, especially in the Norton Critical Edition of *The Double Helix*, edited by G.S. Stent (1980); and in an article by H.F. Judson in *Science Digest*, January 1986.

Also, the Nobel Prize for Physiology or Medicine in 1962 was awarded to Crick, Wilkins, and Watson "for discoveries concerning the structure of nucleic acids and its significance for information transfer in living material." Franklin's name is not mentioned because the prizes are never awarded posthumously. If Franklin had been alive in 1962, there'd have been a real problem, because Nobel Prizes are never split more than three ways. Would she have been named instead of Wilkins? Or not? We'll never know.

After all the issues of ethics in science, and feminism, and who should get Nobel Prizes, there's one other that hasn't been as thoroughly treated. That is the question of style in science. One author went so far as to suggest that the whole problem in the DNA story was that Wilkins and Franklin were unwilling to collaborate with Watson and Crick. While that might have been desirable, it does not take account of the possibility that collaboration could have been extremely difficult because of differences in attitude towards science. We've seen in earlier chapters that science involves both (sometimes wild) speculation and careful measurements. Some people, like Kepler and Newton were good at both. But recall from Chapter 12 that Newton refused to credit Hooke for a mere speculation. Rosalind Franklin was like that, and she felt strongly that Watson and Crick were jumping to conclusions.

They thought so too; but whereas they felt that was an appropriate style, she didn't. The differences in their styles can be seen on two occasions within a week or so at the end of November 1951. In the first, Watson attended a seminar on DNA given at King's College. Historian Robert Olby (*The Path to the Double Helix*, 1974) described it this way:

> Watson came to London expecting to hear Franklin talk about model building, the details of the layer line intensities, the likelihood of a simple solution. Instead he was treated to a sermon in caution, a lecture on the technical difficulties, and a long and boring enumeration of data about the water content of the specimen over a range of humidities. What had this to do with the model? Who cared about phosphate-phosphate interactions *between* molecules. It was the structure of the molecule itself that Watson was after! By his own admission, much of what Franklin said passed over his head. His attention began to wander and in conformity with his normal practice he made no notes. (p. 351)

Watson was after "the structure of the molecule itself," but where did he suppose that would come from? Franklin would tell him that it came from a close analysis of every spot on her X-ray film. Fortunately for Watson, Crick was not averse to wild guesses. Watson recounted as much as he could of Franklin's talk, and Crick started building a model. Within a week they'd produced a three-stranded model with the bases sticking out, based on what Crick understood from the X-ray data he and Watson had heard from Wilkins and Franklin. Immediately they phoned the London group to come to see what they'd produced. Olby describes the meeting:

> [After] Crick's enthusiastic peroration on the power of the helical theory which Cochran, Vand, and he had arrived at four weeks ago...Franklin was positively snorting at all this talk about helices. As the X-ray crystallographer on the job she knew they had a long way to go before the X-ray data was [*sic*] clear enough even to discuss possible structures. The molecule might well be helical but "there was not a shred of evidence" that it was! (*Helix*, 94)...Then the awful truth came out. Watson had got the water content wrong. It was not four molecules per lattice point as Watson had reported it but eight per nucleotide! (Franklin, 1951b) All sorts of structures now became possible. The argument in favour of their particular model no longer held.
>
> Watson and Crick tried to save the day by suggesting a future course of action in which the two groups would collaborate. But Franklin and Gosling [Raymond Gosling was her assistant] very understandably would have nothing to do with such a suggestion. They had witnessed two

> clowns up to pranks. Why should they condone their behavior by joining forces with them? (p. 362)

Franklin would arrive at her answer by the careful analysis of suitable data. Watson and Crick would build model after model until they chanced on one that fit. If Franklin had appreciated the role of speculation in science, she might have been persuaded to collaborate. On the other hand, Crick and Watson clearly did appreciate the role of good data; their correct model depended on it. However, it also depended on a couple of brilliant insights.

If we consider the number of times people generally don't see eye to eye because of differences in intellectual style, we shouldn't be surprised if we find them among scientists. There just are some people who hypothesize an explanation for every new event, and discard it when further evidence points another way. And there are others who reserve judgment until *every* possible piece of information has been carefully sifted and analyzed. Both styles contribute to scientific knowledge, but collaboration between them may be as difficult as that between a capitalist and a communist.

## *Evolution Revisited*

As we noted near the end of Chapter 16, the theory of evolution by natural selection was defective, because Darwin and Wallace lacked an adequate theory of genetics. They had natural selection operating on the variability of characters within a species, but they were ignorant of the source of the variations. The work of Mendel and his discoverers supplied that lack. However, you should not suppose that it was a simple task merely to paste the new genetics onto a blank page of the theory of evolution. The story is more complicated than that because evolution without an adequate genetics was already falling into a decline.

By 1900 Darwin's work on evolution had mainly achieved only one limited result—most biologists agreed with the idea of descent with modification. Yet they were by no means agreed on the mechanism of modification. While some followed Darwin and Wallace in the notion of many small changes brought about by natural selection, there were others still willing to allow Lamarckian inheritance of acquired characteristics. Still others remained skeptical about having achieved any knowledge of the mechanism.

Then, with developments in genetics and the discovery of Mendel's work, the study of biological change took a new turn. The idea soon arose that genetic mutations provided the whole of the explanation. Various studies revealed at least occasional circumstances in which a plant or animal underwent a relatively large change as the result of a single mutation. Some geneticists then proposed that evolution was the result of such "jumps," which they called *saltations*.

Until about 1930 the theory of saltations was presented as an alternative to natural selection. The major difference between the two was that in natural selection the environment determined which of several existing genetic varieties would be more successful, whereas in saltation theory a dramatic mutation would suddenly alter the characteristics of a variety, and if superior, would overwhelm its alternative. Natural selection operated in tiny steps, saltation in more major ways.

After 1930 this opposition between genetics and natural selection began to be seen as a false contrast. Gradually, biologists began to see that many different mutant strains provided the materials on which natural selection could operate. Finding saltation inadequate, they produced a synthesis of genetics and natural selection.

*The Modern Synthesis*

The major contributor to the synthesis of natural selection and genetics was Ronald A. Fisher (1890–1962), an English mathematician. As early as 1916 he developed a statistical analysis to show that the selection of a particular genetic character in a population over a number of generations could significantly increase its presence in the population. Initially, Fisher's analysis was resisted both by geneticists who clung to saltation and by Darwinians whe held a gradualist position. Eventually, Fisher detailed his arguments in a widely influential book, *The Genetical Theory of Natural Selection* (1930). Sir Gavin de Beer has noted that Fisher

> demonstrated that the mechanism of particulate inheritance of Mendelian genes which remained uncontaminated, segregate, re-combine, self-copy, and occasionally mutate, provides exactly what Darwinian selection theory requires to explain the source of variation; that Selection provides exactly what Mendelian theory requires to explain why some genes become dominant, others recessive, and others again suppressed; and that no mechanism other than selection will explain all the facts. [De Beer, Sir Gavin. "Darwin's Notebooks on Transmutation of Species." *Bulletin of the British Museum (Natural History)*, Historical Series 2 (1959–60), p. 28.]

Fisher's work developed into the field of population genetics, which has had notable successes in explaining both stability and change in populations. One neat example involves the gene for sickle-cell anemia in some African human populations (see box). Population genetics provides a satisfactory explanation for the continued occurrence of the sickle-cell gene even though sickle-cell anemia is usually fatal in childhood.

After 1950, with the identification of the genetic material as DNA, mutations were seen as alterations in the sequence of the bases in the steps on the spiral "ladder." Such changes can be produced by external causes such as radiation, or by "copy errors" during the process of DNA duplication. As a matter of course, most such changes are likely to be harmful. Sometimes a change is merely neutral, and very occasionally, it is beneficial.

In the case of sickle-cell anemia, only one amino acid (valine) in its hemoglobin is different from the normal (glutamic acid). The DNA code for valine is CCG, while that for glutamic acid is CAG. The change involves the replacement of only a single nucleic acid base, adenine replaced by cytosine.

### *The Sickle-Cell Gene*

The sickle-cell gene is one of the alternative genes that control the production of hemoglobin in the blood. If we label the ordinary gene *H* (dominant) and the sickle-cell gene *h* (recessive), then individuals can have three possible hemoglobin types, derived from *HH* for normal, *Hh* (mildly anemic), and *hh* (severely anemic). Anyone born with the *hh* character is likely to die in childhood from the anemia. Thus, you might suppose that the frequency of the *h* gene would decline in the population, since the possessors of the *hh* gene combination seldom become parents. However, in some parts of Africa malaria is an ever-present threat to health; and it turns out that persons with the *HH* character are more susceptible to malaria than those with *Hh*. The presence of malaria in the environment provides a selection pressure against *HH*. So, there is a higher than expected frequency of *Hh* in the population of parents. As a result, the children of two *Hh* parents have a one in four chance of inheriting the *hh* factor; one in four the *HH* factor, and two in four the *Hh* factor. With the homozygous (egg with *same* gene from both parents) offspring subject either to malaria (*HH*) or anemia (*hh*), there is hybrid vigor in the heterozygous (egg with *different* genetic contributions from the two parents) offspring. This genetic characteristic is called balanced polymorphism (literally, *many forms*). The frequency of the *h* gene in a number of African populations is about 20%.

In the population of a single species, there are many genetic variations along the DNA chains that direct the manufacture of organic molecules. Variations in these molecules provide the material on which natural selection can operate. Thus, developments in genetics were finally able to provide the key to unlock the mystery that had baffled Darwin and Wallace. The small differences among the members of a population are provided by variant genes, and selected by the conditions encountered in the species' envivonment.

Recently, there has been some controversy about whether evolution by a series of tiny steps is sufficient to account for the evidence of evolution found in the fossil record. A theory of "punctuated equilibrium" has been proposed, which suggests that significant speciation events occur rather rapidly—in a few tens of thousands of years—often in isolated groups at the edges of a species' range. Once a superior species has developed, it spreads over the range of the parent and replaces it. Then, for several millions of years, as long as there is little environmental change, there are no more major alterations in the species' characteristics. According to paleontologists Niles Eldridge and Stephen J. Gould, punctuated equilibrium conforms better to the fossil record than a continuing series of small changes. On this theory, transitional forms are infrequent in the fossil record, because they were never numerous.

Whether punctuated equilibrium should be considered Darwinian or not has been brilliantly discussed by Richard Dawkins in his book *The Blind Watchmaker* (pp. 223–252). Dawkins believes that neo-Darwinians can accommodate punctuated equilibrium quite easily. He suggests that no one ever really claimed that evolution proceeded only by such miniscule steps as to make every transition of organic forms entirely smooth. The argument seems largely to hinge on how jagged the graph of changes should be. Unless you imagine the graph so jagged that species are transformed in a single leap (which takes us back to saltation), then it appears that evolution by natural selection can handle many different degrees of jaggedness.

*Sociobiology*

In the 1960s, some biologists began to question how Darwinian selection could account for what seemed on the surface to be non-adaptive behavior. If, as Darwin urged, selection operates through the survival of the fittest individuals, how can you explain the development of co-operative behavior among the members of a species? If a young antelope at the edge of a herd is killed by a leopard after it issues a cry of warning, that helpful behavior will not be passed on from her genes. How then can such behavior be passed on to future generations? Biologists are dubious about supposing that co-operative behavior is transmitted to the rest of the herd by a process of learning. Still, they want to find *some* explanation for the development of co-operative instincts.

Finding it difficult to explain co-operative behavior at the level of individual survival, some biologists suggested that perhaps selection operates at the level of groups. Groups that exhibit co-operation will leave more offspring than those that don't. However, other biologists felt Darwin had been right to maintain the operation of selection at the level of individuals, and that they should be able to explain co-operation that way. Initially, they argued that while the young antelope contributed no genes to succeeding generations, her close relatives did. Since they have essentially the same genetic make-up, the antelope's surviving siblings and cousins would pass on the gene-based instinct that had prompted her warning cry.

Working out the details of such arguments led biologists to establish a new field of study that has come to be called sociobiology. It applies particularly to understanding the behavior of such social insects as bees and ants, where large groups of neuter workers have no possiblity of contributing genes to succeeding generations. However, they can contribute to the next generation by their careful tending of the queen bee whose fertile eggs can ensure the survival of future generations. Another intriguing example is provided by certain species of fireflies, in which the males in thick vegetation flash in unison to attract females (see box).

This aspect of sociobiology has been very successful in solving some knotty problems of animal social behavior. However, some sociobiologists have proposed extending their principles into the realm of human behavior.

### *Cooperation among Fireflies*

Fireflies at heavily wooded sites in southeast Asia engage in flashing regularly in unison. The males sit on the leaves of the trees in late evening, and flash all together about once a second. Sometimes it appears to a human observer as if whole trees are flashing like a beacon. This beacon provides a signal to the females so that they can find mates.

The hard Darwinian question is: how could such behavior develop and continue generation after generation? For, on the basis of individual selection, a "rogue" male could save himself the trouble and energy of flashing in unison with all the others. He could simply lurk nearby and pick off unsuspecting females lured to the vicinity by his industrious cousins. If this strategy were successful, the instinct of unison flashing would not be passed on, because the uncooperative males would outbreed their cousins.

But, entomologists found by careful observation that firefly vision is inhibited for about half a second after each flash. So the only flashes they can even see are the simultaneous ones. Once the females are close enough to the action they choose the brightest (often nearest) of several simultaneous flashes. A rogue flashing at random or not at all wouldn't even be seen. Over many generations, these male fireflies have survived in the thick foliage by cooperating in a general display to bring the females to the right vicinity. If you think of this behavior as a mating lottery, a male's ticket is drawn by a female only when she is close enough to see his individual flash. But the only way for a male to get his ticket into the bin is to sit there flashing away in unison. The transmission of his particular genetic message can only occur if he first buys a ticket. Rogue males without tickets are out of luck.

Sometimes they seem to have extrapolated their theory far beyond the range of testability. They have suggested that some human social characteristics have a genetic basis, similar to the instincts of other animals. This has aroused considerable opposition, especially from sociologists who believe that many aspects of human culture are learned—transmitted socially, not genetically.

Some social scientists are now trying to apply the principles of population genetics to human cultures—not as an extrapolation of genetic control, but by *analogy* with Darwinian selection. They have constructed mathematical models for the transmission of cultural traits, where offspring learn both from their parents and from others in their environment. The continuity of cultural behavior through the generations can display features that appear to be genetically inherited, while actually not being genetic at all, or only to a very limited extent. When people say they are the third or fourth generation of their family to be nurses or policemen, that doesn't have to mean that nursing or policing are somehow built into their genes. For, besides whatever genetic features they may have inherited from their parents, they have also been subjected to their parents' behavior as well as that of friends of their parents who share the same occupation. Cultural inheritance can have the same kind of continuity and variety as genetic inheritance, while being entirely learned.

An interesting aspect of cultural inheritance is that it contains a component of the inheritance of acquired characteristics, which we identified in Chapter 16 as Lamarckian. We said there that children do not inherit improved muscles genetically as a result of their parents' working out. But children of body-builders might become body-builders through imitation and learning by the process of *cultural inheritance*.

Finally, some applications of sociobiology to human culture contain a very strong undertone of ethical or moral directives: that we *ought* to live in ways that are appropriate to our genetic structure. When the sociobiologists then proceed to tell us they can derive what those ways are from their scientific theories, they have overstepped the bounds of good science. Even if some of their recommendations are useful, we should not suppose that they carry any scientific authority. While some moral teachings may be better than others, their value can only be determined by experience.

### *Conclusion*

After a hundred years of work in many related fields, the message of the genes has been basically decoded. The work involved the efforts of many people—Mendel with his pea plants, microscopists with their cells and chromosomes, organic chemists with their nucleic acids, T.H. Morgan with his fruit flies, Rosalind Franklin and her X-ray analysis of crystals, Watson and Crick with their double helix structure for DNA. A fascinating story of scientific detection.

And now that molecular biologists have unravelled the DNA code, what next? After genetic science comes genetic engineering. Now we can manufacture DNA molecules to order. By a process of gene splicing we can make a bacterium produce the insulin or other serums used in the treatment of diabetes and other diseases. The United States Supreme Court has determined that newly manufactured biological organisms can actually be patented. Brave New World—here we come!

As Aldous Huxley made clear in his novel, *Brave New World*, the possibilities of genetic engineering are exhilarating, and frightening. Do you want to make human beings to order? Grow food in factories so you can get a nutritious sizzle without the steak? Or the steak without the sizzle? Who's to decide? Where will it all lead?

Genetic engineering is out there now, just like nuclear engineering is. Scientists and engineers can make the technical decisions. But their technical vocabulary does not contain words like good and evil. *We* have to make the moral decisions.

*Male fireflies (Lampyridae) flashing in synchronous patterns.*

*The Earth seen from space.*

## CHAPTER TWENTY-TWO

# Resources and the Future

Here we are, all rolling along together on spaceship earth—six billion human souls. Careening through space at half-a-million miles an hour—with no Captain Kirk at the controls. The image of our planet as a spaceship is intended to remind us that the earth and its resources are all we have to work with. Except for the light and heat energy we get from the sun.

So we can begin to ask questions like how many passengers have we got room for? How well can they be fed, clothed, and housed? What are the chances we'll run out of precious non-renewable minerals like aluminum, iron, coal, and petroleum? And can we avoid fouling our nest with more pollution than we can handle?

Technology consists of the machines and techniques we humans use to control our material world. We transport goods and communicate information in quantities and speeds our ancestors could only marvel at. We've conquered the earth!

But our control has been bought at a price. What are the limits on science and technology we've been inclined to overlook?

What about world population growth? In Chapter 16 we talked about the Reverend Thomas Robert Malthus and his Essay on Population. In 1800, Malthus claimed that populations have a natural tendency to expand beyond the limit of the food they can produce, so that populations are often limited by miseries like famine or vices like war. He claimed we could avoid these calamities only by limiting population ourselves—by moral restraint.

Malthus wrote near the beginning of the Industrial Revolution. For a hundred and fifty years, our new technologies and sciences increased food production to levels Malthus had never dreamed possible. The Malthusian crisis never came. Or did it? In 1800, the world population was 900 million. Today it's six billion. And not all of the world's population today is well fed.

We may ask why Malthus's predictions didn't come true? The short answer is technology, particularly agricultural technology. Countries like the United States, Canada, and Argentina have produced food at rates Malthus could not have foreseen. These gains have been made through improved strains of grains and animals, the mechanization of farms, and the widespread use of chemical fertilizers and pesticides. In the process the amount of inanimate energy devoted to food production has increased enormously.

For a more complicated answer to the question, we'd have to look at production, distribution, and consumption statistics in much more detail. If we do that, we might discover that, in fact, the Malthusian limits have been reached in localized areas in the Third World. That certainly would be one way to explain the massive famine relief that's needed from time to time. Just as Malthus could not have foreseen intensive agricultural technology, he might also be staggered by the advances in disease control that have exacerbated population crises.

Since Malthus's time, we've had revolutions in industrial production, agricultural production, and medical technology. With our expanding technology we've achieved a world population more than six times what it was in 1800. But we've done it by placing a tremendous load on those finite resources of our spaceship earth.

We're now realizing that the issue is more complicated than simply attempting to balance food supply with population. That's particularly true because of the life-style we in the industrialized world have adopted. The majority of people in North America and western Europe consume a far larger share of the earth's resources than people in most other countries. So the pressure on those resources doesn't come merely from their large populations, but also from their life-style.

Then, before we run out of the materials we need to live, we have to consider how we use those materials—how much each person needs and how many of us there are. This problem was tackled 30 years ago by a group of international experts called the Club of Rome. They examined the resources of the earth in a new way. They treated the earth, its resources, and its people as a system of interdependent sectors, and they created a computer model of those various interconnected sectors. Their sectors included population, natural resources, agriculture, capital investment in industry, and pollution. You can think of the sectors as boxes, producing those various products. But the boxes are all interconnected.

For example, the size of a population depends on how much food it has to eat. Think of an island that supports a thousand people. Suppose they're using the total food production of the island for a barely adequate diet. If the population increases, each person will get less to eat. That might result in people dying earlier of diseases they aren't strong enough to resist. So the population would decline. Then there'd be more food per person and they'd get healthier again. So the interaction between population and food operates to keep the population around a thousand.

Then suppose some clever farmer discovers that a particular rock on the island can be crushed to powder and spread as fertilizer, which doubles food production. Being better fed, the population increases. Then, some years later, the islanders realize that rock fertilizer resource is running out. What'll they do now? Maybe with more investment in mining they can dig deeper and find more rock. Maybe it'll just run out, and the population will decline back to the level they maintained before their great fertilizer discovery. It all depends.

The factors of our real world economy are even more tightly interrelated. As we dig deeper for minerals like iron and coal, we invest in more machinery to do it. More machines may result in more pollution, like more acid rain, which poses health hazards. Production, pollution, and population are all interconnected.

The Club of Rome investigators built a computer model of all these interconnected sectors of the world economy. Then they had their computer run forecasts of the future. What would happen if current trends and rates continue? Their first result suggested that world population might reach 10 billion before 2050, but that such a population could not be sustained—we'd run out of resources, pollution would increase, world population would decline, and the people still alive would live on a planet much poorer in resources.

A gloomy forecast! But why should we suppose that present trends will continue? All sorts of things could change. The Club of Rome analysts tried various alternatives: they changed certain factors in their model, one at a time, keeping everything else the same, and ran new forecasts. Suppose resources are actually more abundant than we've thought; then pollution would increase even faster, and population would still decline.

Suppose there are large improvements in pollution controls and agricultural technology. Population and industrial production would increase more rapidly, but only temporarily. Although pollution rates are controlled, large increases in population and industry would eventually enlarge total pollution, and death rates would increase again. Other combinations of improved technology of various sorts proved just as unsuccessful.

To get a world economy that could sustain a population at reasonable levels for some centuries, the analysts tried a different solution. Instead of more of this or that, they built zero growth into their computer model—zero growth of population and also zero growth of industrial production. That would mean allowing no new person to be born or new industry to be built, until an old one had been put out of commission. Zero population growth—of both humans and machines.

When the Club of Rome analysts ran their computer model with zero growth, it worked. Graphs of population and pollution that before had risen precipitously and then collapsed, flattened out. Eureka!

We must realize that these were not forecasts of what would actually happen, but of what could happen under various scenarios of how we use the earth's resources. The value of the computer modeling was to point a finger at the two critical factors—industrial growth and population growth.

We can either continue our present economy of growth until we run into the limits Malthus described, or we can impose limits of our own before that happens.

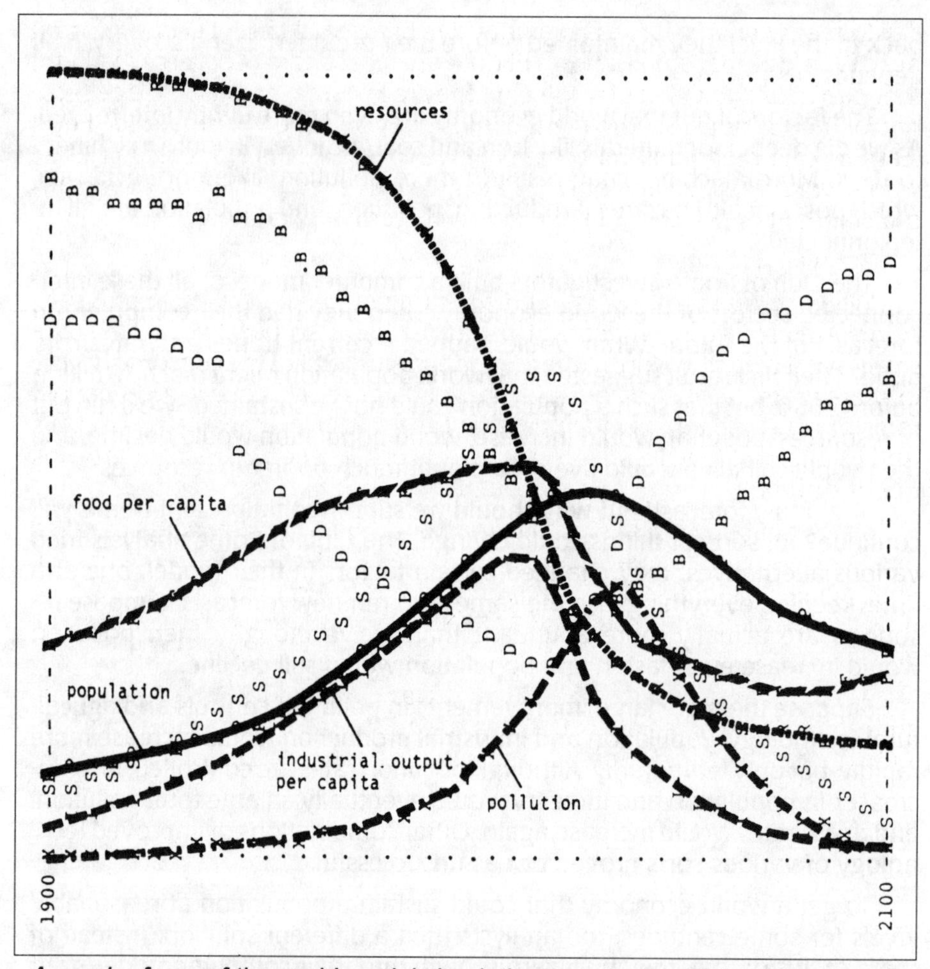


*A graph of one of the world model simulations run by the Club of Rome and reported in* The Limits to Growth *(1972). The model assumes no major change in physical, economic, or social relationships from historical patterns. As the graph shows, this model predicts that resources will decline precipitously from the latter half of the 20th century to the mid-21st century, and that in the first decades of the 21st century lack of resources will precipitate a shortage of food and of industrial output. That, plus rising pollution will bring about a rapid decline in the population of the world. A very grim prediction.*

We've already suggested that people in North America are inclined to think the problem of balancing resources and population lies beyond our shores. Our population has pretty well levelled off, but just look at Asia and Africa. Their populations are growing out of control, it seems. That's the problem that needs to be solved. Why won't they learn the joys of birth control?

One answer, of course, is that they live in a very different economy from ours. With little industrialization, children are an economic resource for them; they can help in the fields by the time they're five and provide old age security for their parents. So, we might think, if they become industrialized, they'll learn to limit their family size the way we've done. Simply industrialize the whole world and the problem'll go away. When everybody in the world has our standard of living, we'll have achieved earthly paradise.

But wait a minute!

While populations in Asia and Africa have been growing explosively, North American consumption of resources has also been growing explosively. A child born in North America today will in its lifetime consume twenty or more times as much as one born in Bangladesh. We have to worry about rates of consumption as much as about the number of people consuming. And for the past few decades, our rates of consumption have been growing out of control.

When anything grows so that it doubles in equal time intervals, it's called exponential growth. If your bank pays seven per cent compound interest, your money'll double every 10 years. Start with $100 today. Ten years from now you'll have $200, in 20 years $400, in 30 years $800, in 40 years $1600, and so on.

Malthus saw that population growth was exponential too. In the twentieth century, the world population has been growing exponentially at about 1.4% a year, doubling every 50 years.

The growth of our North American consumption of resources has been exponential as well. For example, until recently, our consumption of electrical energy grew at about 7% a year, that is, doubling every ten years. If that continued, we'd have to double our capacity to produce electricity every ten years. Suppose we have ten generating stations now. We'd have to build one a year to have ten more in 1999. Then we'd have to build two a year to get us to 2009. And four a year after that. Only one a year now, but four a year just twenty years from now. Where would all the resources come from?

This same kind of exponential growth in production is needed for anything where our consumption doubles every ten years—not only electricity, but also automobiles, refrigerators, TV sets, and so on.

Our industrial growth rates are greatly outpacing third world population growth rates. That's why we're using up the earth's resources faster than they are.

About 1970, books like *The Limits to Growth* by Meadows et al. made some people realize we were depleting our resources too rapidly. And they reacted. They formed lobby groups. They looked at the plans of companies and governments for the future of energy. When they saw how intensive the growth of the energy-producing industry needed to be, they objected, particularly on environmental grounds. If electricity was to be produced by fossil fuels, more production would mean more air pollution and more acid rain. If nuclear generation was planned, that would increase the problem of disposing of radioactive wastes.

Environmentalists also objected to the deterioration of the countryside as corridors had to be built for the power lines. In areas of high population, the ten-year doubling would lead to a high concentration of energy-producing plants. Since these new plants all use steam turbines to turn the generators, they produce a lot of waste heat. That heat must be dissipated, either into a large body of water or into the atmosphere. If into the atmosphere, via cooling or evaporation towers, local weather can be adversely affected. If water at 60°C is discharged into the local lake, temperature changes can drastically alter the balance of aquatic life. Thermal pollution is invisible, but its effects can be as harmful as more visible sorts of pollution.

An interesting alternative is to reduce thermal emission by what's called co-generation. That is, instead of squeezing the last possible joule of energy out of steam by letting it condense, make the production of electricity a little less efficient; use steam at 200°C for space heating in offices and factories. By that scheme, fewer heating furnaces would be required. While the efficiency of electricity production from fuel might fall from 40% to 35%, the overall thermal efficiency would increase.

Now, numerous arguments like the foregoing were presented to businesses and governments in the early 1970s. They mostly fell on deaf ears. At first, environmentalists were treated like a lunatic fringe. Plans had to be made for the real majority of consumers—consumers who were often being encouraged in wasteful ways by the energy companies ("live better electrically" in your "all electric home"). Then, the Arab oil embargo came along, and soon we were paying four times as much for oil as we had only months before.

Energy became much more expensive overnight. The result was that people started using less of it. Instead of energy consumption growing at 7% per year, it fell to 2 or 3%. All those plans for expansion began to look ridiculous. And governments began to pay attention to environmentalists. They needed good arguments to use against the growth enthusiasts who were slower to grasp the new realities.

One lesson to be learned from the energy crisis of the mid-seventies is an old lesson: "money talks." Sound environmental arguments could not achieve what a quadrupling of the price of oil achieved. And yet the price mechanism is always available. In the days of high, wasteful consumption, industrial customers were rewarded for inefficiency; the more they consumed the cheaper it got. That may still be true in some jurisdictions. And it is the ideal way to induce higher consumption. The fact that rate manipulation distorts economies of scale (for example) is not entirely relevant. Prices can be set to achieve objectives. You might find it an eye-opening exercise to enquire into the hidden subsidies or surcharges in the prices you pay for things.

We are now getting a large fraction of our energy from non-renewable resources, from fossil fuels and radioactive minerals. The earth's supply of these materials is not infinite. As long as energy consumption was doubling every ten years, no one could see how new energy finds could possibly keep up. Now, consumption rates are down, but there is still growth in energy consumption. In some areas of life, like improved amenities for poorer people or sufficient food for every person on earth, there should be growth. But all growth pushes against limited resources. Slower growth means only that petroleum reserves might last 100 years instead of 25 years.

Every time we find a way to improve the efficiency of a process—to use one barrel of oil, where we used to use two—we've saved a barrel of oil. One less barrel of oil to be dragged out of the earth, refined, transported, accounted for, advertised, bought and sold (all energy consuming activities in themselves).

That's why the best barrel of oil is the barrel that's saved through conservation, through the increased efficiency of machinery, better insulation of our homes, and so on. And paying attention to alternatives like solar heating. But there are limits to efficiency too. Particularly in the field of transportation.

For example, engines that run on burning fuels produce a lot of waste heat. More than half the heat any engine produces is wasted; it just goes off into the air. Our cars heat up the highways, our electrical generating stations heat up our lakes and rivers.

So, while there's still lots of energy conservation to do, there are also limitations we can't avoid. That's why the limits to growth analysts found that industrial limitations are every bit as important as population limitations. Maybe more important!

The limits to growth! Material resources are not unlimited. But many human resources remain untapped.

Indeed, one of the ways to improve our present situation is to consider redefining what we mean by a resource. Resources are materials or systems that we have uses for. If we consider them not to be useful, then we have to dispose of them—either they're resources or they're garbage. But garbage can be a resource if we take the trouble to separate it into its components: glass and metals can be re-cycled; organic wastes can be composted; paper products can be burned for heat. If we looked at garbage that way, we would realize more quickly that there are certain things that should never be produced in the first place, such as plastic bags and containers which are too dangerous to burn and cannot be easily re-cycled. If you're running out of landfill sites in your area, it may be because you're throwing away valuable resources.

However, there's one kind of waste product that seems to be the very opposite of a resource—the by-products of nuclear reactions. When spent fuel rods are removed from a nuclear reactor, they're still hot, literally in temperature, and also in being still highly radioactive. But they're not hot enough to be useful any more in the reactor. They have to be stored with great care.

Storage of highly radioactive by-products is a serious technical and political problem. While one set of scientists claims to have the problem well in hand, others are throwing various nettles at them. Some of the by-products have half-lives of tens of thousands of years; where do you look for a dump site you can be sure of for several times that period? How do you guarantee the security of such a site? If we expand the number of nuclear power stations, how many such sites will be needed? And who will you find willing to allow the dump site in their backyard? This is not an issue that will be settled easily or soon.

Resources are not only material. They're human too. Until very recently there was little thought given to employment opportunities for women. The political economy of nations defined women's roles very narrowly as housewives and mothers. Now, that's slowly changing, and the resource of human intelligence that used to be chained to the cook stove and the diaper bucket is now ready to be deployed in many ways, some old and some new. Look out!

We can't know what the future holds for us, but we can think of ways of avoiding disaster as long as we pay attention to the realities of our spaceship earth. The message should be clear—to us, to the captains of industry, to our political leaders—less may be more!

*The lamb "Dolly" shown with its surrogate mother. Dolly was cloned from an adult sheep.*

*The first human footprint on the moon. 1969.*

# EPILOGUE

We began with the idea that science and technology are human activities. That should be self-evident since scientists, engineers, inventors, and artisans are human beings. But, we may be worried by a little niggle in our heads that perhaps somehow, some people are more (or less) human than others. That's not intended to raise visions of racism, nor to suggest that some cultural groups are better or worse than others. It is intended to direct your attention to those who might be called the "leading actors" in any given culture, to focus on the "high priests" of your own culture.

Think back over the route we've traced and consider for a moment which occupations were most highly prized from era to era: where would ambitious youth be directed to seek their "fortunes"? For most of recorded history, to be well-born was to have no need to choose. You would move into the seat of power your father held, whether in the Roman senate, the British House of Lords, or the presidency of the Ford Motor Company. Set them aside and think about the rest of us.

Some traditional routes to power have had long staying power, like the army and the church. Bright, ambitious young men have been sent to military academies from Roman times all the way down to today's Royal Military College and West Point; or to the seminaries and monasteries, whether Hindu, Buddhist, Muslim, Hebrew, or Christian. From medieval times, the universities were the "trade schools" for physicians, lawyers, and priests. In some parts of the world today, priests are still powerful; in others, lawyers have more clout. And to be a physician is still a leading ambition of girls and boys in high school.

All these are traditional occupations, prized by cultures for preserving the status quo and easing our pains, whether they are physical, psychological, or social pains. We might not expect to look among these "caring" professions for the innovators in society—for, innovators care for something else.

In fact, during the period 1600–1800, many physicians eagerly participated in the new science—and not just in matters medical. Also, particularly in the seventeenth century, numerous priests also participated. This should not be too surprising when we realize that the revolution in science was built partly on Greek foundations, which these chaps studied in their universities.

The seventeenth century generated a strong role for mathematics in the new science so that names like Newton, Maxwell, and Einstein come to our attention. However, except for a few like Newton, most mathematically based advances in science are made more by those who apply the math than by those who invent it.

Over the past couple of hundred years, mathematics has penetrated deeper and deeper into our culture. Routes to power nowadays lie among accountants and statisticians and computer engineers, alongside the traditional routes that have not faded. Mathematical skills are also found useful in a wide range of sciences, both social and natural. I remember the look of horror on the face of a fellow student—first class in evolutionary biology—when the instructor handed out a math test. He thought he'd chosen a field in historical biology as far as he could possibly get from those hateful numbers. It was not to be.

Our medieval heroes were filled with piety; more and more, our modern heroes are filled with equations. Does that make the world different? Do equations drive out piety?

### *Science*

We can look at the science practised in the twentieth century in two aspects—normal and extraordinary. While every scientist in a university graduate program is required to make some addition to knowledge—it's a requirement for the PhD or the DSc—most of them are modest additions. Most current scientific activity, sometimes called "normal" science, involves mopping up puzzles in some small corner of a field. Only occasionally are scientists lucky or clever enough to become renowned for making a "revolutionary" contribution that transforms a field. What's been called "extraordinary" science comes along once in a while, apparently at random, and usually unpredictably. When such a revolution does occur, it's the task of the "normal" scientists to change their focus, and move their mops into various corners of a new field.

Buonamico was mopping up puzzles in motion philosophically in 1591, while Galileo was getting ready to transform the study of motion. Newton completed the transformation, creating a whole new field for the newly mathematized mops wielded so effectively during the eighteenth century. Between 1890 and 1930, chemists were mopping up Dalton's atomic theory by determining atomic weights with great precision. Meanwhile, Rutherford, Bohr, and others were transforming our picture of atomic structure, and creating a whole new set of puzzles to be mopped, such as the precise determination of energy levels for all the elements.

Of course, sometimes the puzzle-solving mops stumble on something new, as, for example, Fermi's irradiation of all the elements with neutrons, except of course that Fermi didn't recognize the fissioning of uranium when it happened (Chapter

19). That time, it was the quizzical mops of Hahn and Strassmann that made the breakthrough.

Since 99% of science is puzzle-solving, any history (such as this one) that concentrates most of its attention on the 1% is a distortion of the realities of scientific activity. But, then, all history is like that. We watch the strategies of Napoleon and Wellington at Waterloo, or Rommell and Montgomery in North Africa; we follow the tactics of a regiment or a corps; we look at total body counts; but we don't enquire into the activities of every colonel or lieutenant, let alone every corporal or private. Those are the normal distortions of a history that is necessarily selective.

Besides, the inspirations of genius are fascinating—the nine-to-five routines in innumerable laboratories are just dull—however essential they may be to winning the prize. As they say, 99% perspiration (sweat of your brow) and 1% inspiration.

That leads to another distortion worth noting in the history of science. It is that while we may wish to emulate our heroes, we must mostly be content to wield our mops. Some scientists have such deep insights into the workings of nature that they seem to be able to dispense with the mops of experimental verification. Galileo had Salviati say in his *Dialogue* about stones falling vertically even if the earth moved:

> Without experiment, I am sure that the effect will happen as I tell you, because it must happen that way.

Similarly, Einstein was so convinced of the correctness of his special theory of relativity that he knew contrary experimental results must be wrong. Richard Feynman (1918–1988) described this attitude while discussing the time-dilation involved in high speed travel. He described an experiment using mesons:

> ...it would surely be found that the one that went around the circle lasted longer. Although we have not actually carried out an experiment using a complete circle, it is not really necessary of course, because everything fits together all right. This may not satisfy those who insist that every single fact be demonstrated directly, but we confidently predict the result.

"Put down that mop!" Feynman might have said.

But does that mean we should expect to get all our knowledge of the world just by thinking about it? Of course not. The experimental mops that test the grand theories are absolutely essential—how else would we know if the theory was correct? Yet, while theories do require testing, it's an amazing fact about good theories that the tests often appear to have been foregone conclusions. But only "often," not always.

So, while mop-wielding may frequently appear to be unglamorous, it is an absolutely essential part of the scientific enterprise.

## *Science and Society*

In the Foreword, I suggested that many cultures had managed to get along well enough without science as I've defined it. I said that, fully conscious of the fact that many cultures have mythologies and other principles they find perfectly satisfactory for explaining the world they find around them. Those principles may even be said to "work," at least in a psychological sense. (Whether Voodoo works in any other way is something I know nothing about. Don't stick pins in my doll!)

The successes of western science during the past 300 years may actually create a problem for us. Think of what it must mean to say that we live in a "scientific society." From the perspective of scientists, it means that we oughtn't to "believe" anything that hasn't been certified by their mills and mops—inspired guesses that have been tested logically against careful observations and exact measurements. From that point of view we get to modern cosmology—we believe in Evolution, Relativity, the Big Bang, and a myriad other outcomes of scientific activity. By cosmology I mean a "big picture" in our heads that's supposed to represent what we understand about the structure and nature of the world around us. That makes scientists into the modern myth-makers, the high priests of western civilization.

Many scientists, rightly, wear their priestly mantles with great discomfort. They spend their time mopping up the little puzzles of the big theories, and they (mostly) find the theories useful in making sense of their observations. An intriguing example was given in Chapter 21 of entomologists trying to understand the preservation of co-operative flashing among fireflies in southeast Asia. (See box on page 431 and picture on page 433.) Those entomologists found that the theory of natural selection provided a rigorous technique for weeding out invalid hypotheses. Whether they all "believe" in Evolution is an entirely different question.

To extrapolate from successful applications of the theory of evolution to "There is no God" is as wild and unfounded an extrapolation as to deduce from the Bible that the earth is only 6000 years old. The only legitimate cavil scientists may have with fundamentalists is when the latter want to tell them how to do science. No other aspect of the many possible differences among them are genuinely scientific.

Hold on a minute! Are we supposed to see scientists as some select segment of society somehow separate from standard social scruples?

Not at all.

Particularly today, the scientific enterprise is big business and has to be funded, often out of general national revenues. An eighty-kilometer diameter accelerator costing ten billion dollars is not to be treated as a toy to keep particle physicists quiet. Whether to build or not to build is as important a political and social question as deciding to buy a new fleet of nuclear-powered submarines. To keep the admirals quiet?

Neither one is an issue to be solved in this or any other book. Books raise issues; people solve them with intelligence, with compassion, and with common sense. The point about "science and society" is that citizens have a responsibility to inform themselves about matters that affect their lives, and experts have a responsibility to assist them in that.

### *Technology*

Technology penetrates so deeply into so many aspects of modern life that we're in danger of being like the fish who don't realize they're swimming in water. You may choose to give barely a thought to science, but it's hard to avoid technology every time you open your eyes; even your sleep may have been induced by the latest product from the chemist or the electronic engineer.

At the levels of design and production, the keynote of technology is *efficiency.* Do more with less, save time and steps, get more farther, faster, cheaper. James K. Feibleman once suggested, for example, that the aim of aircraft designers appears to be a machine that "will carry an infinite pay-load at an infinite speed while itself weighing nothing at all!"

Another aspect of efficiency is the design dictum that form should follow function. One-piece molded-plastic chairs may be efficient, but they're sure not Chippendales. Of course, beauty is often a personal enough decision that we shouldn't be too quick to pass judgment. And we shouldn't suppose that form from function is an entirely modern concept. Recent research has shown that many decorative features of Gothic cathedrals fitted structural necessities. For example, the pinnacles atop the outer ends of flying buttresses provided a load to hold the beam down. If you need a pile of stone there, why not carve it in the form of a saint? Downspouts let the rain run out through the mouths of gargoyles. Yet the medieval design motivation was different, since the cathedral also had symbolic functions to fill. It was the House of the Lord for people of a particular culture at a particular time. For us moderns, the symbols we value must reside someplace else.

Or compare the elaborately crafted stocks of old hunting muskets with the deadly efficient Sten gun. Maybe we should be grateful to Mr. Sten for making his gun so ugly.

Now, to enquire into the control of technology, paralleling the control of science, is a curious business. For, we seem to control technology every day in the marketplace, choosing this instead of that, or in the wide range of governmental licensing and protection agencies. Any industrialist will tell you that his products must be fine—see how many people buy them. They don't have to, do they?

Of course, at some levels they do have to. We all have to eat, and when agribusiness pumps hormones into cows and chickens for the sake of efficiency and profit, consumers may have little choice. Well, you could all become vegetarians, but have you noticed those beans, lately?

Around 1957, the Ford Motor Company built its advertising campaign on safety. Across the street, the GM cars were sporting curious tail fins. As Ford dealers reported empty showrooms, the company had to revamp its ads in midseason to get the customers back. Fortunately, they didn't challenge their designers to come up with bolt-on fins. That year, apparently, safety didn't sell.

Such stories prompt some observers to think the public doesn't know its own best interests and must be taught to see the light. Good luck to them. Other observers say that technology has a life of its own, that it's running out of control, and no one can stop it. Using phrases like "technological determinism" and "the autonomy of technique," they note that the goals and procedures of technology are often chosen from within technology, without necessarily referring to reasonable goals of the total society. So, our efficiency-driven aircraft designer may pay too little attention to such environmental impacts as sonic booms or atmospheric pollution: "That's not my department," they'll say.

For everyone concerned with the toxic outputs of apparently runaway technology, the simple answer is, make it *your* department. And some do, although they often feel like they're climbing straight up the side of a cliff. If technology is truly autonomous, nothing can be done—we daren't let it be autonomous. Perhaps the scariest part would be if humans are becoming technological; e.g., prizing efficiency above other virtues, with traditional values like sympathy going the way of the dinosaurs. If we care, we should do more than hope not; we should find ways to put our heads where our hearts are.

### *Technology and Society*

The current technology that seems to be impacting our lives the most is the joining of computers to communications networks: everything from airline reservations to your local patrol car with quick access to the data banks of the FBI and the RCMP. The depersonalization of transactions proceeds apace: you can't make a date with the friendly automatic teller on the corner.

Skilled artisans are being replaced by robot arms and secretaries by word processors, as every conceivable task is turned into a replicable, mechanical operation. The meaning of work must be re-defined to fit the realities of burgeoning hi-tech in a post-industrial age. After thousands of years when most of us defined ourselves by family, tribe, or nation, the Industrial Revolution came along and forced us to redefine ourselves by the work we did. For two hundred years it's been our daily work, our skills, that gave us dignity and an income. Now, we're in the middle of a second industrial revolution where new ways to define human worth must be found.

No one should really be much surprised that there's a lot of discontentment or confusion abroad in the land. Nor, I suppose, should we be surprised at the enthusiasm with which folk follow their fortunes through the Roman circuses they call government-operated lotteries. What? I hear Socrates, Shakespeare, and Sartre crying in unison: define your worth by your skill at choosing six numbers? Or, set your life by the whims of a random-number generator to match little balls dropping through a gate? Better you should depend on gladiators and witches.

In the workplace, menials and professionals work too long and too hard for a pittance or a heart attack, while the middle range of industrial skills gets handed over to machines. The only compensation for the de-skilling of our industrial work force is that technology has made some of their skills available to us all: with our touch-tone phones we can call overseas, without either a routing operator, or the telegrapher of a hundred years ago. With our personal computers, we can email chatty letters to distant relatives complete with photos or maybe even video clips as attachments, without the need for letter carriers, post office clerks, or pony express.

In our homes, the upstairs maid is replaced by the vacuum cleaner, the scullery maid by the food-processor and the dishwasher, and the downstairs maid by the washer-dryer set. Though, as Ruth Schwartz Cowan has shown, these "labor-saving" gadgets are likely to make *More Work for Mother* rather than less. She can do it all in 15 hours a week and still manage to work full time downtown.

You might think that all those computers would need lots of programmers, designers, manufacturers, and repairers, and they do, some. But the very nature of the machines means that fewer and fewer hands can do more and more and more. Labor Department forecasts of future employment see many more openings in fast food restaurants and on custodial and cleaning staffs than in the professional echelons.

If you want these gloomy portents not to be realized, then start now to find ways to reinstate human dignity, and seek political and social institutions willing to admit the existence of our changed circumstances without succumbing to them.

### *People and Society and Technology and Science*

How well people handle change is a crucial question today. What ought to startle you is that it is a new question. Seen across the broad span of human history (5000 years of "civilization" preceded by 50,000 years or more of pre-history) handling novelty has not often been a big issue. In many places, for long stretches of time, life in any generation went on pretty much the same as it had in many earlier generations.

Throughout most of history, people have seldom had reason to remark about how different things were in their time compared to that of their grandparents. Even an event apparently so startling as the agricultural revolution (Chapter 2) was a very slow affair.

Think how different things are today. My grandmother washed clothes with a bar of soap on a scrubbing board in a wooden tub. My grandfather drove to church in a horse and buggy. I can't even imagine how different my grandchildren's life will be from mine.

Given this accelerated pace of change, what inner resources do people call upon to handle novelty? Where can they look in the broad sweep of human experience?

Genetically, humans have changed very little in 50,000 years. The hormones that drive us were laid down in a context fantastically different from the one we face today. Adrenalin provided humans with the spurts of energy needed to survive in a hostile world—the "fight or flight" response so loved by psychologists. Can we simply say that's exactly what's needed in today's urban jungles or in the shopping malls and boardrooms of our nation or on the floor of the stock exchange? Do we still live in a savage world that requires savage responses? *The more things change, the more they stay the same* Is it our incapacity to control our hormones that makes us manufacture jungles to keep us human? Or do we use a distorted view of our savage past to justify responses we should be able to control? After having tamed nature, can we not tame ourselves?

Technology and science have provided us with many layers of protection against external savage elements. They protect us from savage climates (heating in winter, air-conditioning in summer) and innumerable other vagaries of human existence, like disease and hunger. But, they've also supplied us with the means to expand our internal savage elements—to destroy ourselves with rocket-delivered nuclear weapons or with aircraft-delivered chemical mind-benders.

We may wonder if our highly-prized techniques have driven us into a world that is more complicated than our genes can handle.

We should not suppose that every human problem has a technical solution. Yet today we are apt to find seriously proposed technical solutions to almost any problem. Here's one we used to hear about from time to time. There are serious shortages of fresh water in some parts of the world. Why not build gigantic nuclear-powered desalination plants, say along African coasts, to make the desert bloom. It might work—but now we're more concerned about radioactive wastes and thermal pollution than we were when that was proposed. Often, technical solutions carry a whole new set of problems with them. You cannot assume that the saving technology will be benign.

Other problems, such as economic and religious strife in the world's hot spots (Northern Ireland, the Middle East, or the Balkans, not to mention Korea and Chechnya), may seem to be entirely independent of technical solution—or any solution at all. One of the problems with technical solutions is that they are devised and conducted by humans—humans who are likely to have personal agendas that are a lot more complicated than the mere alleviation of human suffering.

On the other hand, you may not wish to agree with the Biblical claim, "The poor you have always with you." Without preaching too much, I suggest that a helpful motto could be: Try to leave the world a little better than you found it. The problem is that it is impossible to get world-wide agreement on what is "better." Scientists and technologists can help; but their style of help cannot be assumed to be adequate to every human problem.

These are social issues that should be of great concern to all of us. Science and technology have made great changes in our lives in the past; we've no reason to suppose they won't continue to do so in the future. We ordinary citizens foot the bills (*all* the bills if you take it far enough) so we should know what we're getting for our money, and should have some say, if not the major say, in how the money is spent. It's hard to do that unless you have some sense of the range of possible ways the money can be spent, that is, understanding the nature of the endeavors called technology and science. I hope this book has in some small measure helped you to do that.

How you get to have your say heard is an issue in politics, not in science or technology. You'll have to find another book for that.

This world is the way we've made it—it's the only one we have. Will you help to keep it, to make it better, or will you be part of the problem?

## *In a Mind's Eye*

Ana Seto woke gently under the soft urgings of her mother's voice. Soon, mother, soon, she murmured. The voice gradually became more insistent. All right, all right; look, *I'm awake.*

The voice-activated tape machine turned off. Still her mother's face smiled sweetly from the high-definition integrated monitor in the wall. Ana looked longingly at the face; then, *Good morning mother;* the frame froze and glided smoothly to the bottom right corner of the screen. My mother the icon, Ana smiled weakly.

As Ana dressed, the illumination gradually increased to daylight intensity. She folded the futon into its drawer, rearranged the couch for work, and moved across the room for breakfast. *News headlines,* she called to the nerve center (at least, that's how she'd come to think of it). During her brief meal the nerve center displayed the most recent headlines stored in its memory, programmed in the order of Ana's major interests. As she finished her meal, she saw the bloody aftermath of the latest battle in the Sudan.

Following that final item, the screen displayed the headline titles of items available in greater depth. Ana touched the console to allow a few moments of Jazz-Fest to fill the tiny room, which looked larger because of the great mirror that filled the western wall.

She swayed rhythmically to the beat, dancing with her image in the mirror, then turned it off reluctantly as the next number was announced. There was too much to do today.

At her command, *today,* the screen listed the tasks she entered the night before. After a couple of changes and rearrangements, she settled onto the couch with her console in convenient position. At a touch, the list glided to the right side of the screen, above her mother's smiling icon.

First, the office. They'd found a small glitch in her program; twenty minutes to fix that. Then, the information from the Library of Congress was ready for her. She was using them so much these days, she'd programmed a voice macro for it—at her *LC,* the link was established, and the data were downloaded. In a few days her analysis of communications during World War II would be ready for the historical data bank.

She worked over the keys nimbly for a couple of hours, opening up windows of stored data from time to time to piece the story together.

After a period of intense concentration, she slipped a *kabuki* disk into the DVD and for a while was lost in the formal gestures of her ancestors. She thought back to the teeming Tokyo streets of her childhood. The bittersweet recollection faded; at least, she thought, we don't have to get pushed onto the trains anymore.

The phone icon at bottom left flashed softly, accompanied by the pleasant tinkle of temple bells. That would be Koji in Hamburg. She smiled broadly as his jovial face appeared on the screen. The contract was signed, he'd be back tomorrow. She touched the screen lovingly as the image faded.

Time for workout. She went down to the health club for an hour, bathed, and returned for lunch.

Ana spent the early afternoon reading the accumulated messages from friends and then sent her correspondence out along the network. After that, a couple of hours were spent cleaning up some files.

Finally, Ana could relax. Soon, she'd dress for dinner with friends at the roof garden. Now, she turned toward the mirror and touched a button. The reflection of the room gradually faded.

Through the clear window she looked across at mighty Fujiyama with its proud summit rising above the low-lying clouds, pink in the setting sun.

### *For Further Reading*

Crease Robert P., and Charles C. Mann. *The Second Creation: Makers of the Revolution in Twentieth-Century Physics*. New York: Macmillan, 1986.

Dawkins, Richard. *The Blind Watchmaker: Why the Evidence of Evolution Reveals a Universe Without Design*. New York: Norton, 1987.

Kidder, Tracy. *The Soul of a New Machine*. Boston: Little, Brown, 1981.

Menzies, Heather. *Fast Forward and Out of Control: How Technology is Changing Your Life*. Toronto: Macmillan, 1989.

Vonnegut, Kurt. *Player Piano*. New York: Dell, 1986.

Wall, Byron E., ed. *Science in Society: Classical and Contemporary Readings*. Toronto: Wall & Emerson, 1989.

ubof f, Shoshana. *In the Age of the Smart Machine: The Future of Work and Power.* New York: Basic Books, 1988.

*Edwin Aldren on the Moon. Photo taken by Niel Armstrong.*

*Deep space seen from the Hubble telescope.*

# Index

D



E